Design Data Book

Properties of
Steam, Selected Refrigerants, *n*-Hexane and Brines (SI Units)

Design Data Book

Properties of
Steam, Selected Refrigerants, *n*-Hexane and Brines (SI Units)

Compiled by

B.I. Bhatt

Consulting Engineer,
Aavishkar Consultancy Services,
Ahmedabad

CBS

CBS Publishers & Distributors Pvt. Ltd.

New Delhi • Bengaluru • Chennai • Kochi • Kolkata • Mumbai
Hyderabad • Uttarakhand • Nagpur • Patna • Pune • Jharkhand

ISBN: 81-239-1389-3

First Edition: 2007
Reprint: 2008, 2012, 2019

Published by **Satish Kumar Jain** and produced by **Varun Jain** for
CBS Publishers & Distributors Pvt. Ltd.,
4819/XI Prahlad Street, 24 Ansari Road, Daryaganj, New Delhi - 110002
delhi@cbspd.com, cbspubs@airtelmail.in • www.cbspd.com
Ph.: 23289259, 23266861, 23266867 • Fax: 011-23243014

Corporate Office: 204 FIE, Industrial Area, Patparganj, Delhi - 110 092
Ph: 49344934 • Fax: 011-49344935
E-mail: publishing@cbspd.com • publicity@cbspd.com

Branches:

- ***Bengaluru:*** 2975, 17th Cross, K.R. Road, Bansankari 2nd Stage, Bengaluru - 70 • Ph: +91-80-26771678/79 • Fax: +91-80-26771680 E-mail: cbsbng@gmail.com, bangalore@cbspd.com
- ***Chennai:*** No. 7, Subbaraya Street, Shenoy Nagar, Chennai - 600030 Ph: +91-44-26681266, 26680620 • Fax: +91-44-42032115 E-mail: chennai@cbspd.com
- ***Kochi:*** Ashana House, 39/1904, A.M. Thomas Road, Valanjambalam, Ernakulum, Kochi • Ph: +91-484-4059061-65 Fax: +91-484-4059065 • E-mail: cochin@cbspd.com
- ***Kolkata:*** 6-B, Ground Floor, Rameshwar Shaw Road, Kolkata - 700014 Ph: +91-33-22891126/7/8 • E-mail: kolkata@cbspd.com
- ***Mumbai:*** 83-C, Dr. E. Moses Road, Worli, Mumbai - 400018 Ph: +91-9833017933, 022-24902340/41 • E-mail: mumbai@cbspd.com

Representatives:

- Hyderabad: 0-9885175004
- Patna: 0-9334159340
- Jharkhand: 0-9811541605
- Nagpur: 0-9021734563
- Pune: 0-9623451994
- Uttarakhand: 0-9716462459

Printed at:
India Binding House, Noida, UP (India)

Preface

Thermodynamic properties of steam and refrigerants are important in evaluation of heat changes in cyclic processes. It is needless to emphasize the need for authentic data for accurate calculations.

In this book, an attempt is made to present the authentic thermodynamic data of ice, water, steam and selected refrigerants (most commonly used) from the reliable sources after their careful evaluation. In most property data books, refrigerant properties in superheat region are not presented in tabular form for all refrigerants although in all refrigeration cycles, the refrigerant achieves superheat conditions at different stages. Inclusion of thermodynamic properties of the refrigerants in superheat region should therefore be of interest to the engineers, involved in design of the system.

Water and carbon dioxide are used in supercritical conditions for different applications. Removal of non-biodegradable chemical oxygen demand from aqueous effluents, production of biocrude from organic wastes, etc. are the applications in which water is used in supercritical conditions. Similarly, extraction of oil from various seeds, purification of pharmaceutical products, carrying out certain heterogeneous catalytic reactions with ease, etc. are the applications in which carbon dioxide is used in supercritical conditions. Thermodynamic properties of both these compounds in supercritical conditions should prove useful to the design engineers.

Design of freeze drying system and design of high vacuum system, involving water vapour, require thermodynamic data of ice and water on sublimation curve. These data, covered in the book, should be of special interest to the design engineer.

Liquid nitrogen is used as refrigerant to achieve low temperatures (say below –60°C) in pharmaceutical industry. Also nitrogen is used as a raw material in important synthesis such as ammonia. Property data of nitrogen will be valuable for design of such systems.

Solvent extraction is extensively used in extracting out oil from the oil cake with the help of *n*-hexane. Thermodynamic properties of *n*-hexane are hard to find in literature for carrying out design of *n*-hexane recovery systems. Coverage of these properties in the book will hopefully fill this gap.

Brines are extensively used in chemical, pharmaceutical, plastic and other industries as a secondary coolant. Properties of brines will help design engineers in sizing the chilling equipment and designing its piping.

I wish to sincerely thank Mr. Jinesh K. Vora for his valuable assistance in computer related work.

I welcome suggestions from the users for further improvement.

B. I. Bhatt
August, 2006
aavishkar.consultancy@gmail.com

Contents

Notations

p = Absolute pressure in torr or bar or Pa
t = Temperature in °C
T = Absolute temperature in K
v = Specific volume of superheated gas in m^3/kg
v' = Specific volume of saturated liquid in m^3/kg
v'' = Specific volume of saturated vapour in m^3/kg
h = Specific enthalpy of saturated liquid in kJ/kg
h_s = Specific enthalpy of solid in kJ/kg
H = Specific enthalpy of saturated vapour in kJ/kg
λ_f = Specific latent enthalpy of fusion in kJ/kg
λ_s = Specific latent enthalpy of sublimation in kJ/kg
λ_v = Specific latent enthalpy of evaporation in kJ/kg
i = Specific enthalpy of superheated vapour in kJ/kg
s = Specific entropy of superheated gas in kJ/(kg·K)
s_s = Specific entropy of solid in kJ/(kg·K)
s' = Specific entropy of saturated liquid in kJ/(kg·K)
s'' = Specific entropy of saturated vapour in kJ/(kg·K)
ρ_l = Density of saturated liquid in kg/m^3
ρ_s = Density of solid in kg/m^3

Useful Conversions

Pressure:	1 bar	= 0.1 MPa = 100 kPa = 10^5 Pa
		= 1.019716 kgf/cm^2 **or** at
		= 750.0616 torr
		= 0.986923 atm
		= 14.50377 lbf/in^2
Specific Volume:	1 m^3/kg	= 16.018462 ft^3/lb
Specific Enthalpy:	1 kJ/kg	= 0.238846 kcal/kg
		= 0.429923 Btu/lb
Specific Entropy:	1 kJ/(kg·K)	= 0.238846 kcal/(kg·°C)
		= 0.238846 Btu/(lb·°F)
Thermal Conductivity:	1 W/(m·K)	= 0.859854 kcal/(h·m·°C)
		= 0.577789 Btu/(h·ft·°F)
Heat Transfer Coefficient:	1 W/(m^2·K)	= 0.859854 kcal/(h·m^2·°C)
		= 0.17611 Btu/(h·ft^2·°F)
Density:	1 kg/L	= 62.42795 lb/ft^3
Temperature:	T / K	= t / °C + 273.15
		= 5/9 (T' / °R)
		= 5/9 (t' / °F + 459.67)

Properties of Ice, Water and Steam

TABLE: 1.1 Thermodynamic Properties of Ice and Water Vapour on Sublimation Curve (Temperature Basis)

Temperature t	Temperature T	Saturation pressure, p		Solid density ρ_s	Specific volume of vapour, v''	Specific enthalpy h_s	Specific enthalpy λ_s	Specific enthalpy H	Specific entropy s_s	Specific entropy s''	Temp t
°C	K	Pa	bar	kg/m³	m³/kg	kJ/kg			kJ/(kg·K)		°C
0.01*	**273.66**	**6.1165×10^{2}**	**6.1165×10^{-3}**	**916.7079**	**2.5099×10^{2}**	**–333.39**	**2834.31**	**2500.92**	**–1.22050**	**9.15550**	**0.01**
0	273.15	6.1115×10^{2}	6.1115×10^{-3}	916.708	2.0615×10^{2}	–333.41	2834.31	2500.90	–1.22058	9.15580	0
–1	272.15	5.6266×10^{2}	5.6266×10^{-3}	916.859	2.2311×10^{2}	–335.50	2834.57	2499.07	–1.22825	9.18723	–1
–2	271.15	5.1770×10^{2}	5.1770×10^{-3}	917.002	2.4160×10^{2}	–337.59	2834.83	2497.24	–1.23592	9.21891	–2
–3	270.15	4.7604×10^{2}	4.7604×10^{-3}	917.145	2.6178×10^{2}	–339.67	2835.07	2495.40	–1.24360	9.25084	–3
–4	269.15	4.3746×10^{2}	4.3746×10^{-3}	917.288	2.8382×10^{2}	–341.74	2835.31	2493.57	–1.25127	9.28304	–4
–5	268.15	4.0174×10^{2}	4.0174×10^{-3}	917.440	3.0791×10^{2}	–343.80	2835.54	2491.74	–1.25895	9.31550	–5
–6	267.15	3.6871×10^{2}	3.6871×10^{-3}	917.583	3.3426×10^{2}	–345.85	2835.76	2489.91	–1.26663	9.34823	–6
–7	266.15	3.3817×10^{2}	3.3817×10^{-3}	917.726	3.6309×10^{2}	–347.90	2835.97	2488.07	–1.27431	9.38123	–7
–8	265.15	3.0996×10^{2}	3.0996×10^{-3}	917.869	3.9466×10^{2}	–349.94	2836.18	2486.24	–1.28199	9.41451	–8
–9	264.15	2.8391×10^{2}	2.8391×10^{-3}	918.012	4.2925×10^{2}	–351.98	2836.37	2484.39	–1.28967	9.44806	–9
–10	263.15	2.5988×10^{2}	2.5988×10^{-3}	918.156	4.6718×10^{2}	–354.00	2836.56	2482.56	–1.29736	9.48189	–10
–11	262.15	2.3772×10^{2}	2.3772×10^{-3}	918.291	5.0879×10^{2}	–356.02	2836.74	2480.72	–1.30504	9.51602	–11
–12	261.15	2.1730×10^{2}	2.1730×10^{-3}	918.434	5.5449×10^{2}	–358.03	2836.91	2478.88	–1.31273	9.55042	–12
–13	260.15	1.9850×10^{2}	1.9850×10^{-3}	918.577	6.0470×10^{2}	–360.04	2837.08	2477.04	–1.32042	9.58511	–13
–14	259.15	1.8119×10^{2}	1.8119×10^{-3}	918.721	6.5991×10^{2}	–362.03	2837.23	2475.20	–1.32811	9.62011	–14
–15	258.15	1.6528×10^{2}	1.6528×10^{-3}	918.856	7.2067×10^{2}	–364.02	2837.38	2473.36	–1.33580	9.65540	–15
–16	257.15	1.5066×10^{2}	1.5066×10^{-3}	918.999	7.8756×10^{2}	–366.00	2837.52	2471.52	–1.34349	9.69099	–16
–17	256.15	1.3723×10^{2}	1.3723×10^{-3}	919.135	8.6128×10^{2}	–367.98	2837.65	2469.67	–1.35119	9.72688	–17
–18	255.15	1.2491×10^{2}	1.2491×10^{-3}	919.278	9.4258×10^{2}	–369.95	2837.77	2467.82	–1.35888	9.76310	–18
–19	254.15	1.1360×10^{2}	1.1360×10^{-3}	919.413	1.0323×10^{3}	–371.91	2837.89	2465.98	–1.36658	9.79962	–19
–20	253.15	1.0325×10^{2}	1.0325×10^{-3}	919.549	1.1314×10^{3}	–373.86	2838.00	2464.14	–1.37428	9.83645	–20
–21	252.15	9.3763×10^{1}	9.3763×10^{-4}	919.692	1.2409×10^{3}	–375.80	2838.10	2462.30	–1.38198	9.87361	–21
–22	251.15	8.5085×10^{1}	8.5085×10^{-4}	919.828	1.3621×10^{3}	–377.74	2838.19	2460.45	–1.38968	9.91110	–22
–23	250.15	7.7150×10^{1}	7.7150×10^{-4}	919.963	1.4962×10^{3}	–379.67	2838.28	2458.61	–1.39739	9.94890	–23
–24	249.15	6.9899×10^{1}	6.9899×10^{-4}	920.099	1.6448×10^{3}	–381.60	2838.35	2456.75	–1.40510	9.98704	–24
–25	248.15	6.3280×10^{1}	6.3280×10^{-4}	920.234	1.8096×10^{3}	–383.52	2838.42	2454.90	–1.41280	10.02553	–25
–26	247.15	5.7242×10^{1}	5.7242×10^{-4}	920.370	1.9925×10^{3}	–385.42	2838.48	2453.06	–1.42051	10.06435	–26
–27	246.15	5.1737×10^{1}	5.1737×10^{-4}	920.505	2.1955×10^{3}	–387.33	2838.54	2451.21	–1.42823	10.10351	–27
–28	245.15	4.6723×10^{1}	4.6723×10^{-4}	920.641	2.4213×10^{3}	–389.22	2838.59	2449.37	–1.43594	10.14303	–28
–29	244.15	4.2160×10^{1}	4.2160×10^{-4}	920.768	2.6724×10^{3}	–391.11	2838.62	2447.51	–1.44366	10.18290	–29
–30	243.15	3.8010×10^{1}	3.8010×10^{-4}	920.904	2.9521×10^{3}	–392.99	2838.66	2445.67	–1.45138	10.22313	–30
–31	242.15	3.4239×10^{1}	3.4239×10^{-4}	921.039	3.2637×10^{3}	–394.86	2838.68	2443.82	–1.45910	10.26372	–31
–32	241.15	3.0816×10^{1}	3.0816×10^{-4}	921.167	3.6113×10^{3}	–396.73	2838.70	2441.97	–1.46682	10.30468	–32
–33	240.15	2.7711×10^{1}	2.7711×10^{-4}	921.302	3.9994×10^{3}	–398.59	2838.71	2440.12	–1.47454	10.34602	–33
–34	239.15	2.4896×10^{1}	2.4896×10^{-4}	921.430	4.4330×10^{3}	–400.44	2838.71	2438.27	–1.48227	10.38772	–34
–35	238.15	2.2347×10^{1}	2.2347×10^{-4}	921.566	4.9180×10^{3}	–402.29	2838.70	2436.41	–1.49000	10.42981	–35
–36	237.15	2.0041×10^{1}	2.0041×10^{-4}	921.693	5.4609×10^{3}	–404.12	2838.69	2434.57	–1.49773	10.47229	–36
–37	236.15	1.7957×10^{1}	1.7957×10^{-4}	921.820	6.0691×10^{3}	–405.95	2838.67	2432.72	–1.50546	10.51517	–37
–38	235.15	1.6074×10^{1}	1.6074×10^{-4}	921.948	6.7514×10^{3}	–407.78	2838.64	2430.86	–1.51320	10.55843	–38
–39	234.15	1.4375×10^{1}	1.4375×10^{-4}	922.084	7.5172×10^{3}	–409.59	2838.61	2429.02	–1.52094	10.60210	–39
–40	233.15	1.2843×10^{1}	1.2843×10^{-4}	922.211	8.3778×10^{3}	–411.40	2838.57	2427.17	–1.52868	10.64617	–40

(Contd.)

*Triple point

TABLE: 1.1 Thermodynamic Properties of Ice and Water Vapour on Sublimation Curve (Temperature Basis) [Contd.]

Temperature		Saturation pressure, p		Solid density	Specific volume of	Specific enthalpy			Specific entropy		Temp
t	T			ρ_s	vapour, v''	h_s	λ_s	$\dot{H}$	s_s	s''	t
°C	K	Pa	bar	kg/m³	m³/kg	kJ/kg			kJ/(kg·K)		°C
–41	232.15	1.1464×10^{1}	1.1464×10^{-4}	922.339	9.3458×10^{3}	–413.20	2838.52	2425.32	–1.53642	10.69066	**–41**
–42	231.15	1.0222×10^{1}	1.0222×10^{-4}	922.458	1.0436×10^{4}	–415.00	2838.46	2423.46	–1.54417	10.73557	**–42**
–43	230.15	9.1062	9.1062×10^{-5}	922.586	1.1664×10^{4}	–416.78	2838.40	2421.62	–1.55191	10.78090	**–43**
–44	229.15	8.1038	8.1038×10^{-5}	922.714	1.3050×10^{4}	–418.56	2838.33	2419.77	–1.55966	10.82666	**–44**
–45	228.15	7.2045	7.2045×10^{-5}	922.841	1.4615×10^{4}	–420.33	2838.25	2417.92	–1.56742	10.87285	**–45**
–46	227.15	6.3983	6.3983×10^{-5}	922.960	1.6384×10^{4}	–422.10	2838.16	2416.06	–1.57517	10.91949	**–46**
–47	226.15	5.6764	5.6764×10^{-5}	923.088	1.8387×10^{4}	–423.86	2838.07	2414.21	–1.58293	10.96657	**–47**
–48	225.15	5.0306	5.0306×10^{-5}	923.216	2.0655×10^{4}	–425.61	2837.97	2412.36	–1.59069	11.01411	**–48**
–49	224.15	4.4536	4.4536×10^{-5}	923.335	2.3228×10^{4}	–427.35	2837.86	2410.51	–1.59846	11.06209	**–49**
–50	223.15	3.9384	3.9384×10^{-5}	923.455	2.6149×10^{4}	–429.09	2837.75	2408.66	–1.60622	11.11056	**–50**
–51	222.15	3.4790	3.4790×10^{-5}	923.583	2.9470×10^{4}	–430.82	2837.63	2406.81	–1.61399	11.15948	**–51**
–52	221.15	3.0697	3.0697×10^{-5}	923.702	3.3248×10^{4}	–432.54	2837.50	2404.96	–1.62177	11.20888	**–52**
–53	220.15	2.7055	2.7055×10^{-5}	923.822	3.7553×10^{4}	–434.26	2837.36	2403.10	–1.62954	11.25878	**–53**
–54	219.15	2.3819	2.3819×10^{-5}	923.941	4.2463×10^{4}	–435.97	2837.22	2401.25	–1.63732	11.30916	**–54**
–55	218.15	2.0944	2.0944×10^{-5}	924.061	4.8069×10^{4}	–437.67	2837.07	2399.40	–1.64510	11.36004	**–55**
–56	217.15	1.8396	1.8396×10^{-5}	924.180	5.4479×10^{4}	–439.37	2836.91	2397.54	–1.65289	11.41141	**–56**
–57	216.15	1.6138	1.6138×10^{-5}	924.300	6.1816×10^{4}	–441.05	2836.75	2395.70	–1.66068	11.46331	**–57**
–58	215.15	1.4140	1.4140×10^{-5}	924.419	7.0224×10^{4}	–442.73	2836.58	2393.85	–1.66847	11.51572	**–58**
–59	214.15	1.2374	1.2374×10^{-5}	924.531	7.9872×10^{4}	–444.41	2836.40	2391.99	–1.67627	11.56865	**–59**
–60	213.15	1.0815	1.0815×10^{-5}	924.650	9.0957×10^{4}	–446.07	2836.22	2390.15	–1.68406	11.62213	**–60**
–61	212.15	9.4409×10^{-1}	9.4409×10^{-6}	924.770	1.0371×10^{5}	–447.73	2836.02	2388.29	–1.69187	11.67614	**–61**
–62	211.15	8.2308×10^{-1}	8.2308×10^{-6}	924.881	1.1840×10^{5}	–449.38	2835.82	2386.44	–1.69967	11.73071	**–62**
–63	210.15	7.1664×10^{-1}	7.1664×10^{-6}	925.001	1.3534×10^{5}	–451.03	2835.62	2384.59	–1.70748	11.78582	**–63**
–64	209.15	6.2316×10^{-1}	6.2316×10^{-6}	925.112	1.5490×10^{5}	–452.67	2835.41	2382.74	–1.71530	11.84150	**–64**
–65	208.15	5.4114×10^{-1}	5.4114×10^{-6}	925.223	1.7752×10^{5}	–454.30	2835.18	2380.88	–1.72311	11.89776	**–65**
–66	207.15	4.6929×10^{-1}	4.6929×10^{-6}	925.335	2.0372×10^{5}	–455.92	2834.96	2379.04	–1.73093	11.95460	**–66**
–67	206.15	4.0641×10^{-1}	4.0641×10^{-6}	925.455	2.3410×10^{5}	–457.54	2834.72	2377.18	–1.73876	12.01202	**–67**
–68	205.15	3.5148×10^{-1}	3.5148×10^{-6}	925.566	2.6938×10^{5}	–459.15	2834.48	2375.33	–1.74659	12.07004	**–68**
–69	204.15	3.0354×10^{-1}	3.0354×10^{-6}	925.677	3.1040×10^{5}	–460.75	2834.23	2373.48	–1.75442	12.12867	**–69**
–70	203.15	2.6176×10^{-1}	2.6176×10^{-6}	925.780	3.5818×10^{5}	–462.35	2833.98	2371.63	–1.76226	12.18792	**–70**
–71	202.15	2.2541×10^{-1}	2.2541×10^{-6}	925.892	4.1389×10^{5}	–463.94	2833.72	2369.78	–1.77010	12.24779	**–71**
–72	201.15	1.9382×10^{-1}	1.9382×10^{-6}	926.003	4.7897×10^{5}	–465.52	2833.45	2367.93	–1.77795	12.30829	**–72**
–73	200.15	1.6641×10^{-1}	1.6641×10^{-6}	926.115	5.5510×10^{5}	–467.09	2833.17	2366.08	–1.78580	12.36944	**–73**
–74	199.15	1.4266×10^{-1}	1.4266×10^{-6}	926.218	6.4428×10^{5}	–468.66	2832.89	2364.23	–1.79365	12.43125	**–74**
–75	198.15	1.2211×10^{-1}	1.2211×10^{-6}	926.329	7.4893×10^{5}	–470.22	2832.60	2362.38	–1.80151	12.49372	**–75**
–76	197.15	1.0435×10^{-1}	1.0435×10^{-6}	926.432	8.7191×10^{5}	–471.78	2832.30	2360.52	–1.80937	12.55686	**–76**
–77	196.15	8.9042×10^{-2}	8.9042×10^{-7}	926.535	1.0167×10^{6}	–473.33	2832.00	2358.67	–1.81724	12.62068	**–77**
–78	195.15	7.5854×10^{-2}	7.5854×10^{-7}	926.647	1.1873×10^{6}	–474.87	2831.69	2356.82	–1.82512	12.68520	**–78**
–79	194.15	6.4514×10^{-2}	6.4514×10^{-7}	926.750	1.3889×10^{6}	–476.40	2831.37	2354.97	–1.83299	12.75043	**–79**
–80	193.15	5.4778×10^{-2}	5.4778×10^{-7}	926.853	1.6273×10^{6}	–477.93	2831.05	2353.12	–1.84088	12.81636	**–80**

(Contd.)

TABLE: 1.1 Thermodynamic Properties of Ice and Water Vapour on Sublimation Curve (Temperature Basis) [Contd.]

Temperature		Saturation pressure, p		Solid density	Specific volume of vapour, v''	Specific enthalpy			Specific entropy		Temp
t	T			ρ_s		h_s	λ_s	H	s_s	s''	t
°C	K	Pa	bar	kg/m³	m³/kg	kJ/kg			kJ/(kg·K)		°C
–81	192.15	4.6434×10^{-2}	4.6434×10^{-7}	926.956	1.9098×10^{6}	–479.45	2830.72	2351.27	–1.84877	12.88303	**–81**
–82	191.15	3.9293×10^{-2}	3.9293×10^{-7}	927.059	2.2452×10^{6}	–480.96	2830.38	2349.42	–1.85666	12.95044	**–82**
–83	190.15	3.3192×10^{-2}	3.3192×10^{-7}	927.162	2.6439×10^{6}	–482.46	2830.03	2347.57	–1.86456	13.01860	**–83**
–84	189.15	2.7990×10^{-2}	2.7990×10^{-7}	927.257	3.1188×10^{6}	–483.96	2829.68	2345.72	–1.87246	13.08752	**–84**
–85	188.15	2.3561×10^{-2}	2.3561×10^{-7}	927.360	3.6856×10^{6}	–485.46	2829.32	2343.86	–1.88037	13.15722	**–85**
–86	187.15	1.9796×10^{-2}	1.9796×10^{-7}	927.463	4.3631×10^{6}	–486.94	2828.96	2342.02	–1.88829	13.22770	**–86**
–87	186.15	1.6602×10^{-2}	1.6602×10^{-7}	927.558	5.1746×10^{6}	–488.42	2828.59	2340.17	–1.89621	13.29898	**–87**
–88	185.15	1.3898×10^{-2}	1.3898×10^{-7}	927.652	6.1484×10^{6}	–489.89	2828.21	2338.32	–1.90413	13.37108	**–88**
–89	184.15	1.1612×10^{-2}	1.1612×10^{-7}	927.756	7.3192×10^{6}	–491.36	2827.82	2336.46	–1.91207	13.44400	**–89**
–90	183.15	9.6828×10^{-3}	9.6828×10^{-8}	927.850	8.7296×10^{6}	–492.81	2827.43	2334.62	–1.92000	13.51777	**–90**
–91	182.15	8.0585×10^{-3}	8.0585×10^{-8}	927.945	1.0432×10^{7}	–494.26	2827.03	2332.77	–1.92795	13.59238	**–91**
–92	181.15	6.6933×10^{-3}	6.6933×10^{-8}	928.040	1.2491×10^{7}	–495.71	2826.62	2330.91	–1.93590	13.66786	**–92**
–93	180.15	5.5480×10^{-3}	5.5480×10^{-8}	928.135	1.4986×10^{7}	–497.15	2826.21	2329.06	–1.94386	13.74423	**–93**
–94	179.15	4.5892×10^{-3}	4.5892×10^{-8}	928.229	1.8016×10^{7}	–498.58	2825.79	2327.21	–1.95182	13.82149	**–94**
–95	178.15	3.7882×10^{-3}	3.7882×10^{-8}	928.324	2.1704×10^{7}	–500.00	2825.36	2325.36	–1.95979	13.89967	**–95**
–96	177.15	3.1203×10^{-3}	3.1203×10^{-8}	928.419	2.6202×10^{7}	–501.42	2824.93	2323.51	–1.96777	13.97877	**–96**
–97	176.15	2.5645×10^{-3}	2.5645×10^{-8}	928.505	3.1700×10^{7}	–502.83	2824.49	2321.66	–1.97575	14.05882	**–97**
–98	175.15	2.1031×10^{-3}	2.1031×10^{-8}	928.600	3.8435×10^{7}	–504.23	2824.04	2319.81	–1.98374	14.13982	**–98**
–99	174.15	1.7209×10^{-3}	1.7209×10^{-8}	928.686	4.6705×10^{7}	–505.63	2823.59	2317.96	–1.99174	14.22180	**–99**
–100	173.15	1.4049×10^{-3}	1.4049×10^{-8}	928.772	5.6881×10^{7}	–507.02	2823.13	2316.11	–1.99974	14.30478	**–100**
–101	172.15	1.1443×10^{-3}	1.1443×10^{-8}	928.867	6.9433×10^{7}	–508.40	2822.66	2314.26	–2.00775	14.38877	**–101**
–102	171.15	9.2979×10^{-4}	9.2979×10^{-9}	928.954	8.4954×10^{7}	–509.78	2822.19	2312.41	–2.01577	14.47378	**–102**
–103	170.15	7.5270×10^{-4}	7.5270×10^{-9}	929.040	1.0419×10^{8}	–511.15	2821.71	2310.56	–2.02379	14.55985	**–103**
–104	169.15	6.0946×10^{-4}	6.0946×10^{-9}	929.126	1.2809×10^{8}	–512.51	2821.22	2308.71	–2.03183	14.64697	**–104**
–105	168.15	4.9160×10^{-4}	4.9160×10^{-9}	929.213	1.5786×10^{8}	–513.87	2820.73	2306.86	–2.03987	14.73518	**–105**
–106	167.15	3.9553×10^{-4}	3.9553×10^{-9}	929.290	1.9503×10^{8}	–515.22	2820.23	2305.01	–2.04792	14.82450	**–106**
–107	166.15	3.1742×10^{-4}	3.1742×10^{-9}	929.377	2.4158×10^{8}	–516.56	2819.72	2303.16	–2.05597	14.91494	**–107**
–108	165.15	2.5406×10^{-4}	2.5406×10^{-9}	929.463	3.0000×10^{8}	–517.89	2819.20	2301.31	–2.06404	15.00652	**–108**
–109	164.15	2.0281×10^{-4}	2.0281×10^{-9}	929.541	3.7354×10^{8}	–519.22	2818.68	2299.46	–2.07211	15.09927	**–109**
–110	163.15	1.6146×10^{-4}	1.6146×10^{-9}	929.627	4.6635×10^{8}	–520.55	2818.16	2297.61	–2.08019	15.19321	**–110**
–111	162.15	1.2818×10^{-4}	1.2818×10^{-9}	929.705	5.8382×10^{8}	–521.86	2817.62	2295.76	–2.08828	15.28835	**–111**
–112	161.15	1.0148×10^{-4}	1.0148×10^{-9}	929.783	7.3291×10^{8}	–523.17	2817.08	2293.91	–2.09638	15.38472	**–112**
–113	160.15	8.0106×10^{-5}	8.0106×10^{-10}	929.861	9.2268×10^{8}	–524.47	2816.53	2292.06	–2.10449	15.48235	**–113**
–114	159.15	6.3050×10^{-5}	6.3050×10^{-10}	929.938	1.1650×10^{9}	–525.77	2815.98	2290.21	–2.11260	15.58126	**–114**
–115	158.15	4.9479×10^{-5}	4.9479×10^{-10}	930.016	1.4752×10^{9}	–527.06	2815.42	2288.36	–2.12073	15.68146	**–115**
–116	157.15	3.8710×10^{-5}	3.8710×10^{-10}	930.094	1.8736×10^{9}	–528.34	2814.85	2286.51	–2.12886	15.78300	**–116**
–117	156.15	3.0192×10^{-5}	3.0192×10^{-10}	930.172	2.3869×10^{9}	–529.62	2814.28	2284.66	–2.13701	15.88589	**–117**
–118	155.15	2.3474×10^{-5}	2.3474×10^{-10}	930.241	3.0503×10^{9}	–530.88	2813.69	2282.81	–2.14516	15.99016	**–118**
–119	154.15	1.8193×10^{-5}	1.8193×10^{-10}	930.319	3.9105×10^{9}	–532.15	2813.11	2280.96	–2.15332	16.09583	**–119**
–120	153.15	1.4053×10^{-5}	1.4053×10^{-10}	930.388	5.0295×10^{9}	–533.40	2812.51	2279.11	–2.16149	16.20294	**–120**

(Contd.)

TABLE: 1.1 Thermodynamic Properties of Ice and Water Vapour on Sublimation Curve (Temperature Basis) [Contd.]

Temperature t	Temperature T	Saturation pressure, p		Solid density ρ_s	Specific volume of vapour, v''	Specific enthalpy h_s	Specific enthalpy λ_s	Specific enthalpy H	Specific entropy s_s	Specific entropy s''	Temp t
°C	K	Pa	bar	kg/m³	m³/kg	kJ/kg			kJ/(kg·K)		°C
–121	152.15	1.0820×10^{-5}	1.0820×10^{-10}	930.466	6.4900×10^{9}	–534.65	2811.91	2277.26	–2.16967	16.31151	**–121**
–122	151.15	8.3017×10^{-6}	8.3017×10^{-11}	930.536	8.4029×10^{9}	–535.89	2811.30	2275.41	–2.17786	16.42157	**–122**
–123	150.15	6.3477×10^{-6}	6.3477×10^{-11}	930.605	1.0917×10^{10}	–537.13	2810.69	2273.56	–2.18606	16.53315	**–123**
–124	149.15	4.8364×10^{-6}	4.8364×10^{-11}	930.674	1.4233×10^{10}	–538.36	2810.07	2271.71	–2.19427	16.64628	**–124**
–125	148.15	3.6717×10^{-6}	3.6717×10^{-11}	930.743	1.8622×10^{10}	–539.58	2809.44	2269.86	–2.20249	16.76100	**–125**
–126	147.15	2.7772×10^{-6}	2.7772×10^{-11}	930.813	2.4453×10^{10}	–540.79	2808.81	2268.02	–2.21073	16.87732	**–126**
–127	146.15	2.0928×10^{-6}	2.0928×10^{-11}	930.873	3.2230×10^{10}	–542.00	2808.17	2266.17	–2.21897	16.99530	**–127**
–128	145.15	1.5710×10^{-6}	1.5710×10^{-11}	930.943	4.2642×10^{10}	–543.20	2807.52	2264.32	–2.22722	17.11496	**–128**
–129	144.15	1.1747×10^{-6}	1.1747×10^{-11}	931.012	5.6635×10^{10}	–544.40	2806.86	2262.46	–2.23548	17.23635	**–129**
–130	143.15	8.7485×10^{-7}	8.7485×10^{-12}	931.073	7.5517×10^{10}	–545.59	2806.20	2260.61	–2.24376	17.35947	**–130**
–131	142.15	6.4890×10^{-7}	6.4890×10^{-12}	931.133	1.0110×10^{11}	–546.77	2805.54	2258.77	–2.25204	17.48440	**–131**
–132	141.15	4.7931×10^{-7}	4.7931×10^{-12}	931.203	1.3591×10^{11}	–547.95	2804.86	2256.91	–2.26034	17.61115	**–132**
–133	140.15	3.5254×10^{-7}	3.5254×10^{-12}	931.263	1.8347×10^{11}	–549.11	2804.18	2255.07	–2.26864	17.73977	**–133**
–134	139.15	2.5818×10^{-7}	2.5818×10^{-12}	931.324	2.4874×10^{11}	–550.28	2803.49	2253.21	–2.27696	17.87030	**–134**
–135	138.15	1.8824×10^{-7}	1.8824×10^{-12}	931.385	3.3872×10^{11}	–551.43	2802.80	2251.37	–2.28529	18.00278	**–135**
–136	137.15	1.3662×10^{-7}	1.3662×10^{-12}	931.437	4.6330×10^{11}	–552.58	2802.10	2249.52	–2.29363	18.13725	**–136**
–137	136.15	9.8702×10^{-8}	9.8702×10^{-13}	931.498	6.3662×10^{11}	–553.72	2801.39	2247.67	–2.30198	18.27376	**–137**
–138	135.15	7.0971×10^{-8}	7.0971×10^{-13}	931.558	8.7887×10^{11}	–554.85	2800.67	2245.82	–2.31035	18.41234	**–138**
–139	134.15	5.0785×10^{-8}	5.0785×10^{-13}	931.610	1.2191×10^{12}	–555.98	2799.95	2243.97	–2.31872	18.55307	**–139**
–140	133.15	3.6162×10^{-8}	3.6162×10^{-13}	931.671	1.6993×10^{12}	–557.10	2799.22	2242.12	–2.32711	18.69596	**–140**

TABLE: 1.2 Thermodynamic Properties of Ice and Water Vapour on Sublimation Curve (Pressure Basis)

Pressure, p		Saturation temperature		Solid density	Specific volume of vapour, v''	Specific enthalpy			Specific entropy		Press.
		t	T	ρ_s		h_s	λ_s	H	s_s	s''	p
Pa	bar	°C	K	kg/m³	m³/kg	kJ/kg			kJ/(kg·K)		Pa
6.1165×10^2	**6.1165×10^{-3}**	**0.01***	**273.16**	**916.7079**	**2.5099×10^2**	**–333.39**	**2834.31**	**2500.92**	**–1.22050**	**9.15550**	**6.1165×10^2**
6.0×10^2	6.0×10^{-3}	–0.22	272.93	916.742	2.0981×10^2	–333.88	2834.37	2500.49	–1.22229	9.16280	**6.0×10^2**
5.0×10^2	5.0×10^{-3}	–2.42	270.73	917.061	2.4977×10^2	–338.45	2834.93	2496.48	–1.23911	9.23215	**5.0×10^2**
4.0×10^2	4.0×10^{-3}	–5.05	268.10	917.440	3.0920×10^2	–343.90	2835.55	2491.65	–1.25934	9.31716	**4.0×10^2**
3.0×10^2	3.0×10^{-3}	–8.37	264.78	917.920	4.0719×10^2	–350.70	2836.26	2485.56	–1.28485	9.42699	**3.0×10^2**
2.0×10^2	2.0×10^{-3}	–12.92	260.23	918.569	6.0035×10^2	–359.87	2837.07	2477.20	–1.31978	9.58222	**2.0×10^2**
1.0×10^2	1.0×10^{-3}	–20.33	252.82	919.599	1.1666×10^3	–374.51	2838.03	2463.52	–1.37684	9.84877	**1.0×10^2**
9.0×10^1	9.0×10^{-4}	–21.42	251.73	919.743	1.2907×10^3	–376.63	2838.14	2461.51	–1.38524	9.88941	**9.0×10^1**
8.0×10^1	8.0×10^{-4}	–22.63	250.52	919.912	1.4450×10^3	–378.96	2838.24	2459.28	–1.39454	9.93489	**8.0×10^1**
7.0×10^1	7.0×10^{-4}	–23.98	249.17	920.099	1.6426×10^3	–381.57	2838.35	2456.78	–1.40498	9.98649	**7.0×10^1**
6.0×10^1	6.0×10^{-4}	–25.53	247.62	920.302	1.9045×10^3	–384.53	2838.46	2453.93	–1.41690	10.04613	**6.0×10^1**
5.0×10^1	5.0×10^{-4}	–27.34	245.81	920.548	2.2687×10^3	–387.96	2838.56	2450.60	–1.43082	10.11675	**5.0×10^1**
4.0×10^1	4.0×10^{-4}	–29.51	243.64	920.836	2.8109×10^3	–392.07	2838.64	2446.57	–1.44758	10.20332	**4.0×10^1**
3.0×10^1	3.0×10^{-4}	–32.25	240.90	921.201	3.7057×10^3	–397.20	2838.70	2441.50	–1.46878	10.31512	**3.0×10^1**
2.0×10^1	2.0×10^{-4}	–36.02	237.13	921.693	5.4717×10^3	–404.16	2838.69	2434.53	–1.49788	10.47310	**2.0×10^1**
1.0×10^1	1.0×10^{-4}	–42.19	230.96	922.484	1.0659×10^4	–415.34	2838.45	2423.11	–1.54564	10.74419	**1.0×10^1**
9.0	9.0×10^{-5}	–43.10	230.05	922.603	1.1797×10^4	–416.96	2838.39	2421.43	–1.55270	10.78550	**9.0**
8.0	8.0×10^{-5}	–44.11	229.04	922.731	1.3213×10^4	–418.76	2838.32	2419.56	–1.56052	10.83172	**8.0**
7.0	7.0×10^{-5}	–45.24	227.91	922.867	1.5026×10^4	–420.77	2838.23	2417.46	–1.56930	10.88417	**7.0**
6.0	6.0×10^{-5}	–46.54	226.61	923.029	1.7431×10^4	–423.05	2838.11	2415.06	–1.57935	10.94476	**6.0**
5.0	5.0×10^{-5}	–48.05	225.10	923.216	2.0777×10^4	–425.70	2837.96	2412.26	–1.59108	11.01651	**5.0**
4.0	4.0×10^{-5}	–49.87	223.28	923.446	2.5761×10^4	–428.87	2837.76	2408.89	–1.60525	11.10443	**4.0**
3.0	3.0×10^{-5}	–52.18	220.97	923.719	3.3993×10^4	–432.86	2837.47	2404.61	–1.62319	11.21796	**3.0**
2.0	2.0×10^{-5}	–55.36	217.79	924.103	5.0257×10^4	–438.27	2837.00	2398.73	–1.64788	11.37830	**2.0**
1.0	1.0×10^{-5}	–60.58	212.57	924.719	9.8105×10^4	–447.03	2836.11	2389.08	–1.68857	11.65327	**1.0**
9.0×10^{-1}	9.0×10^{-6}	–61.35	211.80	924.804	1.0861×10^5	–448.31	2835.95	2387.64	–1.69460	11.69516	**9.0×10^{-1}**
8.0×10^{-1}	8.0×10^{-6}	–62.21	210.94	924.907	1.2169×10^5	–449.72	2835.78	2386.06	–1.70128	11.74202	**8.0×10^{-1}**
7.0×10^{-1}	7.0×10^{-6}	–63.17	209.98	925.018	1.3844×10^5	–451.31	2835.58	2384.27	–1.70880	11.79518	**7.0×10^{-1}**
6.0×10^{-1}	6.0×10^{-6}	–64.27	208.88	925.146	1.6067×10^5	–453.11	2835.35	2382.24	–1.71740	11.85660	**6.0×10^{-1}**
5.0×10^{-1}	5.0×10^{-6}	–65.56	207.59	925.292	1.9162×10^5	–455.20	2835.06	2379.86	–1.72746	11.92930	**5.0×10^{-1}**
4.0×10^{-1}	4.0×10^{-6}	–67.11	206.04	925.463	2.3773×10^5	–457.72	2834.70	2376.98	–1.73962	12.01837	**4.0×10^{-1}**
3.0×10^{-1}	3.0×10^{-6}	–69.08	204.07	925.686	3.1394×10^5	–460.88	2834.21	2373.33	–1.75504	12.13337	**3.0×10^{-1}**
2.0×10^{-1}	2.0×10^{-6}	–71.79	201.36	925.977	4.6465×10^5	–465.19	2833.49	2368.30	–1.77632	12.29572	**2.0×10^{-1}**
1.0×10^{-1}	1.0×10^{-6}	–76.27	196.88	926.458	9.0864×10^5	–472.20	2832.22	2360.02	–1.81149	12.57400	**1.0×10^{-1}**
9.0×10^{-2}	9.0×10^{-7}	–76.93	196.22	926.535	1.0062×10^6	–473.22	2832.02	2358.80	–1.81671	12.61638	**9.0×10^{-2}**
8.0×10^{-2}	8.0×10^{-7}	–77.67	195.48	926.612	1.1277×10^6	–474.36	2831.79	2357.43	–1.82251	12.66378	**8.0×10^{-2}**
7.0×10^{-2}	7.0×10^{-7}	–78.50	194.65	926.698	1.2834×10^6	–475.63	2831.53	2355.90	–1.82903	12.71755	**7.0×10^{-2}**
6.0×10^{-2}	6.0×10^{-7}	–79.44	193.71	926.793	1.4900×10^6	–477.08	2831.23	2354.15	–1.83650	12.77966	**6.0×10^{-2}**
5.0×10^{-2}	5.0×10^{-7}	–80.55	192.60	926.913	1.7777×10^6	–478.77	2830.86	2352.09	–1.84524	12.85318	**5.0×10^{-2}**
4.0×10^{-2}	4.0×10^{-7}	–81.89	191.26	927.050	2.2067×10^6	–480.80	2830.41	2349.61	–1.85582	12.94324	**4.0×10^{-2}**
3.0×10^{-2}	3.0×10^{-7}	–83.59	189.56	927.222	2.9161×10^6	–483.36	2829.82	2346.46	–1.86926	13.05947	**3.0×10^{-2}**
2.0×10^{-2}	2.0×10^{-7}	–85.94	187.21	927.455	4.3200×10^6	–486.85	2828.96	2342.11	–1.88782	13.22356	**2.0×10^{-2}**
1.0×10^{-2}	1.0×10^{-7}	–89.82	183.33	927.833	8.4609×10^6	–492.56	2827.50	2334.94	–1.91860	13.50468	**1.0×10^{-2}**

*Triple point

(Contd.)

TABLE: 1.2 Thermodynamic Properties of Ice and Water Vapour on Sublimation Curve (Pressure Basis) [Contd.]

Pressure, p		Saturation temperature		Solid density	Specific volume of vapour, v''	Specific enthalpy			Specific entropy		Press. p
		t	T	ρ_s		h_s	λ_s	H	s_s	s''	
Pa	bar	°C	K	kg/m^3	m^3/kg	kJ/kg			kJ/(kg·K)		Pa
9.0×10^{-3}	9.0×10^{-8}	–90.40	182.75	927.885	9.3714×10^{6}	–493.39	2827.27	2333.88	–1.92318	13.54747	**9.0×10^{-3}**
8.0×10^{-3}	8.0×10^{-8}	–91.04	182.11	927.954	1.0506×10^{7}	–494.32	2827.01	2332.69	–1.92826	13.59534	**8.0×10^{-3}**
7.0×10^{-3}	7.0×10^{-8}	–91.76	181.39	928.023	1.1959×10^{7}	–495.36	2826.72	2331.36	–1.93399	13.64963	**7.0×10^{-3}**
6.0×10^{-3}	6.0×10^{-8}	–92.58	180.57	928.100	1.3889×10^{7}	–496.55	2826.38	2329.83	–1.94055	13.71235	**6.0×10^{-3}**
5.0×10^{-3}	5.0×10^{-8}	–93.55	179.60	928.186	1.6578×10^{7}	–497.93	2825.98	2328.05	–1.94823	13.78658	**5.0×10^{-3}**
4.0×10^{-3}	4.0×10^{-8}	–94.72	178.43	928.298	2.0587×10^{7}	–499.60	2825.48	2325.88	–1.95754	13.87748	**4.0×10^{-3}**
3.0×10^{-3}	3.0×10^{-8}	–96.20	176.95	928.436	2.7222×10^{7}	–501.70	2824.84	2323.14	–1.96937	13.99481	**3.0×10^{-3}**
2.0×10^{-3}	2.0×10^{-8}	–98.25	174.90	928.617	4.0360×10^{7}	–504.58	2823.91	2319.33	–1.98575	14.16037	**2.0×10^{-3}**
1.0×10^{-3}	1.0×10^{-8}	–101.65	171.50	928.919	7.9150×10^{7}	–509.30	2822.35	2313.05	–2.01297	14.44396	**1.0×10^{-3}**
9.0×10^{-4}	9.0×10^{-9}	–102.16	170.99	928.962	8.7685×10^{7}	–509.99	2822.11	2312.12	–2.01702	14.48713	**9.0×10^{-4}**
8.0×10^{-4}	8.0×10^{-9}	–102.72	170.43	929.014	9.8322×10^{7}	–510.76	2821.84	2311.08	–2.02152	14.53541	**8.0×10^{-4}**
7.0×10^{-4}	7.0×10^{-9}	–103.35	169.80	929.066	1.1195×10^{8}	–511.62	2821.54	2309.92	–2.02660	14.59015	**7.0×10^{-4}**
6.0×10^{-4}	6.0×10^{-9}	–104.07	169.08	929.135	1.3005×10^{8}	–512.61	2821.18	2308.57	–2.03242	14.65339	**6.0×10^{-4}**
5.0×10^{-4}	5.0×10^{-9}	–104.92	168.23	929.204	1.5528×10^{8}	–513.76	2820.76	2307.00	–2.03924	14.72823	**5.0×10^{-4}**
4.0×10^{-4}	4.0×10^{-9}	–105.95	167.20	929.290	1.9292×10^{8}	–515.15	2820.25	2305.10	–2.04750	14.81989	**4.0×10^{-4}**
3.0×10^{-4}	3.0×10^{-9}	–107.25	165.90	929.403	2.5521×10^{8}	–516.90	2819.58	2302.68	–2.05803	14.93815	**3.0×10^{-4}**
2.0×10^{-4}	2.0×10^{-9}	–109.06	164.09	929.549	3.7865×10^{8}	–519.30	2818.64	2299.34	–2.07261	15.10502	**2.0×10^{-4}**
1.0×10^{-4}	1.0×10^{-9}	–112.06	161.09	929.791	7.4345×10^{8}	–523.25	2817.05	2293.80	–2.09689	15.39077	**1.0×10^{-4}**
9.0×10^{-5}	9.0×10^{-10}	–112.51	160.64	929.826	8.2376×10^{8}	–523.83	2816.80	2292.97	–2.10051	15.43426	**9.0×10^{-5}**
8.0×10^{-5}	8.0×10^{-10}	–113.01	160.14	929.861	9.2387×10^{8}	–524.48	2816.53	2292.05	–2.10453	15.48290	**8.0×10^{-5}**
7.0×10^{-5}	7.0×10^{-10}	–113.56	159.59	929.904	1.0522×10^{9}	–525.21	2816.22	2291.01	–2.10907	15.53805	**7.0×10^{-5}**
6.0×10^{-5}	6.0×10^{-10}	–114.21	158.94	929.956	1.2226×10^{9}	–526.03	2815.86	2289.83	–2.11427	15.60175	**6.0×10^{-5}**
5.0×10^{-5}	5.0×10^{-10}	–114.96	158.19	930.016	1.4602×10^{9}	–527.00	2815.44	2288.44	–2.12038	15.67713	**5.0×10^{-5}**
4.0×10^{-5}	4.0×10^{-10}	–115.87	157.28	930.085	1.8147×10^{9}	–528.17	2814.92	2286.75	–2.12778	15.76944	**4.0×10^{-5}**
3.0×10^{-5}	3.0×10^{-10}	–117.03	156.12	930.172	2.4018×10^{9}	–529.65	2814.26	2284.61	–2.13721	15.88854	**3.0×10^{-5}**
2.0×10^{-5}	2.0×10^{-10}	–118.63	154.52	930.293	3.5657×10^{9}	–531.68	2813.31	2281.63	–2.15030	16.05656	**2.0×10^{-5}**
1.0×10^{-5}	1.0×10^{-10}	–121.30	151.85	930.484	7.0082×10^{9}	–535.02	2811.73	2276.71	–2.17212	16.34423	**1.0×10^{-5}**
9.0×10^{-6}	9.0×10^{-11}	–121.70	151.45	930.510	7.7665×10^{9}	–535.52	2811.49	2275.97	–2.17538	16.38800	**9.0×10^{-6}**
8.0×10^{-6}	8.0×10^{-11}	–122.14	151.01	930.544	8.7118×10^{9}	–536.06	2811.22	2275.16	–2.17900	16.43695	**8.0×10^{-6}**
7.0×10^{-6}	7.0×10^{-11}	–122.64	150.51	930.579	9.9235×10^{9}	–536.68	2810.91	2274.23	–2.18309	16.49246	**7.0×10^{-6}**
6.0×10^{-6}	6.0×10^{-11}	–123.21	149.94	930.622	1.1533×10^{10}	–537.38	2810.56	2273.18	–2.18777	16.55658	**6.0×10^{-6}**
5.0×10^{-6}	5.0×10^{-11}	–123.88	149.27	930.665	1.3778×10^{10}	–538.21	2810.14	2271.93	–2.19327	16.63244	**5.0×10^{-6}**
4.0×10^{-6}	4.0×10^{-11}	–124.69	148.46	930.717	1.7129×10^{10}	–539.20	2809.63	2270.43	–2.19995	16.72533	**4.0×10^{-6}**
3.0×10^{-6}	3.0×10^{-11}	–125.72	147.43	930.795	2.2680×10^{10}	–540.46	2808.98	2268.52	–2.20846	16.84517	**3.0×10^{-6}**
2.0×10^{-6}	2.0×10^{-11}	–127.16	145.99	930.891	3.3689×10^{10}	–542.19	2808.05	2265.86	–2.22028	17.01422	**2.0×10^{-6}**
1.0×10^{-6}	1.0×10^{-11}	–129.55	143.60	931.047	6.6275×10^{10}	–545.05	2806.50	2261.45	–2.24002	17.30360	**1.0×10^{-6}**
9.0×10^{-7}	9.0×10^{-12}	–129.90	143.25	931.064	7.3456×10^{10}	–545.48	2806.27	2260.79	–2.24297	17.34763	**9.0×10^{-7}**
8.0×10^{-7}	8.0×10^{-12}	–130.30	142.85	931.090	8.2409×10^{10}	–545.94	2806.00	2260.06	–2.24625	17.39686	**8.0×10^{-7}**
7.0×10^{-7}	7.0×10^{-12}	–130.75	142.40	931.116	9.3888×10^{10}	–546.47	2805.70	2259.23	–2.24995	17.45270	**7.0×10^{-7}**
6.0×10^{-7}	6.0×10^{-12}	–131.26	141.89	931.151	1.0914×10^{11}	–547.08	2805.36	2258.28	–2.25420	17.51717	**6.0×10^{-7}**
5.0×10^{-7}	5.0×10^{-12}	–131.86	141.29	931.194	1.3041×10^{11}	–547.78	2804.95	2257.17	–2.25919	17.59346	**5.0×10^{-7}**
4.0×10^{-7}	4.0×10^{-12}	–132.59	140.56	931.237	1.6218×10^{11}	–548.64	2804.46	2255.82	–2.26524	17.68688	**4.0×10^{-7}**
3.0×10^{-7}	3.0×10^{-12}	–133.52	139.63	931.289	2.1481×10^{11}	–549.72	2803.82	2254.10	–2.27297	17.80737	**3.0×10^{-7}**
2.0×10^{-7}	2.0×10^{-12}	–134.81	138.34	931.368	3.1924×10^{11}	–551.21	2802.92	2251.71	–2.28370	17.97735	**2.0×10^{-7}**
1.0×10^{-7}	1.0×10^{-12}	–136.96	136.19	931.498	6.2854×10^{11}	–553.67	2801.42	2247.75	–2.30165	18.26827	**1.0×10^{-7}**

(Contd.)

TABLE: 1.2 Thermodynamic Properties of Ice and Water Vapour on Sublimation Curve (Pressure Basis) [Contd.]

Pressure, p		Saturation temperature		Solid density	Specific volume of vapour, v''	Specific enthalpy			Specific entropy		Press.
		t	T	ρ_s		h_s	λ_s	H	s_s	s''	p
Pa	bar	°C	K	kg/m³	m³/kg	kJ/kg			kJ/(kg·K)		Pa
9.0 × 10⁻⁸	9.0 × 10⁻¹³	–137.28	135.87	931.515	6.9673×10^{11}	–554.04	2801.19	2247.15	–2.30433	18.31253	**9.0 × 10⁻⁸**
8.0 × 10⁻⁸	8.0 × 10⁻¹³	–137.64	135.51	931.532	7.8176×10^{11}	–554.44	2800.93	2246.49	–2.30732	18.36202	**8.0 × 10⁻⁸**
7.0 × 10⁻⁸	7.0 × 10⁻¹³	–138.04	135.11	931.558	8.9079×10^{11}	–554.90	2800.64	2245.74	–2.31069	18.41814	**7.0 × 10⁻⁸**
6.0 × 10⁻⁸	6.0 × 10⁻¹³	–138.50	134.65	931.584	1.0357×10^{12}	–555.42	2800.31	2244.89	–2.31456	18.48294	**6.0 × 10⁻⁸**
5.0 × 10⁻⁸	5.0 × 10⁻¹³	–139.05	134.10	931.619	1.2378×10^{12}	–556.03	2799.92	2243.89	–2.31911	18.55962	**5.0 × 10⁻⁸**
4.0 × 10⁻⁸	4.0 × 10⁻¹³	–139.70	133.45	931.654	1.5397×10^{12}	–556.77	2799.44	2242.67	–2.32463	18.65350	**4.0 × 10⁻⁸**
3.0 × 10⁻⁸	3.0 × 10⁻¹³	–140.54	132.61	931.697	2.0400×10^{12}	–557.71	2798.82	2241.11	–2.33167	18.77461	**3.0 × 10⁻⁸**
2.0 × 10⁻⁸	2.0 × 10⁻¹³	–141.71	131.44	931.758	3.0331×10^{12}	–559.00	2797.95	2238.95	–2.34147	18.94541	**2.0 × 10⁻⁸**
1.0 × 10⁻⁸	1.0 × 10⁻¹³	–143.66	129.49	931.862	5.9765×10^{12}	–561.14	2796.47	2235.33	–2.35787	19.23772	**1.0 × 10⁻⁸**

TABLE: 1.3 Thermodynamic Properties of Ice and Water on Melting Curve

Melting Temperature		Saturation pressure	Solid density	Liquid density,	Specific enthalpy			Specific entropy		Temp.
t	T	p	ρ_s	ρ_l	h_s	λ_f	h	s_s	s'	t
°C	K	bar	kg/m³	kg/m³	kJ/kg			kJ/(kgK)		°C
0.01*	**273.16**	**0.0061165**	**916.7079**	**999.78**	**–333.39**	**333.39**	**0.00**	**–1.22050**	**0.00000**	**0.01**
0	273.15	1.352	916.725	999.86	–333.27	333.37	0.10	–1.22060	–0.00015	**0**
–1	272.15	132.277	918.282	1006.40	–321.71	330.85	9.14	–1.23053	–0.01484	**–1**
–2	271.15	256.727**	919.752	1012.48	–310.82	328.23	17.41	–1.24029	–0.02979	**–2**
–3	270.15	375.661	921.158	1018.23	–300.51	325.52	25.01	–1.24992	–0.04496	**–3**
–4	269.15	489.809	922.501	1023.65	–290.69	322.74	32.05	–1.25943	–0.06031	**–4**
–5	268.15	599.746	923.788	1028.79	–281.30	319.92	38.62	–1.26884	–0.07579	**–5**
–6	267.15	705.932	925.027	1033.68	–272.31	317.04	44.73	–1.27815	–0.09139	**–6**
–7	266.15	808.745	926.218	1038.35	–263.66	314.14	50.48	–1.28737	–0.10706	**–7**
–8	265.15	908.498	927.377	1042.83	–255.33	311.21	55.88	–1.29651	–0.12279	**–8**
–9	264.15	1005.458	928.496	1047.13	–247.28	308.27	60.99	–1.30559	–0.13857	**–9**
–10	263.15	1099.851	929.575	1051.26	–239.49	305.32	65.83	–1.31459	–0.15435	**–10**
–11	262.15	1191.875	930.631	1055.24	–231.95	302.36	70.41	–1.32354	–0.17015	**–11**
–12	261.15	1281.701	931.654	1059.08	–224.63	299.41	74.78	–1.33243	–0.18591	**–12**
–13	260.15	1369.482	932.653	1062.79	–217.51	296.48	78.97	–1.34126	–0.20162	**–13**
–14	259.15	1455.355	933.628	1066.37	–210.59	293.56	82.97	–1.35005	–0.21728	**–14**
–15	258.15	1539.441	934.571	1069.83	–203.85	290.66	86.81	–1.35879	–0.23284	**–15**
–16	257.15	1621.851	935.497	1073.18	–197.28	287.80	90.52	–1.36748	–0.24829	**–16**
–17	256.15	1702.689	936.400	1076.43	–190.86	284.97	94.11	–1.37613	–0.2636	**–17**
–18	255.15	1782.047	937.286	1079.58	–184.59	282.19	97.60	–1.38475	–0.27876	**–18**
–19	254.15	1860.014	938.148	1082.63	–178.46	279.46	101.00	–1.39332	–0.29373	**–19**
–20	253.15	1936.672	938.994	1085.59	–172.46	276.79	104.33	–1.40186	–0.30848	**–20**
–21	252.15	2012.099	939.823	1088.46	–166.58	274.18	107.60	–1.41037	–0.32300	**–21**
–22	251.15	2086.369	940.636	1091.25	–160.81	271.64	110.83	–1.41885	–0.33725	**–22**
–23	250.15	2159.554	941.433	1093.95	–155.15	269.19	114.04	–1.42729	–0.35118	**–23**
–24	249.15	2231.723	942.223	1096.59	–149.59	266.82	117.23	–1.43571	–0.36479	**–24**
–25	248.15	2302.943	942.987	1099.12	–144.12	264.55	120.43	–1.44410	–0.37801	**–25**

*Triple point

**Data above 221.2 bar refer to the special fluid state

Reference State:

Specific enthalpy of saturated liquid (water) at 0.01°C (273.16 K) = 0.0 kJ/kg

Specific enthalpy of saturated liquid (water) at 0.01°C (273.16 K) = 0.00000 kJ/(kg·K)

Note:

Data, contained in Tables 1.1, 1.2 and 1.3, are expressed on the ITS–90 temperature scale and are consistent with the Standard IAPWS–95.

(Data obtained by Private Communication with Dr. Rainer Feistel, Sea Research Institute, Warnemünde, Germany and are gratefully acknowledged.)

TABLE: 1.4 Thermodynamic Properties of Steam up to 1 Atmosphere (Temperature Basis)

Temperature		Saturation pressure, p		Liquid density	Specific volume of vapour, v''	Specific enthalpy			Specific entropy		Temp.
t	T			ρ_l		h	λ_v	H	s'	s''	t
°C	K	bar	torr	kg/m³	m³/kg	kJ/kg			kJ/(kg·K)		°C
0.01*	**273.16**	**0.006112**	**4.584**	**999.780**	**206.163**	**0.0**	**2501.6**	**2501.6**	**0.00000**	**9.15746**	**0.01**
1	274.15	0.006566	4.925	999.850	192.607	4.2	2499.2	2503.4	0.01525	9.13107	**1**
2	275.15	0.007055	5.292	999.910	179.923	8.4	2496.8	2505.2	0.03059	9.10467	**2**
3	276.15	0.007575	5.682	999.950	168.169	12.6	2494.5	2507.1	0.04586	9.07850	**3**
4	277.15	0.008129	6.097	999.970	157.272	16.8	2492.1	2508.9	0.06106	9.05258	**4**
5	278.15	0.008718	6.539	999.970	147.163	21.0	2489.7	2510.7	0.07621	9.0269	**5**
6	279.15	0.009345	7.009	999.960	137.779	25.2	2487.4	2512.6	0.09128	9.00145	**6**
7	280.15	0.010012	7.510	999.930	129.064	29.4	2485.0	2514.4	0.10630	8.97624	**7**
8	281.15	0.010720	8.041	999.880	120.966	33.6	2482.6	2516.2	0.12126	8.95125	**8**
9	282.15	0.011472	8.605	999.820	113.435	37.8	2480.3	2518.1	0.13615	8.92649	**9**
10	283.15	0.012270	9.203	999.750	106.430	42.0	2477.9	2519.9	0.15099	8.90196	**10**
11	284.15	0.013116	9.838	999.660	99.9087	46.2	2475.5	2521.7	0.16577	8.87765	**11**
12	285.15	0.014014	10.51	999.560	93.8354	50.4	2473.2	2523.6	0.18049	8.85355	**12**
13	286.15	0.014965	11.22	999.440	88.1761	54.6	2470.8	2525.4	0.19515	8.82968	**13**
14	287.15	0.015973	11.98	999.310	82.8997	58.8	2468.5	2527.3	0.20976	8.80602	**14**
15	288.15	0.017039	12.78	999.171	77.9779	62.9	2466.1	2529.0	0.22432	8.78257	**15**
16	289.15	0.018168	13.63	999.011	73.3842	67.1	2463.8	2530.9	0.23882	8.75933	**16**
17	290.15	0.019362	14.52	998.851	69.0948	71.3	2461.4	2532.7	0.25327	8.73629	**17**
18	291.15	0.020624	15.47	998.672	65.0873	75.5	2459.0	2534.5	0.26766	8.71346	**18**
19	292.15	0.021957	16.47	998.482	61.3414	79.7	2456.7	2536.4	0.28201	8.69083	**19**
20	293.15	0.023366	17.53	998.283	57.8383	83.9	2454.3	2538.2	0.29630	8.66840	**20**
21	294.15	0.024853	18.64	998.064	54.5606	88.0	2452.0	2540.0	0.31054	8.64617	**21**
22	295.15	0.026422	19.82	997.845	51.4923	92.2	2449.6	2541.8	0.32473	8.62413	**22**
23	296.15	0.028076	21.06	997.616	48.6186	96.4	2447.2	2543.6	0.33887	8.60228	**23**
24	297.15	0.029821	22.37	997.367	45.9260	100.6	2444.9	2545.5	0.35296	8.58062	**24**
25	298.15	0.031660	23.75	997.118	43.4017	104.8	2442.5	2547.3	0.36701	8.55916	**25**
26	299.15	0.033597	25.20	996.850	41.0343	108.9	2440.2	2549.1	0.38100	8.53787	**26**
27	300.15	0.035636	26.73	996.582	38.8129	113.1	2437.8	2550.9	0.39495	8.51678	**27**
28	301.15	0.037782	28.34	996.304	36.7276	117.3	2435.4	2552.7	0.40885	8.49586	**28**
29	302.15	0.040040	30.03	996.006	34.7691	121.5	2433.1	2554.6	0.42270	8.47512	**29**
30	303.15	0.042415	31.81	995.708	32.9289	125.7	2430.7	2556.4	0.43651	8.45456	**30**
31	304.15	0.044911	33.69	995.401	31.1991	129.8	2428.3	2558.1	0.45027	8.43417	**31**
32	305.15	0.047534	35.65	995.084	29.5724	134.0	2425.9	2559.9	0.46399	8.41396	**32**
33	306.15	0.050288	37.72	994.758	28.0419	138.2	2423.6	2561.8	0.47766	8.39392	**33**
34	307.15	0.053180	39.89	994.421	26.6013	142.4	2421.2	2563.6	0.49128	8.37405	**34**
35	308.15	0.056216	42.17	994.085	25.2449	146.6	2418.8	2565.4	0.50486	8.35434	**35**
36	309.15	0.059400	44.55	993.730	23.9671	150.7	2416.4	2567.1	0.51840	8.33480	**36**
37	310.15	0.062739	47.06	993.374	22.7628	154.9	2414.1	2569.0	0.53190	8.31542	**37**
38	311.15	0.066240	49.68	993.009	21.6275	159.1	2411.7	2570.8	0.54535	8.29621	**38**
39	312.15	0.069908	52.44	992.635	20.5566	163.3	2409.3	2572.6	0.55876	8.27715	**39**
40	313.15	0.073750	55.32	992.251	19.5461	167.5	2406.9	2574.4	0.57212	8.25826	**40**
41	314.15	0.077773	58.33	991.857	18.5923	171.6	2404.5	2576.1	0.58545	8.23952	**41**
42	315.15	0.081985	61.49	991.463	17.6915	175.8	2402.1	2577.9	0.59873	8.22093	**42**
43	316.15	0.086391	64.80	991.061	16.8407	180.0	2399.7	2579.7	0.61197	8.20249	**43**
44	317.15	0.091001	68.26	990.648	16.0635	184.2	2397.3	2581.5	0.62517	8.18421	**44**
45	318.15	0.095820	71.87	990.226	15.2762	188.4	2394.9	2583.3	0.63832	8.16607	**45**

*Triple point

(Contd.)

TABLE: 1.4 Thermodynamic Properties of Steam up to 1 Atmosphere (Temperature Basis) [Contd.]

Temperature		Saturation pressure, p		Liquid density	Specific volume of vapour, v''	Specific enthalpy			Specific entropy		Temp.
t	T			ρ_l		h	λ_v	H	s'	s''	t
°C	K	bar	torr	kg/m³	m³/kg	kJ/kg			kJ/(kg·K)		°C
46	319.15	0.10086	75.65	989.805	14.5572	192.5	2392.5	2585.0	0.65144	8.14809	**46**
47	320.15	0.10612	79.60	989.374	13.8768	196.7	2390.1	2586.8	0.66452	8.13024	**47**
48	321.15	0.11162	83.72	988.934	13.2329	200.9	2387.7	2588.6	0.67756	8.11255	**48**
49	322.15	0.11736	88.03	988.484	12.6232	205.1	2385.3	2590.4	0.69055	8.09499	**49**
50	323.15	0.12335	92.52	988.035	12.0457	209.3	2382.9	2592.2	0.70351	8.07757	**50**
51	324.15	0.12961	97.22	987.567	11.4985	213.4	2380.5	2593.9	0.71643	8.06030	**51**
52	325.15	0.13613	102.11	987.108	10.9798	217.6	2378.1	2595.7	0.72931	8.04316	**52**
53	326.15	0.14293	107.21	986.631	10.4880	221.8	2375.7	2597.5	0.74215	8.02615	**53**
54	327.15	0.15002	112.52	986.154	10.0215	226.0	2373.2	2599.2	0.75496	8.00929	**54**
55	328.15	0.15741	118.07	985.668	9.57887	230.2	2370.8	2601.0	0.76772	7.99255	**55**
56	329.15	0.16511	123.84	985.173	9.15871	234.4	2368.4	2602.8	0.78045	7.97595	**56**
57	330.15	0.17313	129.86	984.678	8.75976	238.5	2365.9	2604.4	0.79314	7.95947	**57**
58	331.15	0.18147	136.11	984.174	8.38082	242.7	2363.5	2606.2	0.80579	7.94312	**58**
59	332.15	0.19016	142.63	983.661	8.02076	246.9	2361.1	2608.0	0.81841	7.92690	**59**
60	333.15	0.19920	149.41	983.149	7.67853	251.1	2358.6	2609.7	0.83099	7.91081	**60**
61	334.15	0.20861	156.47	982.627	7.35315	255.3	2356.2	2611.5	0.84353	7.89484	**61**
62	335.15	0.21838	163.80	982.096	7.04368	259.5	2353.7	2613.2	0.85604	7.87899	**62**
63	336.15	0.22855	171.43	981.557	6.74925	263.7	2351.3	2615.0	0.86851	7.86327	**63**
64	337.15	0.23912	179.35	981.017	6.46904	267.8	2348.8	2616.6	0.88094	7.84766	**64**
65	338.15	0.25009	187.58	980.479	6.20228	272.0	2346.3	2618.3	0.89334	7.83217	**65**
66	339.15	0.26150	196.14	979.921	5.94824	276.2	2343.9	2620.1	0.90571	7.81680	**66**
67	340.15	0.27334	205.02	979.365	5.70624	280.4	2341.4	2621.8	0.91804	7.80155	**67**
68	341.15	0.28563	214.24	978.799	5.47564	284.6	2338.9	2623.5	0.93033	7.78641	**68**
69	342.15	0.29838	223.80	978.234	5.25584	288.8	2336.4	2625.2	0.94259	7.77138	**69**
70	343.15	0.31162	233.73	977.660	5.04627	293.0	2334.0	2627.0	0.95482	7.75647	**70**
71	344.15	0.32535	244.03	977.078	4.84638	297.2	2331.5	2628.6	0.96701	7.74166	**71**
72	345.15	0.33958	254.71	976.496	4.65568	301.4	2329.0	2630.4	0.97917	7.72697	**72**
73	346.15	0.35434	265.78	975.905	4.47370	305.5	2326.5	2632.0	0.99129	7.71239	**73**
74	347.15	0.36964	277.25	975.315	4.29998	309.7	2324.0	2633.7	1.00338	7.69791	**74**
75	348.15	0.38549	289.14	974.716	4.13410	313.9	2321.5	2635.4	1.01544	7.68353	**75**
76	349.15	0.40191	301.46	974.108	3.97566	318.1	2318.9	2637.0	1.02747	7.66926	**76**
77	350.15	0.41891	314.21	973.501	3.82428	322.3	2316.4	2638.7	1.03946	7.65510	**77**
78	351.15	0.43652	327.42	972.886	3.67962	326.5	2313.9	2640.4	1.05142	7.64104	**78**
79	352.15	0.45474	341.08	972.261	3.54132	330.7	2311.4	2642.1	1.06335	7.62708	**79**
80	353.15	0.47360	355.23	971.638	3.40909	334.9	2308.8	2643.7	1.07525	7.61322	**80**
81	354.15	0.49311	369.86	971.006	3.28261	339.1	2306.3	2645.4	1.08712	7.59945	**81**
82	355.15	0.51329	385.00	970.374	3.16160	343.3	2303.8	2647.1	1.09895	7.58579	**82**
83	356.15	0.53416	400.65	969.725	3.04580	347.5	2301.2	2648.7	1.11076	7.57222	**83**
84	357.15	0.55573	416.83	969.086	2.93495	351.7	2298.6	2650.3	1.12253	7.55875	**84**
85	358.15	0.57803	433.56	968.439	2.82881	355.9	2296.1	2652.0	1.13427	7.54537	**85**
86	359.15	0.60108	450.85	967.783	2.72716	360.1	2293.5	2653.6	1.14598	7.53209	**86**
87	360.15	0.62489	468.71	967.118	2.62979	364.3	2290.9	2655.2	1.15766	7.51890	**87**
88	361.15	0.64948	487.15	966.454	2.53647	368.5	2288.4	2656.9	1.16932	7.50580	**88**
89	362.15	0.67487	506.19	965.782	2.44704	372.7	2285.8	2658.5	1.18094	7.49279	**89**
90	363.15	0.70109	525.86	965.111	2.36130	376.9	2283.2	2660.1	1.19253	7.47987	**90**

(Contd.)

TABLE: 1.4 Thermodynamic Properties of Steam up to 1 Atmosphere (Temperature Basis) [Contd.]

Temperature		Saturation pressure, p		Liquid density	Specific volume of vapour, v''	Specific enthalpy			Specific entropy		Temp.
t	T			ρ_l		h	λ_v	H	s'	s''	t
°C	K	bar	torr	kg/m^3	m^3/kg	kJ/kg			kJ/(kg·K)		°C
91	364.15	0.72815	546.16	964.432	2.27907	381.1	2280.6	2661.7	1.20409	7.46704	**91**
92	365.15	0.75607	567.10	963.753	2.20021	385.4	2278.0	2663.4	1.21562	7.45430	**92**
93	366.15	0.78489	588.72	963.066	2.12454	389.6	2275.4	2665.0	1.22713	7.44164	**93**
94	367.15	0.81461	611.01	962.371	2.05192	393.8	2272.8	2666.6	1.23861	7.42907	**94**
95	368.15	0.84526	634.00	961.677	1.98222	398.0	2270.2	2668.2	1.25005	7.41658	**95**
96	369.15	0.87686	657.70	960.975	1.91530	402.2	2267.5	2669.7	1.26147	7.40418	**96**
97	370.15	0.90944	682.14	960.264	1.85104	406.4	2264.9	2671.3	1.27286	7.39185	**97**
98	371.15	0.94301	707.32	959.555	1.78931	410.6	2262.2	2672.9	1.28423	7.37962	**98**
99	372.15	0.97761	733.27	958.846	1.72999	414.8	2259.6	2674.4	1.29556	7.36746	**99**
100*	**373.15**	**1.01325**	**760.00**	**958.121**	**1.67300**	**419.1**	**2256.9**	**2676.0**	**1.30687**	**7.35538**	**100**

*Normal boiling point

Reference State:

Specific enthalpy of saturated water at 0.01°C (273.16 K) = 0.0 kJ/kg

Specific entropy of saturated water at 0.01°C (273.16 K) = 0.00000 kJ/(kg·K)

TABLE: 1.5 Thermodynamic Properties of Steam above 1 Atmosphere (Temperature Basis)

Temperature		Saturation pressure,	Liquid density	Specific volume of vapour, v''	Specific enthalpy			Specific entropy		Temp.
t	T	p	ρ_l		h	λ_v	H	s'	s''	t
°C	K	bar	kg/m³	m³/kg	kJ/kg			kJ/(kg·K)		°C
101	374.15	1.0500	957.396	1.61821	423.3	2254.3	2677.6	1.31815	7.34338	**101**
102	375.15	1.0878	956.672	1.56553	427.5	2251.6	2679.1	1.32940	7.33146	**102**
103	376.15	1.1267	955.941	1.51488	431.7	2248.9	2680.6	1.34063	7.31961	**103**
104	377.15	1.1668	955.201	1.46615	435.9	2246.3	2682.2	1.35183	7.30785	**104**
105	378.15	1.2080	954.463	1.41928	440.2	2243.6	2683.8	1.36301	7.29615	**105**
106	379.15	1.2504	953.716	1.37417	444.4	2240.9	2685.3	1.37416	7.28454	**106**
107	380.15	1.2941	952.962	1.33075	448.6	2238.2	2686.8	1.38528	7.27299	**107**
108	381.15	1.3390	952.209	1.28895	452.9	2235.4	2688.3	1.39637	7.26152	**108**
109	382.15	1.3852	951.457	1.24870	457.1	2232.7	2689.8	1.40744	7.25012	**109**
110	383.15	1.4327	950.688	1.20994	461.3	2230.0	2691.3	1.41849	7.23880	**110**
111	384.15	1.4815	949.920	1.17259	465.5	2227.3	2692.8	1.42951	7.22754	**111**
112	385.15	1.5316	949.154	1.13661	469.8	2224.5	2694.3	1.44051	7.21636	**112**
113	386.15	1.5832	948.380	1.10194	474.0	2221.8	2695.8	1.45148	7.20524	**113**
114	387.15	1.6362	947.598	1.06852	478.3	2219.0	2697.3	1.46242	7.19419	**114**
115	388.15	1.6906	946.817	1.03629	482.5	2216.2	2698.7	1.47334	7.18321	**115**
116	389.15	1.7465	946.029	1.00522	486.7	2213.4	2700.1	1.48424	7.17229	**116**
117	390.15	1.8039	945.233	0.975251	491.0	2210.7	2701.7	1.49511	7.16144	**117**
118	391.15	1.8628	944.439	0.946340	495.2	2207.9	2703.1	1.50596	7.15066	**118**
119	392.15	1.9233	943.637	0.918445	499.5	2205.1	2704.6	1.51679	7.13994	**119**
120	393.15	1.9854	942.836	0.891524	503.7	2202.2	2705.9	1.52759	7.12928	**120**
121	394.15	2.0492	942.028	0.865539	508.0	2199.4	2707.4	1.53837	7.11869	**121**
122	395.15	2.1145	941.212	0.840452	512.2	2196.6	2708.8	1.54913	7.10816	**122**
123	396.15	2.1816	940.398	0.816228	516.5	2193.7	2710.2	1.55986	7.09769	**123**
124	397.15	2.2504	939.576	0.792833	520.7	2190.9	2711.6	1.57057	7.08728	**124**
125	398.15	2.3210	938.747	0.770233	525.0	2188.0	2713.0	1.58126	7.07693	**125**
126	399.15	2.3933	937.919	0.748399	529.2	2185.2	2714.4	1.59192	7.06664	**126**
127	400.15	2.4675	937.084	0.727300	533.5	2182.3	2715.8	1.60257	7.05641	**127**
128	401.15	2.5435	936.251	0.706908	537.8	2179.4	2717.2	1.61319	7.04624	**128**
129	402.15	2.6215	935.410	0.687195	542.0	2176.5	2718.5	1.62379	7.03612	**129**
130	403.15	2.7013	934.562	0.668136	546.3	2173.6	2719.9	1.63436	7.02606	**130**
131	404.15	2.7831	933.707	0.649706	550.6	2170.7	2721.3	1.64492	7.01606	**131**
132	405.15	2.8670	932.853	0.631880	554.8	2167.8	2722.6	1.65546	7.00611	**132**
133	406.15	2.9528	932.001	0.614636	559.1	2164.8	2723.9	1.66597	6.99622	**133**
134	407.15	3.0407	931.133	0.597952	563.4	2161.9	2725.3	1.67646	6.98638	**134**
135	408.15	3.1307	930.267	0.581807	567.7	2158.9	2726.6	1.68693	6.97659	**135**
136	409.15	3.2229	929.403	0.566181	572.0	2155.9	2727.9	1.69739	6.96686	**136**
137	410.15	3.3173	928.522	0.551054	576.2	2153.0	2729.2	1.70782	6.95718	**137**
138	411.15	3.4138	927.644	0.536409	580.5	2150.0	2730.5	1.71823	6.94755	**138**
139	412.15	3.5127	926.758	0.522228	584.8	2147.0	2731.8	1.72862	6.93797	**139**
140	413.15	3.6138	925.874	0.508493	589.1	2144.0	2733.1	1.73899	6.92844	**140**
141	414.15	3.7172	924.984	0.495189	593.4	2140.9	2734.3	1.74934	6.91896	**141**
142	415.15	3.8231	924.086	0.482300	597.7	2137.9	2735.6	1.75967	6.90953	**142**
143	416.15	3.9313	923.191	0.469811	602.0	2134.9	2736.9	1.76998	6.90015	**143**
144	417.15	4.0420	922.280	0.457707	606.3	2131.8	2738.1	1.78028	6.89081	**144**
145	418.15	4.1552	921.379	0.445975	610.6	2128.7	2739.3	1.79055	6.88153	**145**

(Contd.)

TABLE: 1.5 Thermodynamic Properties of Steam above 1 Atmosphere (Temperature Basis) [Contd.]

Temperature		Saturation pressure,	Liquid density	Specific volume of vapour, v''	Specific enthalpy			Specific entropy		Temp.
t	T	p	ρ_l		h	λ_v	H	s'	s''	t
°C	K	bar	kg/m³	m³/kg	kJ/kg			kJ/(kg·K)		°C
146	419.15	4.2709	920.463	0.434601	614.9	2125.7	2740.6	1.80081	6.87228	**146**
147	420.15	4.3892	919.549	0.423573	619.2	2122.6	2741.8	1.81104	6.86309	**147**
148	421.15	4.5101	918.628	0.412880	623.5	2119.5	2743.0	1.82126	6.85394	**148**
149	422.15	4.6337	917.701	0.402508	627.8	2116.3	2744.1	1.83146	6.84484	**149**
150	423.15	4.7600	916.775	0.392447	632.1	2113.2	2745.3	1.84164	6.83578	**150**
151	424.15	4.8890	915.843	0.382686	636.5	2110.1	2746.6	1.85180	6.82676	**151**
152	425.15	5.0208	914.905	0.373215	640.8	2106.9	2747.7	1.86194	6.81779	**152**
153	426.15	5.1554	913.960	0.364024	645.1	2103.8	2748.9	1.87207	6.80886	**153**
154	427.15	5.2929	913.017	0.355103	649.4	2100.6	2750.0	1.88218	6.79997	**154**
155	428.15	5.4333	912.068	0.346442	653.8	2097.4	2751.2	1.89227	6.79112	**155**
156	429.15	5.5767	911.112	0.338034	658.1	2094.2	2752.3	1.90234	6.78231	**156**
157	430.15	5.7230	910.150	0.329869	662.4	2091.0	2753.4	1.91240	6.77355	**157**
158	431.15	5.8725	909.190	0.321939	666.8	2087.7	2754.5	1.92244	6.76482	**158**
159	432.15	6.0250	908.224	0.314238	671.1	2084.5	2755.6	1.93246	6.75613	**159**
160	433.15	6.1806	907.252	0.306756	675.5	2081.3	2756.8	1.94247	6.74749	**160**
161	434.15	6.3395	906.273	0.299486	679.8	2078.0	2757.8	1.95246	6.73808	**161**
162	435.15	6.5016	905.297	0.292423	684.2	2074.7	2758.9	1.96243	6.73031	**162**
163	436.15	6.6669	904.314	0.285558	688.5	2071.4	2759.9	1.97239	6.72177	**163**
164	437.15	6.8356	903.326	0.278885	692.9	2068.1	2761.0	1.98233	6.71327	**164**
165	438.15	7.0077	902.332	0.272399	697.3	2064.8	2762.1	1.99225	6.70481	**165**
166	439.15	7.1831	901.380	0.266092	701.6	2061.4	2763.0	2.00216	6.69638	**166**
167	440.15	7.3621	900.333	0.259960	706.0	2058.1	2764.1	2.01205	6.68799	**167**
168	441.15	7.5445	899.321	0.253996	710.4	2054.7	2765.1	2.02193	6.67964	**168**
169	442.15	7.7306	898.311	0.248196	714.7	2051.3	2766.0	2.03179	6.67131	**169**
170	443.15	7.9202	897.296	0.242553	719.1	2047.9	2767.1	2.04164	6.66303	**170**
171	444.15	8.1135	896.282	0.237064	723.5	2044.5	2768.0	2.05147	6.65477	**171**
172	445.15	8.3106	895.255	0.231723	727.9	2041.1	2769.0	2.06128	6.64655	**172**
173	446.15	8.5114	894.230	0.226525	732.3	2037.7	2770.0	2.07108	6.63836	**173**
174	447.15	8.7160	893.192	0.221466	736.7	2034.2	2770.9	2.08087	6.63020	**174**
175	448.15	8.9244	892.156	0.216542	741.1	2030.7	2771.8	2.09064	6.62207	**175**
176	449.15	9.1368	891.115	0.211748	745.5	2027.3	2772.8	2.10040	6.61398	**176**
177	450.15	9.3532	890.068	0.207081	749.9	2023.8	2773.7	2.11014	6.60591	**177**
178	451.15	9.5736	889.015	0.202537	754.3	2020.2	2774.5	2.11987	6.59787	**178**
179	452.15	9.7980	887.958	0.198111	758.7	2016.7	2775.4	2.12959	6.58987	**179**
180	453.15	10.027	886.902	0.193800	763.1	2013.1	2776.2	2.13929	6.58189	**180**
181	454.15	10.259	885.834	0.189602	767.5	2009.6	2777.1	2.14898	6.57394	**181**
182	455.15	10.496	884.768	0.185511	772.0	2006.0	2778.0	2.15865	6.56602	**182**
183	456.15	10.738	883.689	0.181525	776.4	2002.4	2778.8	2.16831	6.55813	**183**
184	457.15	10.983	882.613	0.177642	780.8	1998.8	2779.6	2.17796	6.55026	**184**
185	458.15	11.233	881.531	0.173857	785.3	1995.2	2780.5	2.18760	6.54242	**185**
186	459.15	11.488	880.437	0.170168	789.7	1991.5	2781.2	2.19722	6.53461	**186**
187	460.15	11.747	879.345	0.166572	794.2	1987.8	2782.0	2.20683	6.52682	**187**
188	461.15	12.010	878.248	0.163066	798.6	1984.2	2782.8	2.21643	6.51906	**188**
189	462.15	12.278	877.147	0.159648	803.1	1980.5	2783.6	2.22601	6.51133	**189**
190	463.15	12.551	876.033	0.156316	807.5	1976.7	2784.2	2.23558	6.50361	**190**

(Contd.)

TABLE: 1.5 Thermodynamic Properties of Steam above 1 Atmosphere (Temperature Basis) [Contd.]

Temperature		Saturation pressure,	Liquid density	Specific volume of	Specific enthalpy			Specific entropy		Temp.
t	T	p	ρ_l	vapour, v''	h	λ_v	H	s'	s''	t
°C	K	bar	kg/m³	m³/kg	kJ/kg			kJ/(kg·K)		°C
191	464.15	12.829	874.921	0.153065	812.0	1973.0	2785.0	2.24514	6.49593	**191**
192	465.15	13.111	873.805	0.149895	816.5	1969.3	2785.8	2.25469	6.48826	**192**
193	466.15	13.398	872.684	0.146803	820.9	1965.5	2786.4	2.26423	6.48062	**193**
194	467.15	13.690	871.558	0.143787	825.4	1961.7	2787.1	2.27375	6.47301	**194**
195	468.15	13.987	870.421	0.140844	829.9	1957.9	2787.8	2.28326	6.46541	**195**
196	469.15	14.289	869.286	0.137972	834.4	1954.1	2788.5	2.29276	6.45784	**196**
197	470.15	14.596	868.138	0.135170	838.9	1950.2	2789.1	2.30225	6.45029	**197**
198	471.15	14.909	866.994	0.132435	843.4	1946.4	2789.8	2.31173	6.44276	**198**
199	472.15	15.226	865.838	0.129766	847.9	1942.5	2790.4	2.32120	6.43525	**199**
200	473.15	15.549	864.678	0.127160	852.4	1938.6	2791.0	2.33066	6.42776	**200**
201	474.15	15.877	863.521	0.124616	856.9	1934.6	2791.5	2.34011	6.42030	**201**
202	475.15	16.210	862.351	0.122133	861.4	1930.7	2792.1	2.34954	6.41285	**202**
203	476.15	16.549	861.171	0.119707	865.9	1926.7	2792.6	2.35897	6.40542	**203**
204	477.15	16.893	859.993	0.117339	870.5	1922.8	2793.3	2.36838	6.39801	**204**
205	478.15	17.243	858.811	0.115026	875.0	1918.8	2793.8	2.37779	6.39062	**205**
206	479.15	17.598	857.618	0.112767	879.5	1914.7	2794.2	2.38719	6.38325	**206**
207	480.15	17.959	856.421	0.110560	884.1	1910.7	2794.8	2.39657	6.37589	**207**
208	481.15	18.326	855.220	0.108404	888.6	1906.6	2795.2	2.40595	6.36855	**208**
209	482.15	18.699	854.015	0.106297	893.2	1902.6	2795.8	2.41532	6.36123	**209**
210	483.15	19.077	852.806	0.104239	897.7	1898.5	2796.2	2.42467	6.35393	**210**
211	484.15	19.462	851.586	0.102227	902.3	1894.3	2796.6	2.43402	6.34664	**211**
212	485.15	19.852	850.362	0.100261	906.9	1890.2	2797.1	2.44336	6.33937	**212**
213	486.15	20.249	849.134	0.0983397	911.5	1886.0	2797.5	2.45270	6.33211	**213**
214	487.15	20.651	847.896	0.0964615	916.0	1881.8	2797.8	2.46202	6.32487	**214**
215	488.15	21.060	846.654	0.0946253	920.6	1877.6	2798.2	2.47133	6.31765	**215**
216	489.15	21.475	845.409	0.0928301	925.2	1873.4	2798.6	2.48064	6.31043	**216**
217	490.15	21.896	844.160	0.0910747	929.8	1869.1	2798.9	2.48994	6.30324	**217**
218	491.15	22.324	842.900	0.0893583	934.4	1864.9	2799.3	2.49923	6.29605	**218**
219	492.15	22.758	841.652	0.0876796	939.1	1860.6	2799.7	2.50851	6.28888	**219**
220	493.15	23.198	840.364	0.0860378	943.7	1856.2	2799.9	2.51779	6.28172	**220**
221	494.15	23.645	839.088	0.0844319	948.3	1851.9	2800.2	2.52706	6.27457	**221**
222	495.15	24.099	837.809	0.0828609	952.9	1847.5	2800.4	2.53632	6.26744	**222**
223	496.15	24.560	836.519	0.0813240	957.6	1843.1	2800.7	2.54557	6.26031	**223**
224	497.15	25.027	835.227	0.0798204	962.2	1838.7	2800.9	2.55482	6.25320	**224**
225	498.15	25.501	833.924	0.0783491	966.9	1834.3	2801.2	2.56406	6.24610	**225**
226	499.15	25.982	832.612	0.0769093	971.5	1829.8	2801.3	2.57329	6.23901	**226**
227	500.15	26.470	831.304	0.0755002	976.2	1825.3	2801.5	2.58252	6.23193	**227**
228	501.15	26.965	829.979	0.0741211	980.9	1820.8	2801.7	2.59174	6.22486	**228**
229	502.15	27.467	828.651	0.0727712	985.6	1816.3	2801.9	2.60096	6.21779	**229**
230	503.15	27.976	827.321	0.0714498	990.3	1811.7	2802.0	2.61017	6.21074	**230**
231	504.15	28.493	825.982	0.0701561	995.0	1807.1	2802.1	2.61937	6.20370	**231**
232	505.15	29.016	824.633	0.0688895	999.7	1802.5	2802.2	2.62857	6.19666	**232**
233	506.15	29.547	823.282	0.0676493	1004.4	1797.9	2002.3	2.63777	6.18963	**233**
234	507.15	30.086	821.922	0.0664349	1009.1	1793.2	2802.3	2.64696	6.18260	**234**
235	508.15	30.632	820.553	0.0652454	1013.8	1788.5	2802.3	2.65614	6.17559	**235**

(Contd.)

TABLE: 1.5 Thermodynamic Properties of Steam above 1 Atmosphere (Temperature Basis) [Contd.]

Temperature		Saturation pressure,	Liquid density	Specific volume of vapour, v''	Specific enthalpy			Specific entropy		Temp.
t	T	p	ρ_l		h	λ_v	H	s'	s''	t
°C	K	bar	kg/m³	m³/kg	kJ/kg			kJ/(kg·K)		°C
236	509.15	31.186	819.182	0.0640807	1018.6	1783.8	2802.4	2.66532	6.16858	**236**
237	510.15	31.747	817.802	0.0629396	1023.3	1779.0	2802.3	2.67450	6.16157	**237**
238	511.15	32.317	816.413	0.0618220	1028.1	1774.2	2802.3	2.68367	6.15457	**238**
239	512.15	32.893	815.016	0.0607271	1032.8	1769.4	2802.2	2.69284	6.14758	**239**
240	513.15	33.478	813.617	0.0596544	1037.6	1764.6	2802.2	2.70200	6.14059	**240**
241	514.15	34.071	812.203	0.0586034	1042.4	1759.7	2802.1	2.71116	6.13360	**241**
242	515.15	34.672	810.787	0.0575735	1047.2	1754.9	2802.1	2.72032	6.12662	**242**
243	516.15	35.281	809.363	0.0565643	1052.0	1749.9	2801.9	2.72947	6.11964	**243**
244	517.15	35.898	807.931	0.0555751	1056.8	1745.0	2801.8	2.73862	6.11266	**244**
245	518.15	36.523	806.491	0.0546057	1061.6	1740.0	2801.6	2.74777	6.10569	**245**
246	519.15	37.157	805.043	0.0536554	1066.4	1735.0	2801.4	2.75692	6.09871	**246**
247	520.15	37.799	803.587	0.0527239	1071.2	1730.0	2801.2	2.76606	6.09174	**247**
248	521.15	38.449	802.124	0.0518107	1076.1	1724.9	2801.0	2.77520	6.08477	**248**
249	522.15	39.108	800.653	0.0509153	1080.9	1719.8	2800.7	2.78435	6.07780	**249**
250	523.15	39.776	799.175	0.0500374	1085.8	1714.7	2800.5	2.79348	6.07083	**250**
251	524.15	40.452	797.690	0.0491766	1090.7	1709.5	2800.2	2.80262	6.06385	**251**
252	525.15	41.137	796.191	0.0483324	1095.5	1704.3	2799.8	2.81176	6.05688	**252**
253	526.15	41.831	794.691	0.0475045	1100.4	1699.1	2799.5	2.82090	6.04990	**253**
254	527.15	42.534	793.179	0.0466924	1105.3	1693.8	2799.1	2.83003	6.04292	**254**
255	528.15	43.246	791.659	0.0458938	1110.2	1688.5	2798.7	2.83917	6.03594	**255**
256	529.15	43.967	790.127	0.0451145	1115.2	1683.2	2798.4	2.84830	6.02896	**256**
257	530.15	44.697	788.594	0.0443479	1120.1	1677.8	2797.9	2.85744	6.02197	**257**
258	531.15	45.437	787.042	0.0435958	1125.0	1672.4	2797.4	2.86657	6.01497	**258**
259	532.15	46.185	785.490	0.0428579	1130.0	1666.9	2796.9	2.87571	6.00797	**259**
260	533.15	46.943	783.926	0.0421338	1134.9	1661.5	2796.4	2.88485	6.00097	**260**
261	534.15	47.711	782.350	0.0414232	1139.9	1656.0	2795.9	2.89399	5.99395	**261**
262	535.15	48.488	780.768	0.0407258	1144.9	1650.4	2795.3	2.90313	5.98693	**262**
263	536.15	49.275	779.174	0.0400413	1149.9	1644.8	2794.7	2.91227	5.97991	**263**
264	537.15	50.071	777.575	0.0393694	1154.9	1639.2	2794.1	2.92142	5.97287	**264**
265	538.15	50.877	775.958	0.0387099	1159.9	1633.5	2793.4	2.93056	5.96583	**265**
266	539.15	51.693	774.341	0.0380625	1165.0	1627.8	2792.8	2.93971	5.95877	**266**
267	540.15	52.519	772.708	0.0374268	1170.0	1622.1	2792.1	2.94887	5.95171	**267**
268	541.15	53.355	771.064	0.0368027	1175.1	1616.3	2791.4	2.95802	5.94463	**268**
269	542.15	54.202	769.414	0.0361898	1180.2	1610.5	2790.7	2.96718	5.93754	**269**
270	543.15	55.050	767.754	0.0355880	1185.2	1604.6	2789.8	2.97635	5.93045	**270**
271	544.15	55.925	766.078	0.0349969	1190.3	1598.7	2789.0	2.98552	5.92333	**271**
272	545.15	56.802	764.397	0.0344165	1195.4	1592.8	2788.2	2.99469	5.91621	**272**
273	546.15	57.689	762.701	0.0338463	1200.6	1586.8	2787.4	3.00387	5.90907	**273**
274	547.15	58.587	761.000	0.0332863	1205.7	1580.8	2786.5	3.01306	5.90191	**274**
275	548.15	59.496	759.284	0.0327361	1210.9	1574.7	2785.6	3.02225	5.89474	**275**
276	549.15	60.415	757.553	0.0321956	1216.0	1568.5	2784.5	3.03144	5.88755	**276**
277	550.15	61.346	755.818	0.0316645	1221.2	1562.4	2783.6	3.04065	5.88035	**277**
278	551.15	62.287	754.068	0.0311427	1226.4	1556.2	2782.6	3.04986	5.87313	**278**
279	552.15	63.239	752.304	0.0306299	1231.6	1549.9	2781.5	3.05907	5.86589	**279**
280	553.15	64.202	750.531	0.0301260	1236.8	1543.6	2780.4	3.06830	5.85863	**280**

(Contd.)

TABLE: 1.5 Thermodynamic Properties of Steam above 1 Atmosphere (Temperature Basis) [Contd.]

Temperature		Saturation pressure,	Liquid density	Specific volume of	Specific enthalpy			Specific entropy		Temp.
t	T	p	ρ_l	vapour, v''	h	λ_v	H	s'	s''	t
°C	K	bar	kg/m³	m³/kg	kJ/kg			kJ/(kg·K)		°C
281	554.15	65.176	748.744	0.0296308	1242.1	1537.2	2779.3	3.07753	5.85134	**281**
282	555.15	66.161	746.949	0.0291441	1247.3	1530.8	2778.1	3.08678	5.84404	**282**
283	556.15	67.158	745.134	0.0286656	1252.6	1524.3	2776.9	3.09603	5.83672	**283**
284	557.15	68.167	743.312	0.0281954	1257.9	1517.8	2775.7	3.10529	5.82937	**284**
285	558.15	69.186	741.477	0.0277330	1263.2	1511.3	2774.5	3.11456	5.82200	**285**
286	559.15	70.218	739.629	0.0272785	1268.5	1504.6	2773.1	3.12384	5.81461	**286**
287	560.15	71.261	737.762	0.0268316	1273.9	1498.0	2771.9	3.13313	5.80719	**287**
288	561.15	72.315	735.889	0.0263922	1279.2	1491.2	2770.4	3.14244	5.79975	**288**
289	562.15	73.382	733.999	0.0259601	1284.6	1484.5	2769.1	3.15175	5.79228	**289**
290	563.15	74.461	732.091	0.0255351	1290.0	1477.6	2767.6	3.16108	5.78478	**290**
291	564.15	75.551	730.172	0.0251172	1295.4	1470.7	2766.1	3.17042	5.77725	**291**
292	565.15	76.654	728.242	0.0247061	1300.9	1463.8	2764.7	3.17978	5.76970	**292**
293	566.15	77.769	726.296	0.0243018	1306.3	1456.8	2763.1	3.18914	5.76211	**293**
294	567.15	78.897	724.328	0.0239041	1311.8	1449.7	2761.5	3.19853	5.75450	**294**
295	568.15	80.037	722.350	0.0235128	1317.3	1442.6	2759.9	3.20792	5.74685	**295**
296	569.15	81.189	720.357	0.0231279	1322.8	1435.4	2758.2	3.21734	5.73917	**296**
297	570.15	82.355	718.350	0.0227491	1328.3	1428.1	2756.4	3.22677	5.73146	**297**
298	571.15	83.532	716.322	0.0223764	1333.9	1420.8	2754.7	3.23622	5.72371	**298**
299	572.15	84.723	714.281	0.0220096	1339.5	1413.4	2752.9	3.24568	5.71593	**299**
300	573.15	85.927	712.220	0.0216487	1345.1	1406.0	2751.0	3.25317	5.70812	**300**
301	574.15	87.144	710.142	0.0212935	1350.7	1398.5	2749.2	3.26467	5.70026	**301**
302	575.15	88.374	708.050	0.0209438	1356.3	1390.9	2747.2	3.27419	5.69237	**302**
303	576.15	89.617	705.936	0.0205996	1362.0	1383.2	2745.2	3.28374	5.68444	**303**
304	577.15	90.873	703.804	0.0202608	1367.7	1375.5	2743.2	3.29330	5.67647	**304**
305	578.15	92.144	701.656	0.0199272	1373.4	1367.7	2741.1	3.30289	5.66846	**305**
306	579.15	93.427	699.491	0.0195987	1379.2	1359.8	2739.0	3.31250	5.66041	**306**
307	580.15	94.725	697.301	0.0192752	1384.9	1351.9	2736.8	3.32213	5.65232	**307**
308	581.15	96.036	695.096	0.0189567	1390.7	1343.9	2734.6	3.33179	5.64418	**308**
309	582.15	97.361	692.866	0.0186429	1396.5	1335.8	2732.3	3.34148	5.63599	**309**
310	583.15	98.700	690.622	0.0183339	1402.4	1327.6	2730.0	3.35119	5.62776	**310**
311	584.15	100.05	688.350	0.0180295	1408.3	1319.4	2727.7	3.36093	5.61948	**311**
312	585.15	101.42	686.059	0.0177295	1414.2	1311.0	2725.2	3.37070	5.61114	**312**
313	586.15	102.80	683.747	0.0174339	1420.1	1302.6	2722.7	3.38049	5.60275	**313**
314	587.15	104.20	681.412	0.0171427	1426.1	1294.1	2720.2	3.39032	5.59431	**314**
315	588.15	105.61	679.053	0.0168556	1432.1	1285.5	2717.6	3.40018	5.58581	**315**
316	589.15	107.04	676.668	0.0165726	1438.1	1276.8	2714.9	3.41007	5.57725	**316**
317	590.15	108.48	674.263	0.0162937	1444.2	1268.0	2712.2	3.42000	5.56862	**317**
318	591.15	109.93	671.831	0.0160186	1450.3	1259.1	2709.4	3.42996	5.55993	**318**
319	592.15	111.40	669.371	0.0157473	1456.4	1250.2	2706.6	3.43996	5.55117	**319**
320	593.15	112.89	666.889	0.0154798	1462.6	1241.1	2703.7	3.45000	5.54233	**320**
321	594.15	114.39	664.377	0.0152159	1468.8	1231.9	2700.7	3.46008	5.53342	**321**
322	595.15	115.91	661.835	0.0149555	1475.1	1222.6	2697.7	3.47020	5.52443	**322**
323	596.15	117.44	659.270	0.0146986	1481.3	1213.2	2694.5	3.48036	5.51535	**323**
324	597.15	118.99	656.672	0.0144450	1487.7	1203.6	2691.3	3.49056	5.50619	**324**
325	598.15	120.56	654.044	0.0141947	1494.0	1194.0	2688.0	3.50082	5.49693	**325**

(Contd.)

TABLE: 1.5 Thermodynamic Properties of Steam above 1 Atmosphere (Temperature Basis) [Contd.]

Temperature		Saturation pressure,	Liquid density	Specific volume of vapour, v''	Specific enthalpy			Specific entropy		Temp.
t	T	p	ρ_l		h	λ_v	H	s'	s''	t
°C	K	bar	kg/m³	m³/kg	kJ/kg			kJ/(kg·K)		°C
326	599.15	122.14	651.385	0.0139476	1500.4	1184.2	2684.6	3.51112	5.48756	**326**
327	600.15	123.73	648.698	0.0137036	1506.9	1174.2	2681.1	3.52147	5.47810	**327**
328	601.15	125.35	645.974	0.0134626	1513.4	1164.2	2677.6	3.53187	5.46852	**328**
329	602.15	126.98	643.215	0.0132245	1519.9	1154.0	2673.9	3.54232	5.45883	**329**
330	603.15	128.63	640.422	0.0129994	1526.5	1143.6	2670.1	3.55283	5.44901	**330**
331	604.15	130.29	637.592	0.0127570	1533.2	1133.1	2666.4	3.56340	5.43907	**331**
332	605.15	131.97	634.723	0.0125273	1539.9	1122.5	2662.3	3.57403	5.42899	**332**
333	606.15	133.67	631.816	0.0123003	1546.6	1111.7	2658.3	3.58472	5.41877	**333**
334	607.15	135.38	628.871	0.0120759	1553.4	1100.7	2654.1	3.59548	5.40841	**334**
335	608.15	137.12	625.880	0.0118540	1560.3	1089.5	2649.8	3.60631	5.39789	**335**
336	609.15	138.87	622.847	0.0116345	1567.2	1078.1	2645.3	3.61721	5.38720	**336**
337	610.15	140.64	619.767	0.0114175	1574.2	1066.6	2640.8	3.62819	5.37635	**337**
338	611.15	142.42	616.641	0.0112029	1581.2	1054.8	2636.0	3.63925	5.36533	**338**
339	612.15	144.23	613.463	0.0109905	1588.3	1042.9	2631.2	3.65039	5.35413	**339**
340	613.15	146.05	610.232	0.0107804	1595.5	1030.7	2626.2	3.66162	5.34274	**340**
341	614.15	147.89	606.951	0.0105726	1602.7	1018.3	2621.0	3.67294	5.33115	**341**
342	615.15	149.76	603.610	0.0103670	1610.1	1005.7	2615.8	3.68435	5.31938	**342**
343	616.15	151.64	600.208	0.0101635	1617.5	992.8	2610.3	3.69588	5.30740	**343**
344	617.15	153.54	596.741	0.0099622	1624.9	979.7	2604.6	3.70751	5.29521	**344**
345	618.15	155.45	593.204	0.0097631	1632.5	966.4	2598.9	3.71925	5.28282	**345**
346	619.15	157.39	589.598	0.0095660	1640.2	952.8	2593.0	3.73112	5.27022	**346**
347	620.15	159.35	585.916	0.0093711	1648.0	938.9	2586.9	3.74313	5.25740	**347**
348	621.15	161.33	582.150	0.0091783	1655.8	924.8	2580.6	3.75527	5.24436	**348**
349	622.15	163.33	578.295	0.0089876	1663.8	910.4	2574.2	3.76757	5.23112	**349**
350	623.15	165.35	574.343	0.0087991	1671.9	895.7	2567.6	3.78004	5.21766	**350**
351	624.15	167.39	570.386	0.0086094	1680.4	880.4	2560.8	3.79334	5.20363	**351**
352	625.15	169.45	566.219	0.0084205	1689.3	864.2	2553.5	3.80707	5.18929	**352**
353	626.15	171.54	561.924	0.0082324	1698.4	847.7	2546.1	3.82094	5.17462	**353**
354	627.15	173.64	557.507	0.0080453	1707.5	830.9	2538.4	3.83487	5.15959	**354**
355	628.15	175.77	552.944	0.0078592	1716.6	813.8	2530.4	3.84886	5.14417	**355**
356	629.15	177.92	548.216	0.0076741	1725.9	796.2	2522.1	3.86295	5.12835	**356**
357	630.15	180.09	543.301	0.0074897	1735.2	778.3	2513.5	3.87716	5.11206	**357**
358	631.15	182.29	538.213	0.0073061	1744.7	759.9	2504.6	3.89155	5.09529	**358**
359	632.15	184.51	532.935	0.0071229	1754.3	740.9	2495.2	3.90615	5.07796	**359**
360	633.15	186.75	527.454	0.0069398	1764.2	721.3	2485.5	3.92102	5.06003	**360**
361	634.15	189.02	521.730	0.0067566	1774.2	701.0	2475.2	3.93622	5.04141	**361**
362	635.15	191.31	515.783	0.0065727	1784.6	679.8	2464.4	3.95182	5.02202	**362**
363	636.15	193.62	509.528	0.0063877	1795.3	657.8	2453.1	3.96792	5.00173	**363**
364	637.15	195.96	502.968	0.0062010	1806.4	634.6	2441.0	3.98462	4.98042	**364**
365	638.15	198.33	496.032	0.0060116	1818.0	610.0	2428.0	4.00208	4.95788	**365**
366	639.15	200.72	488.663	0.0058186	1830.2	583.9	2414.1	4.02048	4.93389	**366**
367	640.15	203.13	480.723	0.0056206	1843.2	555.7	2398.9	4.04009	4.90812	**367**
368	641.15	205.57	472.121	0.0054157	1857.3	525.1	2382.4	4.06127	4.88012	**368**
369	642.15	208.04	462.577	0.0052011	1872.8	491.1	2363.9	4.08456	4.84922	**369**
370	643.15	210.54	451.753	0.0049728	1890.2	452.6	2342.8	4.11080	4.81439	**370**

(Contd.)

TABLE: 1.5 Thermodynamic Properties of Steam above 1 Atmosphere (Temperature Basis) [Contd.]

Temperature		Saturation pressure,	Liquid density	Specific volume of vapour, v''	Specific enthalpy			Specific entropy		Temp.
t	T	p	ρ_l		h	λ_v	H	s'	s''	t
°C	K	bar	kg/m³	m³/kg	kJ/kg			kJ/(kg·K)		°C
371	644.15	213.06	439.020	0.0047233	1910.5	407.4	2317.9	4.14144	4.77384	**371**
372	645.15	215.62	423.083	0.0044389	1935.6	351.4	2287.0	4.17942	4.72403	**372**
373	646.15	218.20	400.593	0.0040840	1970.5	273.5	2244.0	4.23255	4.65588	**373**
374	647.15	220.81	351.778	0.0034659	2046.7	109.5	2156.2	4.34934	4.51853	**374**
374.15*	**647.30**	**221.20**	**315.457**	**0.00317**	**2107.4**	**0.0**	**2107.4**	**4.44286**	**4.44286**	**374.15**

*Critical point

TABLE: 1.6 Thermodynamic Properties of Steam up to 1 Atmosphere (Pressure Basis)

Pressure, p		Saturation temperature		Liquid density	Specific volume of vapour, v'	Specific enthalpy			Specific entropy		Press. p
		t	T	ρ_l		h	λ_v	H	s'	s''	
bar	torr	°C	K	kg/m³	m³/kg	kJ/kg			kJ/(kg·K)		bar
0.006112	**4.584**	**0.01***	**273.16**	**999.780**	**206.163**	**0.00**	**2501.6**	**2501.6**	**0.00000**	**9.15746**	**0.006112**
0.010	7.50	6.98	280.13	999.930	129.209	29.3	2485.0	2514.4	0.10604	8.97667	**0.010**
0.015	11.25	13.04	286.19	999.430	87.9821	54.7	2470.8	2525.5	0.19567	8.82883	**0.015**
0.020	15.00	17.51	290.66	998.762	67.0061	73.5	2460.1	2533.6	0.26065	8.72456	**0.020**
0.025	18.75	21.10	294.25	998.044	54.2562	88.4	2451.8	2540.2	0.31191	8.64403	**0.025**
0.030	22.50	24.10	297.25	997.347	45.6673	101.0	2444.6	2545.6	0.35436	8.57848	**0.030**
0.035	26.25	26.69	299.84	996.671	39.4787	111.8	2438.6	2550.4	0.39068	8.52322	**0.035**
0.040	30.00	28.98	302.13	996.016	34.8022	121.4	2433.1	2554.5	0.42246	8.47548	**0.040**
0.045	33.75	31.03	304.18	995.391	31.1408	130.0	2428.2	2558.2	0.45075	8.43347	**0.045**
0.050	37.50	32.90	306.05	994.797	28.1944	137.8	2423.8	2561.6	0.47626	8.39596	**0.050**
0.055	41.25	34.61	307.76	994.214	25.7707	144.9	2419.8	2564.7	0.49951	8.36210	**0.055**
0.060	45.00	36.18	309.33	993.670	23.7410	151.5	2416.0	2567.5	0.52088	8.33124	**0.060**
0.065	48.75	37.65	310.80	993.137	22.0159	157.6	2412.6	2578.2	0.54066	8.30289	**0.065**
0.070	52.50	39.02	312.17	992.625	20.5310	163.4	2409.2	2572.6	0.55909	8.27669	**0.070**
0.075	56.25	40.32	313.47	992.132	19.2391	168.8	2406.1	2574.9	0.57633	8.25233	**0.075**
0.080	60.00	41.53	314.68	991.650	18.1046	173.9	2403.2	2577.1	0.59255	8.22956	**0.080**
0.085	63.76	42.69	315.84	991.188	17.1001	178.7	2400.5	2579.2	0.60786	8.20821	**0.085**
0.090	67.51	43.79	316.94	990.737	16.2043	183.3	2397.9	2581.1	0.62235	8.18810	**0.090**
0.095	71.26	44.83	317.98	990.295	15.4003	187.7	2395.3	2583.0	0.63613	8.16909	**0.095**
0.100	75.01	45.83	318.98	989.874	14.6746	191.8	2393.0	2584.8	0.64925	8.15108	**0.100**
0.110	82.51	47.71	320.86	989.061	13.4161	199.7	2388.4	2588.1	0.67378	8.11766	**0.110**
0.120	90.01	49.45	322.60	988.279	12.3619	206.9	2304.3	2591.2	-	8.08721	**0.120**
0.130	97.51	51.06	324.21	987.537	11.4657	213.7	2380.3	2594.0	0.71723	8.05924	**0.130**
0.140	105.01	52.57	325.72	986.836	10.6942	220.0	2376.7	2596.7	0.73669	8.03338	**0.140**
0.150	112.51	54.00	327.15	986.154	10.0228	226.0	2373.2	2599.2	0.75492	8.00933	**0.150**
0.160	120.01	55.34	328.49	985.503	9.43314	231.6	2370.0	2601.6	0.77207	7.98607	**0.160**
0.170	127.51	56.61	329.76	984.872	8.91095	236.9	2366.9	2603.8	0.78826	7.96580	**0.170**
0.180	135.01	57.83	330.98	984.262	8.44521	242.0	2363.9	2605.9	0.80360	7.94595	**0.180**
0.190	142.51	58.98	332.13	983.671	8.02716	246.8	2361.1	2607.9	0.81818	7.92720	**0.190**
0.200	150.01	60.09	333.24	983.101	7.64977	251.5	2358.4	2609.9	0.83207	7.90943	**0.200**
0.21	157.51	61.14	334.29	982.550	7.30732	255.9	2355.8	2611.7	0.84535	7.89254	**0.21**
0.22	165.01	62.16	335.31	982.010	6.99514	260.1	2353.4	2613.5	0.85805	7.87645	**0.22**
0.23	172.51	63.14	336.29	981.489	6.70934	264.2	2351.0	2615.2	0.87024	7.86108	**0.23**
0.24	180.01	64.08	337.23	980.979	6.44669	268.2	2348.6	2616.8	0.88196	7.84639	**0.24**
0.25	187.52	64.99*	338.14	980.479	6.20447	272.0	2346.3	2618.3	0.89324	7.83230	**0.25**
0.26	195.02	65.87	339.02	979.998	5.98034	275.7	2344.2	2619.9	0.90411	7.81878	**0.26**
0.27	202.52	66.72	339.87	979.518	5.77235	279.2	2342.1	2621.3	0.91481	7.80578	**0.27**
0.28	210.02	67.55	340.70	979.058	5.57879	282.7	2340.0	2622.7	0.92476	7.79326	**0.28**
0.29	217.52	68.35	341.50	978.608	5.39820	286.0	2338.1	2624.1	0.93459	7.78118	**0.29**
0.30	225.02	69.12	342.27	978.167	5.22930	289.3	2336.1	2625.4	0.94411	7.76953	**0.30**
0.31	232.52	69.88	343.03	977.727	5.07098	292.5	2334.2	2626.7	0.95335	7.75526	**0.31**
0.32	240.02	70.61	343.76	977.307	4.92227	295.5	2332.4	2628.0	0.96232	7.74736	**0.32**
0.33	247.52	71.33	344.48	976.887	4.78232	298.6	2330.6	2429.2	0.97103	7.73679	**0.33**
0.34	255.02	72.03	345.18	976.486	4.65036	301.5	2328.9	2630.4	0.97952	7.72655	**0.34**
0.35	262.52	72.71	345.86	976.076	4.52571	304.3	2327.2	2631.5	0.98777	7.71661	**0.35**

*Triple point

(Contd.)

TABLE: 1.6 Thermodynamic Properties of Steam up to 1 Atmosphere (Pressure Basis) [Contd.]

Pressure, p		Saturation temperature		Liquid density	Specific volume of	Specific enthalpy			Specific entropy		Press.
		t	T	ρ_l	vapour, v'	h	λ_v	H	s'	s''	p
bar	torr	°C	K	kg/m^3	m^3/kg	kJ/kg			kJ/(kg·K)		bar
0.36	270.02	73.37	346.52	975.686	4.40779	307.1	2325.5	2632.6	0.99582	7.70696	**0.36**
0.37	277.52	74.02	347.17	975.296	4.29605	309.8	2323.9	2633.7	1.00366	7.69757	**0.37**
0.38	285.02	74.66	347.81	974.915	4.19003	312.5	2322.3	2634.8	1.01132	7.68844	**0.38**
0.39	292.52	75.28	348.43	974.545	4.08928	315.1	2320.8	2635.9	1.01879	7.67955	**0.39**
0.40	300.02	75.89	349.04	974.175	3.99342	317.7	2319.2	2636.9	1.02610	7.67089	**0.40**
0.41	307.53	76.48	349.63	973.814	3.90210	320.1	2317.8	2637.9	1.03323	7.66245	**0.41**
0.42	315.03	77.06	350.21	973.463	3.81500	322.6	2316.3	2638.9	1.04022	7.65421	**0.42**
0.43	322.53	77.63	350.78	973.103	3.73183	325.0	2314.8	2639.8	1.04705	7.64618	**0.43**
0.44	330.03	78.19	351.34	972.763	3.65232	327.3	2313.4	2640.7	1.05374	7.63832	**0.44**
0.45	337.53	78.74	351.89	972.422	3.57625	329.6	2312.1	2641.7	1.06029	7.63065	**0.45**
0.46	345.03	79.28	352.43	972.082	3.50338	331.9	2310.7	2642.6	1.06672	7.62315	**0.46**
0.47	352.53	79.81	352.96	971.751	3.43352	334.1	2309.3	2643.4	1.07302	7.61582	**0.47**
0.48	360.03	80.33	353.48	971.430	3.36649	336.3	2308.0	2644.3	1.07919	7.60864	**0.48**
0.49	367.53	80.84	353.99	971.100	3.30210	338.5	2306.7	2645.2	1.08526	7.60161	**0.49**
0.50	375.03	81.35	354.50	970.789	3.24022	340.6	2305.4	2646.0	1.09121	7.59472	**0.50**
0.52	390.03	82.33	355.48	970.158	3.12338	344.7	2302.9	2647.6	1.10279	7.58137	**0.52**
0.54	405.03	83.27	356.42	969.556	3.01494	348.7	2300.5	2649.2	1.11399	7.56852	**0.54**
0.56	420.03	84.19	357.34	968.955	2.91401	352.5	2298.2	2450.7	1.12481	7.55615	**0.56**
0.58	435.04	85.09	358.24	968.373	2.81984	356.3	2295.8	2652.1	1.13529	7.54422	**0.58**
0.60	450.04	85.95	359.10	967.811	2.73175	359.9	2293.7	2653.6	1.14544	7.53270	**0.60**
0.62	465.04	86.80	359.95	967.249	2.64918	363.5	2291.4	2654.9	1.15530	7.52156	**0.62**
0.64	480.04	67.62	340.77	966.707	2.57162	366.9	2289.4	2656.3	1.16487	7.51079	**0.64**
0.66	495.04	88.42	361.57	966.174	2.49861	370.3	2287.3	2657.6	1.17418	7.50035	**0.66**
0.68	510.04	89.20	362.35	965.652	2.42978	373.6	2285.2	2658.8	1.18324	7.49022	**0.68**
0.70	525.04	89.96	363.11	965.139	2.36473	376.8	2283.3	2680.1	1.19205	7.48040	**0.70**
0.72	540.04	90.70	363.85	964.636	2.30320	379.9	2281.4	2661.3	1.20065	7.47085	**0.72**
0.74	555.05	91.43	364.58	964.144	2.24490	382.9	2279.5	2662.4	1.20903	7.46157	**0.74**
0.76	570.05	92.14	365.29	963.660	2.18957	385.9	2277.7	2663.6	1.21721	7.45254	**0.76**
0.78	585.05	92.83	365.98	963.178	2.13699	388.9	2275.8	2664.7	1.22520	7.44375	**0.78**
0.80	600.05	93.51	366.66	962.705	2.08696	391.7	2274.1	2665.8	1.23301	7.43519	**0.80**
0.82	615.05	93.51	366.66	962.251	2.08696	394.5	2272.3	2666.8	1.24065	7.42684	**0.82**
0.84	630.05	94.83	367.98	961.788	1.99383	397.3	2270.6	2667.9	1.24812	7.41869	**0.84**
0.86	645.05	95.47	368.62	961.344	1.95041	400.0	2268.9	2668.9	1.25543	7.41074	**0.86**
0.88	660.05	96.10	369.25	960.901	1.90891	402.6	2267.3	2669.9	1.26259	7.40297	**0.88**
0.90	675.06	96.71	369.86	960.467	1.86919	403.2	2265.7	2670.9	1.24960	7.39538	**0.90**
0.92	690.06	97.32	370.47	960.043	1.83115	407.8	2264.0	2671.8	1.27648	7.38798	**0.92**
0.94	705.06	97.91	371.06	959.619	1.79467	410.3	2262.4	2672.7	1.28322	7.38070	**0.94**
0.96	720.06	98.49	371.64	959.205	1.75967	412.7	2261.0	2673.7	1.28984	7.37359	**0.96**
0.98	735.06	99.07	372.22	958.791	1.72605	415.1	2259.5	2674.6	1.29633	7.36663	**0.98**
1.00	750.06	99.63	372.78	958.387	1.69373	417.5	2257.9	2675.4	1.30271	7.35982	**1.00**
1.01325	**760.00**	**100.00***	**373.15**	**958.121**	**1.67300**	**419.1**	**2256.9**	**2676.0**	**1.30687**	**7.35538**	**1.01325**

*Normal boiling point

Reference State:

Specific enthalpy of saturated water at 0.01°C (273.16 K) = 0.0 kJ/kg

Specific entropy of saturated water at 0.01°C (273.16 K) = 0.00000 kJ/(kg·K)

TABLE: 1.7 Thermodynamic Properties of Steam above 1 Atmosphere (Pressure Basis)

Pressure, p bar	Saturation Temperature t °C	Saturation Temperature T K	Liquid density, ρ_l kg/m³	Specific volume of vapour, v'' m³/kg	Specific enthalpy h kJ/kg	Specific enthalpy λ_v kJ/kg	Specific enthalpy H kJ/kg	Specific entropy s' kJ/(kg·K)	Specific entropy s'' kJ/(kg·K)	Press. p bar
1.05	101.00	374.15	957.396	1.61816	423.3	2254.3	2677.6	1.31816	7.34337	**1.05**
1.10	102.32	375.47	956.444	1.54924	428.8	2250.8	2679.6	1.33297	7.32769	**1.10**
1.15	103.59	376.74	955.511	1.48614	434.2	2247.4	2681.6	1.34719	7.31272	**1.15**
1.20	104.81	377.96	954.581	1.42813	439.4	2244.0	2683.4	1.36087	7.29839	**1.20**
1.25	105.99	379.14	953.725	1.37461	444.4	2240.8	2685.2	1.37404	7.28465	**1.25**
1.30	107.13	380.28	952.862	1.32589	449.2	2237.8	2687.0	1.38676	7.27146	**1.30**
1.35	108.24	381.39	952.027	1.27912	453.9	2234.8	2688.7	1.39904	7.25877	**1.35**
1.40	109.32	382.47	951.212	1.23633	458.4	2231.9	2690.3	1.41093	7.24655	**1.40**
1.45	110.36	383.51	950.417	1.19639	462.8	2229.0	2691.8	1.42244	7.23476	**1.45**
1.50	111.37	384.52	949.641	1.15904	467.1	2226.3	2693.4	1.43361	7.22337	**1.50**
1.55	112.36	385.51	948.875	1.12401	471.3	2223.5	2694.8	1.44445	7.21235	**1.55**
1.60	113.32	386.47	948.128	1.09111	475.4	2220.8	2696.2	1.45498	7.20169	**1.60**
1.65	114.26	387.41	947.400	1.06014	479.3	2218.3	2697.6	1.46523	7.19138	**1.65**
1.70	115.17	388.32	946.683	1.03093	483.2	2215.8	2699.0	1.47520	7.18134	**1.70**
1.75	116.06	389.21	945.975	1.00334	487.0	2213.3	2700.3	1.48491	7.17162	**1.75**
1.80	116.93	390.08	945.287	0.977227	490.7	2210.8	2701.5	1.49439	7.16217	**1.80**
1.85	117.70	390.85	944.608	0.952483	494.3	2208.5	2702.8	1.50363	7.15298	**1.85**
1.90	118.62	391.77	943.948	0.928999	497.8	2206.2	2704.0	1.51265	7.14403	**1.90**
1.95	119.43	392.58	943.289	0.906680	501.3	2203.8	2705.1	1.52147	7.13532	**1.95**
2.00	120.23	393.38	942.649	0.885441	504.7	2201.6	2706.3	1.53008	7.12683	**2.00**
2.05	121.01	394.16	942.019	0.865205	508.0	2199.4	2707.4	1.53851	7.11855	**2.05**
2.10	121.78	394.93	941.389	0.845900	511.3	2197.2	2708.5	1.54675	7.11047	**2.10**
2.15	122.53	395.68	940.778	0.827464	514.5	2195.1	2709.6	1.55484	7.10258	**2.15**
2.20	123.27	396.42	940.177	0.809839	517.6	2193.0	2710.6	1.56275	7.09487	**2.20**
2.25	123.99	397.14	939.576	0.792972	520.7	2190.9	2711.6	1.57050	7.08734	**2.25**
2.30	124.71	397.86	938.994	0.776813	523.7	2188.9	2717.3	1.57811	7.07997	**2.30**
2.35	125.40	398.55	938.412	0.761320	526.7	2186.9	2713.6	1.58557	7.07276	**2.35**
2.40	126.09	399.24	937.840	0.746431	529.6	2184.9	2714.5	1.59289	7.06571	**2.40**
2.45	126.77	399.92	937.277	0.732169	532.5	2183.0	2715.5	1.60000	7.05879	**2.45**
2.50	127.43	400.58	936.698	0.718439	535.3	2181.1	2716.4	1.60714	7.05202	**2.50**
2.55	128.08	401.23	936.181	0.705230	538.1	2179.2	2717.3	1.61408	7.04538	**2.55**
2.60	128.73	401.88	935.637	0.692512	540.9	2177.3	2718.2	1.62089	7.03888	**2.60**
2.65	129.36	402.51	935.104	0.680258	543.6	2175.4	2719.0	1.62760	7.03249	**2.65**
2.70	129.98	403.13	934.571	0.668443	546.2	2173.7	2719.9	1.63419	7.02622	**2.70**
2.75	130.60	403.75	934.056	0.657043	548.9	2171.8	2720.7	1.64068	7.02007	**2.75**
2.80	131.20	404.35	933.541	0.646037	551.4	2170.1	2721.5	1.64706	7.01403	**2.80**
2.85	131.80	404.95	933.097	0.635404	554.0	2168.3	2722.3	1.65335	7.00810	**2.85**
2.90	132.39	405.54	932.523	0.625125	556.5	2166.6	2723.1	1.65953	7.00227	**2.90**
2.95	132.97	406.12	932.027	0.615184	559.0	2164.9	2723.9	1.66563	6.99653	**2.95**
3.00	133.54	406.69	931.532	0.605362	561.4	2163.3	2724.7	1.67164	6.99090	**3.00**
3.1	134.66	407.81	930.562	0.587219	566.2	2159.9	2726.1	1.68339	6.97990	**3.1**
3.2	135.75	408.90	929.610	0.569985	570.9	2156.7	2727.6	1.69481	6.96925	**3.2**
3.3	136.82	409.97	928.686	0.553761	575.5	2153.5	2729.0	1.70593	6.95893	**3.3**
3.4	137.86	411.01	927.773	0.538459	579.9	2150.4	2730.3	1.71675	6.94891	**3.4**
3.5	138.87	412.02	926.870	0.524003	584.3	2147.3	2731.6	1.72730	6.93918	**3.5**

(Contd.)

TABLE: 1.7 Thermodynamic Properties of Steam above 1 Atmosphere (Pressure Basis) [Contd.]

Pressure, p bar	Saturation Temperature t °C	Saturation Temperature T K	Liquid density, ρ_l kg/m³	Specific volume of vapour, v'' m³/kg	Specific enthalpy h kJ/kg	Specific enthalpy λ_v kJ/kg	Specific enthalpy H kJ/kg	Specific entropy s' kJ/(kg·K)	Specific entropy s'' kJ/(kg·K)	Press. p bar
3.6	139.86	413.01	925.995	0.510323	588.5	2144.4	2732.9	1.73739	6.92972	**3.6**
3.7	140.83	413.98	925.129	0.497357	592.7	2141.4	2734.1	1.74763	6.92052	**3.7**
3.8	141.78	414.93	924.283	0.485051	596.8	2138.5	2735.3	1.75744	6.91156	**3.8**
3.9	142.71	415.86	923.446	0.473355	600.8	2135.7	2736.5	1.76703	6.90283	**3.9**
4.0	143.62	416.77	922.620	0.462224	604.7	2132.9	2737.6	1.77640	6.89433	**4.0**
4.1	144.52	417.67	921.812	0.451618	608.5	2130.2	2738.7	1.78557	6.88602	**4.1**
4.2	145.39	418.54	921.022	0.441500	612.3	2127.5	2739.8	1.79455	6.87792	**4.2**
4.3	146.25	419.40	920.234	0.431836	616.0	2124.9	2740.9	1.80334	6.87000	**4.3**
4.4	147.09	420.24	919.464	0.422597	619.6	2122.3	2741.9	1.81196	6.86226	**4.4**
4.5	147.92	421.07	918.704	0.413754	623.2	2119.7	2742.9	1.82841	6.85470	**4.5**
4.6	148.73	421.88	917.953	0.405283	626.7	2117.2	2743.9	1.82870	6.84730	**4.6**
4.7	149.53	422.68	917.212	0.397160	630.1	2114.7	2744.8	1.83683	6.84005	**4.7**
4.8	150.31	423.46	916.481	0.389364	633.5	2112.2	2745.7	1.84482	6.83295	**4.8**
4.9	151.08	424.23	915.759	0.381875	636.8	2109.8	2746.6	1.85266	6.82600	**4.9**
5.0	151.84	424.99	915.047	0.374676	640.1	2107.4	2747.5	1.86036	6.81919	**5.0**
5.2	153.33	426.48	913.651	0.361000	646.5	2102.8	2749.3	1.87537	6.80595	**5.2**
5.4	154.76	427.91	912.292	0.348457	652.8	2098.1	2750.9	1.88990	6.79320	**5.4**
5.6	156.16	429.31	910.954	0.336705	658.8	2093.7	2752.5	1.90396	6.78090	**5.6**
5.8	157.52	430.67	909.653	0.325736	664.7	2089.3	2754.0	1.91760	6.76903	**5.8**
6.0	158.84	431.99	909.041	0.315474	670.4	2085.1	2755.5	1.93083	6.75754	**6.0**
6.2	160.12	433.27	907.128	0.305851	676.0	2080.9	2756.9	1.94370	6.74643	**6.2**
6.4	161.38	434.53	905.912	0.296809	681.5	2076.7	2758.2	1.95621	6.73565	**6.4**
6.6	162.60	435.75	867.077	0.288296	686.8	2072.7	2759.5	1.96838	6.72520	**6.6**
6.8	163.79	436.94	903.530	0.280267	692.0	2068.8	2760.8	1.98025	6.71505	**6.8**
7.0	164.96	438.11	902.372	0.272681	697.1	2064.9	2762.0	1.99181	6.70518	**7.0**
7.2	166.10	439.25	901.234	0.265502	702.0	2061.1	2763.1	2.00310	6.69558	**7.2**
7.4	167.21	440.36	900.122	0.258697	706.9	2057.4	2764.3	2.01412	6.68624	**7.4**
7.6	168.30	441.45	899.022	0.252239	711.7	2053.7	2765.4	2.02489	6.67714	**7.6**
7.8	169.37	442.52	897.940	0.246100	716.3	2050.1	2766.4	2.03542	6.66826	**7.8**
8.0	170.41	443.56	896.877	0.240257	720.9	2046.6	2767.5	2.04572	6.65960	**8.0**
8.2	171.44	444.59	895.825	0.234690	725.4	2043.1	2768.5	2.05580	6.65114	**8.2**
8.4	172.45	445.60	894.799	0.229378	729.9	2039.5	2769.4	2.06567	6.64288	**8.4**
8.6	173.44	446.59	893.775	0.224305	734.2	2036.2	2770.4	2.07535	6.63480	**8.6**
8.8	174.41	447.56	892.777	0.219454	738.5	2032.8	2771.3	2.08483	6.62690	**8.8**
9.0	175.36	448.51	891.782	0.214812	742.6	2029.5	2772.1	2.09419	6.61917	**9.0**
9.2	176.29	449.44	890.869	0.210364	746.8	2026.2	2773.0	2.10327	6.61160	**9.2**
9.4	177.21	450.36	889.846	0.206099	750.8	2023.0	2773.8	2.11223	6.60419	**9.4**
9.6	178.12	451.27	888.889	0.202006	754.8	2019.8	2774.6	2.12103	6.59692	**9.6**
9.8	179.01	452.16	887.950	0.198073	758.7	2016.7	2775.4	2.12967	6.58980	**9.8**
10.0	179.88	453.03	887.020	0.194293	762.6	2013.6	2776.2	2.13817	6.58281	**10.0**
10.2	180.75	453.90	886.108	0.190655	766.4	2010.5	2776.9	2.14653	6.57595	**10.2**
10.4	181.60	454.75	885.199	0.187153	770.2	2007.4	2777.6	2.15474	6.56922	**10.4**
10.6	182.43	455.58	884.306	0.183778	773.9	2004.4	2778.3	2.16283	6.56261	**10.6**
10.8	183.26	456.41	883.416	0.180524	777.5	2001.5	2779.0	2.17078	6.55612	**10.8**
11.0	184.07	457.22	882.542	0.177384	781.1	1998.6	2779.7	2.17861	6.54973	**11.0**

(Contd.)

TABLE: 1.7 Thermodynamic Properties of Steam above 1 Atmosphere (Pressure Basis) [Contd.]

Pressure, p bar	Saturation Temperature t °C	Saturation Temperature T K	Liquid density, ρ_l kg/m³	Specific volume of vapour, v'' m³/kg	Specific enthalpy h kJ/kg	Specific enthalpy λ_v kJ/kg	Specific enthalpy H kJ/kg	Specific entropy s' kJ/(kg·K)	Specific entropy s'' kJ/(kg·K)	Press. p bar
11.2	184.87	458.02	881.671	0.174352	784.7	1995.6	2780.3	2.18633	6.54346	**11.2**
11.4	185.66	458.81	880.817	0.171422	788.2	1992.8	2780.9	2.19392	6.53729	**11.4**
11.6	186.44	459.59	879.964	0.168591	791.6	1990.0	2781.6	2.20141	6.53122	**11.6**
11.8	187.20	460.35	879.121	0.165851	795.1	1987.1	2782.2	2.20878	6.52524	**11.8**
12.0	187.96	461.11	878.287	0.163200	798.4	1984.3	2782.7	2.21606	6.51936	**12.0**
12.2	188.71	461.86	877.462	0.160633	801.8	1981.5	2783.3	2.22323	6.51357	**12.2**
12.4	189.45	462.60	876.647	0.158146	805.1	1978.8	2783.9	2.23030	6.50787	**12.4**
12.6	190.18	463.33	875.841	0.155735	808.3	1976.1	2784.4	2.23728	6.50225	**12.6**
12.8	190.90	464.05	875.036	0.153396	811.5	1973.4	2784.9	2.24416	6.49672	**12.8**
13.0	191.61	464.76	874.241	0.151127	814.7	1970.7	2785.4	2.25095	6.49126	**13.0**
13.2	192.31	465.46	873.454	0.148924	817.8	1968.1	2785.9	2.25766	6.48588	**13.2**
13.4	193.01	466.16	872.676	0.146785	821.0	1965.4	2786.4	2.26428	6.48058	**13.4**
13.6	193.69	466.84	871.900	0.144706	824.0	1962.9	2786.9	2.27082	6.47535	**13.6**
13.8	194.37	467.52	871.133	0.142606	827.1	1960.2	2787.3	2.27728	6.47018	**13.8**
14.0	195.04	468.19	870.375	0.140721	830.1	1957.7	2787.8	2.28366	6.46509	**14.0**
14.2	195.71	468.86	869.618	0.138810	833.1	1955.1	2788.2	2.28997	6.46006	**14.2**
14.4	196.36	469.51	868.870	0.136949	836.0	1952.7	2788.7	2.29620	6.45510	**14.4**
14.6	197.01	470.16	868.131	0.135138	838.9	1950.2	2789.1	2.30236	6.45020	**14.6**
14.8	197.65	470.80	867.393	0.133375	841.8	1947.7	2789.5	2.30845	6.44536	**14.8**
15.0	198.29	471.44	866.656	0.131656	844.7	1945.2	2789.9	2.31447	6.44059	**15.0**
15.2	198.92	472.07	865.936	0.129981	847.5	1942.8	2790.3	2.32043	6.43586	**15.2**
15.4	199.54	472.69	865.217	0.128349	850.3	1940.4	2790.7	2.32632	6.43120	**15.4**
15.6	200.16	473.31	864.499	0.126756	853.1	1937.9	2791.0	2.33214	6.42659	**15.6**
15.8	200.77	473.92	863.789	0.125203	855.8	1935.6	2791.4	2.33791	6.42203	**15.8**
16.0	201.37	474.52	863.081	0.123686	858.6	1933.1	2791.7	2.34361	6.41753	**16.0**
16.2	201.97	475.12	862.381	0.122207	861.3	1930.8	2792.1	2.34926	6.41307	**16.2**
16.4	202.56	475.71	861.690	0.120761	863.9	1928.5	2792.4	2.35485	6.40867	**16.4**
16.6	203.15	476.30	861.000	0.119350	866.6	1926.2	2792.8	2.36038	6.40431	**16.6**
16.8	203.73	476.88	860.311	0.117971	869.2	1923.8	2793.1	2.36585	6.40000	**16.8**
17.0	204.31	477.46	859.631	0.116623	871.8	1921.6	2793.4	2.37127	6.39574	**17.0**
17.2	204.88	478.03	858.952	0.115105	874.4	1919.3	2793.7	2.37664	6.39152	**17.2**
17.4	205.44	478.59	858.281	0.114017	877.0	1917.0	2794.0	2.38196	6.38735	**17.4**
17.6	206.00	479.15	857.611	0.112757	879.5	1914.8	2794.3	2.38723	6.38321	**17.6**
17.8	206.56	479.71	856.949	0.111524	882.1	1912.5	2794.6	2.39245	6.37912	**17.8**
18.0	207.11	480.26	856.289	0.110317	884.6	1910.2	2794.8	2.39762	6.37507	**18.0**
18.2	207.66	480.81	855.637	0.109136	887.1	1908.0	2795.1	2.40274	6.37107	**18.2**
18.4	208.20	481.35	854.986	0.107980	889.5	1905.9	2795.4	2.40781	6.36710	**18.4**
18.6	208.74	481.89	854.336	0.106848	892.0	1903.6	2795.6	2.41284	6.36316	**18.6**
18.8	209.27	482.42	853.694	0.105739	894.4	1901.5	2795.9	2.41783	6.35927	**18.8**
19.0	209.80	482.95	853.053	0.104653	896.8	1899.3	2796.1	2.42277	6.35541	**19.0**
19.2	210.32	483.47	852.413	0.103589	899.2	1897.1	2796.3	2.42767	6.35159	**19.2**
19.4	210.84	483.99	851.782	0.102545	901.6	1895.0	2796.6	2.43253	6.34781	**19.4**
19.6	211.36	484.51	851.151	0.101523	903.9	1892.9	2796.8	2.43735	6.34405	**19.6**
19.8	211.87	485.02	850.528	0.100520	906.3	1890.7	2797.0	2.44212	6.34033	**19.8**
20.0	212.37	485.52	849.899	0.0995361	908.6	1888.6	2797.2	2.44686	6.33665	**20.0**

(Contd.)

TABLE: 1.7 Thermodynamic Properties of Steam above 1 Atmosphere (Pressure Basis) [Contd.]

Pressure, p bar	Saturation Temperature t °C	Saturation Temperature T K	Liquid density, ρ_l kg/m³	Specific volume of vapour, v'' m³/kg	Specific enthalpy h kJ/kg	Specific enthalpy λ_v kJ/kg	Specific enthalpy H kJ/kg	Specific entropy s' kJ/(kg·K)	Specific entropy s'' kJ/(kg·K)	Press. p bar
20.5	213.63	486.78	848.356	0.0971583	914.3	1883.4	2797.7	2.45854	6.32758	**20.5**
21.0	214.85	488.00	846.840	0.0948898	920.0	1878.2	2798.2	2.46998	6.31870	**21.0**
21.5	216.06	489.21	845.337	0.0927232	925.5	1873.1	2798.6	2.48120	6.31000	**21.5**
22.0	217.24	490.39	843.853	0.0906516	931.0	1868.1	2799.1	2.49221	6.30148	**22.0**
22.5	218.41	491.56	842.382	0.0886688	936.3	1863.1	2799.4	2.50302	6.29312	**22.5**
23.0	219.55	492.70	840.937	0.0867692	941.6	1858.2	2799.8	2.51363	6.28493	**23.0**
23.5	220.68	493.83	839.504	0.0849477	946.8	1853.3	2800.1	2.52406	6.27688	**23.5**
24.0	221.78	494.93	838.082	0.0831994	951.9	1848.5	2800.4	2.53430	6.26899	**24.0**
24.5	222.87	496.02	836.687	0.0815200	957.0	1843.7	2800.7	2.54438	6.26123	**24.5**
25.0	223.94	497.09	835.296	0.0799053	962.0	1838.9	2800.9	2.55429	6.25361	**25.0**
25.5	225.00	498.15	833.924	0.0783518	966.9	1834.3	2801.2	2.56404	6.24611	**25.5**
26.0	226.04	499.19	832.563	0.0768560	971.7	1829.7	2801.4	2.57364	6.23874	**26.0**
26.5	227.06	500.21	831.221	0.0754146	976.5	1825.1	2801.6	2.58309	6.23149	**26.5**
27.0	228.07	501.22	829.889	0.0740247	981.2	1820.5	2801.7	2.59239	6.22436	**27.0**
27.5	229.07	502.22	828.569	0.0726836	985.9	1816.0	2801.9	2.60156	6.21733	**27.5**
28.0	230.05	503.20	827.260	0.0713887	990.5	1811.5	2802.0	2.61060	6.21041	**28.0**
28.5	231.01	504.16	825.962	0.0701377	995.0	1807.1	2802.1	2.61950	6.20359	**28.5**
29.0	231.97	505.12	824.674	0.0689282	999.5	1802.7	2802.2	2.62829	6.19687	**29.0**
29.5	232.91	506.06	823.404	0.0677584	1004.0	1798.2	2802.2	2.63695	6.19025	**29.5**
30.0	233.84	506.99	822.139	0.0666261	1008.4	1793.9	2802.3	2.64550	6.18372	**30.0**
30.5	234.76	507.91	820.883	0.0655297	1012.7	1789.6	2802.3	2.65393	6.17728	**30.5**
31.0	235.67	508.82	819.639	0.0644674	1017.0	1785.3	2802.3	2.66225	6.17092	**31.0**
31.5	236.56	509.71	818.404	0.0634376	1021.2	1781.1	2802.3	2.67047	6.16465	**31.5**
32.0	237.45	510.60	817.180	0.0624389	1025.4	1776.9	2802.3	2.67858	6.15845	**32.0**
32.5	238.32	511.47	815.967	0.0614698	1029.6	1772.7	2802.3	2.68660	6.15234	**32.5**
33.0	239.18	512.33	814.764	0.0605290	1033.7	1768.6	2802.3	2.69451	6.14630	**33.0**
33.5	240.04	513.19	813.564	0.0596154	1037.8	1764.4	2802.2	2.70234	6.14033	**33.5**
34.0	240.88	514.03	812.374	0.0587276	1041.8	1760.3	2802.1	2.71007	6.13444	**34.0**
34.5	241.72	514.87	811.194	0.0578646	1045.8	1756.3	2802.1	2.71771	6.12861	**34.5**
35.0	242.54	515.69	810.018	0.0570255	1049.7	1752.3	2802.0	2.72527	6.12285	**35.0**
35.5	243.36	516.51	808.852	0.0562091	1053.7	1748.1	2801.8	2.73274	6.11715	**35.5**
36.0	244.16	517.31	807.696	0.0554146	1057.6	1744.1	2801.7	2.74013	6.11152	**36.0**
36.5	244.96	518.11	806.543	0.0546410	1061.4	1740.2	2801.6	2.74744	6.10594	**36.5**
37.0	245.75	518.90	805.399	0.0538877	1065.2	1736.2	2801.4	2.75467	6.10043	**37.0**
37.5	246.54	519.69	804.266	0.0531537	1069.0	1732.3	2801.3	2.76182	6.09498	**37.5**
38.0	247.31	520.46	803.135	0.0524383	1072.7	1728.4	2801.1	2.76890	6.08958	**38.0**
38.5	248.08	521.23	802.015	0.0517408	1076.5	1724.5	2800.9	2.77591	6.08423	**38.5**
39.0	248.84	521.99	800.897	0.0510606	1080.1	17207.7	2800.8	2.78285	6.07894	**39.0**
39.5	249.59	522.74	799.789	0.0503969	1083.8	1716.7	2800.5	2.78972	6.07370	**39.5**
40.0	250.33	523.48	798.684	0.0497493	1087.4	1712.9	2800.3	2.79652	6.06851	**40.0**
40.5	251.07	524.22	797.582	0.0491170	1091.0	1709.1	2800.1	2.80326	6.06337	**40.5**
41.0	251.80	524.95	796.495	0.0484996	1094.6	1705.3	2799.9	2.80993	6.05827	**41.0**
41.5	252.52	525.67	795.406	0.0478965	1098.1	1701.5	2799.6	2.81655	6.05322	**41.5**
42.0	253.24	526.39	794.325	0.0473073	1101.6	1697.8	2799.4	2.82310	6.04822	**42.0**
42.5	253.95	527.10	793.254	0.0467314	1105.1	1694.0	2799.1	2.82959	6.04326	**42.5**

(Contd.)

TABLE: 1.7 Thermodynamic Properties of Steam above 1 Atmosphere (Pressure Basis) [Contd.]

Pressure, p	Saturation Temperature		Liquid density, ρ_l	Specific volume of vapour, v''	Specific enthalpy			Specific entropy		Press. p
	t	T			h	λ_v	H	s'	s''	
bar	°C	K	kg/m³	m³/kg	kJ/kg			kJ/(kg·K)		bar
43.0	254.66	527.81	792.180	0.0461685	1108.5	1690.4	2798.9	2.83602	6.03835	**43.0**
43.5	255.35	528.50	791.120	0.0456179	1112.0	1686.6	2798.6	2.84239	6.03347	**43.5**
44.0	256.05	529.20	790.058	0.0450795	1115.4	1682.9	2798.3	2.84871	6.02864	**44.0**
44.5	256.73	529.88	789.004	0.0445527	1118.8	1679.2	2798.0	2.85498	6.02365	**44.5**
45.0	257.41	530.56	787.954	0.0440371	1122.1	1675.6	2797.7	2.86119	6.01909	**45.0**
45.5	258.08	531.23	786.912	0.0435324	1125.5	1671.9	2797.4	2.86735	6.01437	**45.5**
46.0	258.75	531.90	785.873	0.0430383	1128.8	1668.2	2797.0	2.87346	6.00969	**46.0**
46.5	259.42	532.57	784.837	0.0425545	1132.0	1664.7	2796.7	2.87951	6.00505	**46.5**
47.0	260.07	533.22	783.810	0.0420805	1135.3	1661.1	2796.4	2.88552	6.00044	**47.0**
47.5	260.73	533.88	782.779	0.0416162	1138.6	1657.4	2796.0	2.89149	5.99587	**47.5**
48.0	261.37	534.52	781.757	0.0411611	1141.8	1653.9	2795.7	2.89740	5.99133	**48.0**
48.5	262.02	535.17	780.744	0.0407151	1145.0	1650.3	2795.3	2.90327	5.98682	**48.5**
49.0	262.65	535.80	779.727	0.0402778	1148.2	1646.8	2794.9	2.90909	5.98235	**49.0**
49.5	263.28	536.43	778.719	0.0398491	1151.3	1643.3	2794.6	2.91487	5.97791	**49.5**
50	263.91	537.06	777.714	0.0394285	1154.5	1639.7	2794.2	2.92060	5.97349	**50**
51	265.15	538.30	775.717	0.0386113	1160.7	1632.7	2793.4	2.93194	5.96476	**51**
52	266.37	539.52	773.730	0.0378242	1166.9	1625.7	2792.6	2.94312	5.95614	**52**
53	267.58	540.73	771.766	0.0370657	1172.9	1618.8	2791.7	2.95414	5.94763	**53**
54	268.76	541.91	769.805	0.0363342	1175.9	1611.9	2790.8	2.96501	5.93923	**54**
55	269.93	543.08	767.866	0.0356282	1184.9	1605.0	2789.9	2.97573	5.93092	**55**
56	271.09	544.24	765.937	0.0349465	1190.8	1598.2	2789.0	2.98631	5.92272	**56**
57	272.22	545.37	764.018	0.0342876	1196.6	1591.4	2788.0	2.99675	5.91461	**57**
58	273.35	546.50	762.114	0.0336506	1202.4	1584.6	2787.0	3.00706	5.90658	**58**
59	274.46	547.61	760.219	0.0330343	1208.1	1578.0	2786.0	3.01724	5.89865	**59**
60	275.55	548.70	758.334	0.0324378	1213.7	1571.3	2785.0	3.02730	5.89079	**60**
61	276.63	549.78	756.464	0.0318600	1219.3	1564.7	2783.9	3.03724	5.88302	**61**
62	277.70	550.85	754.597	0.0313000	1224.8	1558.1	2782.9	3.04706	5.87532	**62**
63	278.75	551.90	752.746	0.0307571	1230.3	1551.5	2781.8	3.05677	5.86769	**63**
64	279.79	552.94	750.903	0.0302304	1235.8	1544.9	2780.6	3.06637	5.86014	**64**
65	280.82	553.97	749.069	0.0297192	1241.1	1538.4	2779.5	3.07587	5.85265	**65**
66	281.84	554.99	747.239	0.0292229	1246.5	1531.8	2778.3	3.08527	5.84523	**66**
67	282.84	555.99	745.423	0.0287407	1251.8	1525.3	2777.1	3.09456	5.83788	**67**
68	283.84	556.99	743.616	0.0282721	1257.0	1518.9	2775.9	3.10376	5.83058	**68**
69	284.82	557.97	741.812	0.0278165	1262.2	1512.5	2774.7	3.11287	5.82334	**69**
70	285.79	558.94	740.017	0.0273733	1267.4	1506.0	2773.5	3.12189	5.81616	**70**
71	286.75	559.90	738.231	0.0269420	1272.6	1499.6	2772.2	3.13082	5.80904	**71**
72	287.70	560.85	736.448	0.0265223	1277.6	1493.3	2770.9	3.13966	5.80197	**72**
73	288.64	561.79	734.673	0.0261134	1282.7	1486.9	2769.6	3.14843	5.79494	**73**
74	289.57	562.72	732.907	0.0257152	1287.7	1480.6	2768.3	3.15711	5.78797	**74**
75	290.50	563.65	731.144	0.0253270	1292.7	1474.2	2766.9	3.16571	5.78105	**75**
76	291.41	564.56	729.389	0.0249486	1297.6	1467.9	2765.5	3.17424	5.77417	**76**
77	292.31	565.46	727.638	0.0245796	1302.6	1461.6	2764.2	3.18269	5.76734	**77**
78	293.21	566.36	725.890	0.0242196	1307.4	1455.4	2762.8	3.19107	5.76055	**78**
79	294.09	567.24	724.150	0.0238683	1312.3	1449.0	2761.3	3.19938	5.75380	**79**
80	294.97	568.12	722.418	0.0235253	1317.1	1442.8	2759.9	3.20762	5.74710	**80**

(Contd.)

TABLE: 1.7 Thermodynamic Properties of Steam above 1 Atmosphere (Pressure Basis) [Contd.]

Pressure, p	Saturation Temperature		Liquid density, ρ_l	Specific volume of vapour, v''	Specific enthalpy			Specific entropy		Press. p
	t	T			h	λ_v	H	s'	s''	
bar	°C	K	kg/m³	m³/kg	kJ/kg			kJ/(kg·K)		bar
81	295.84	568.99	720.684	0.0231904	1321.9	1436.5	2758.4	3.21580	5.74043	**81**
82	296.70	569.85	718.959	0.0228633	1326.6	1430.4	2757.0	3.22391	5.73380	**82**
83	297.55	570.70	717.237	0.0225436	1331.4	1424.1	2755.5	3.23196	5.72721	**83**
84	298.39	571.54	715.517	0.0222312	1336.1	1417.9	2754.0	3.23994	5.72065	**84**
85	299.23	572.38	713.806	0.0219258	1340.7	1411.8	2752.5	3.24787	5.71413	**85**
86	300.06	573.21	712.093	0.0216271	1345.4	1405.5	2750.9	3.25574	5.70764	**86**
87	300.88	574.03	710.389	0.0213349	1350.0	1399.4	2749.4	3.26355	5.70119	**87**
88	301.70	574.85	708.687	0.0210490	1354.6	1393.2	2747.8	3.27131	5.69476	**88**
89	302.51	575.66	706.984	0.0207693	1359.2	1387.0	2746.2	3.27901	5.68837	**89**
90	303.31	576.46	705.288	0.0204953	1363.7	1380.9	2744.6	3.28666	5.68201	**90**
91	304.10	577.25	703.591	0.0202271	1368.3	1374.7	2743.0	3.29426	5.67567	**91**
92	304.89	578.04	701.902	0.0199644	1372.8	1368.5	2741.3	3.30181	5.66937	**92**
93	305.67	578.82	700.211	0.0197070	1377.2	1362.5	2739.7	3.30931	5.66309	**93**
94	306.44	579.59	698.524	0.0194548	1381.7	1356.3	2738.0	3.31676	5.65683	**94**
95	307.21	580.36	696.840	0.0192076	1386.1	1350.3	2736.4	3.32417	5.65060	**95**
96	307.97	581.12	695.154	0.0189653	1390.6	1344.1	2734.7	3.33153	5.64440	**96**
97	308.73	581.88	693.476	0.0187276	1395.0	1338.0	2733.0	3.33885	5.63822	**97**
98	309.48	582.63	691.793	0.0184945	1399.3	1331.9	2731.2	3.34612	5.63206	**98**
99	310.22	583.37	690.117	0.0182658	1403.7	1325.8	2729.5	3.35335	5.62592	**99**
100	310.96	584.11	688.440	0.0180413	1408.0	1319.7	2727.7	3.36055	5.61980	**100**
102	312.42	585.57	685.091	0.0176047	1416.7	1307.5	2724.2	3.37481	5.60762	**102**
104	313.86	587.01	681.742	0.0171839	1425.2	1295.4	2720.6	3.38892	5.59551	**104**
106	315.27	588.42	678.403	0.0167777	1433.7	1283.2	2716.9	3.40289	5.58347	**106**
108	316.67	589.82	675.060	0.0163854	1442.2	1270.9	2713.1	3.41672	5.57147	**108**
110	318.05	591.20	671.718	0.0160062	1450.6	1258.7	2709.3	3.43042	5.55953	**110**
112	319.40	592.55	668.378	0.0156394	1458.9	1246.5	2705.4	3.44399	5.54763	**112**
114	320.74	593.89	665.035	0.0152843	1467.2	1234.4	2701.5	5.45745	5.53575	**114**
116	322.06	595.21	661.686	0.0149402	1475.4	1222.0	2697.4	3.47079	5.52389	**116**
118	323.36	596.51	658.337	0.0146067	1483.6	1209.7	2693.3	3.48404	5.51205	**118**
120	324.65	597.80	654.982	0.0142830	1491.8	1197.4	2689.2	3.49718	5.50022	**120**
122	325.91	599.06	651.619	0.0139688	1499.9	1185.0	2684.9	3.51022	5.48837	**122**
124	327.17	600.32	648.248	0.0136635	1508.0	1172.6	2680.6	3.52318	5.47652	**124**
126	328.40	601.55	644.870	0.0133667	1516.0	1160.1	2676.1	3.53605	5.46465	**126**
128	329.62	602.77	641.482	0.0130780	1524.0	1147.6	2671.6	3.54884	5.45274	**128**
130	330.83	603.98	638.085	0.0127970	1532.0	1135.0	2667.0	3.56157	5.44080	**130**
132	332.02	605.17	634.675	0.0125232	1540.0	1122.3	2662.3	3.57422	5.42881	**132**
134	333.19	606.34	631.249	0.0122564	1547.9	1109.5	2657.4	3.58681	5.41677	**134**
136	334.63	607.78	627.809	0.0119963	1555.8	1096.7	2652.5	3.59934	5.40466	**136**
138	335.51	608.66	624.352	0.0117427	1563.8	1083.8	2647.5	3.61182	5.39250	**138**
140	336.64	609.79	620.875	0.0114950	1571.6	1070.8	2642.4	3.62424	5.38026	**140**
142	337.76	610.91	617.383	0.0112532	1579.5	1057.6	2637.1	3.63663	5.36794	**142**
144	338.87	612.02	613.866	0.0110171	1587.4	1044.4	2631.8	3.64898	5.35554	**144**
146	339.97	613.12	610.325	0.0107863	1595.3	1031.0	2626.3	3.66130	5.34306	**146**
148	341.06	614.21	606.759	0.0105608	1603.1	1017.6	2620.7	3.67359	5.33048	**148**
150	342.13	615.28	603.169	0.0103402	1611.0	1004.0	2615.0	3.68585	5.31728	**150**

(Contd.)

TABLE: 1.7 Thermodynamic Properties of Steam above 1 Atmosphere (Pressure Basis) [Contd.]

Pressure, p	Saturation Temperature		Liquid density,	Specific volume of vapour, v''	Specific enthalpy			Specific entropy		Press. p
	t	T	ρ_l		h	λ_v	H	s'	s''	
bar	°C	K	kg/m³	m³/kg	kJ/kg			kJ/(kg·K)		bar
152	343.19	616.34	599.546	0.0101245	1618.9	990.3	2609.2	3.69810	5.30506	**152**
154	344.24	617.39	595.887	0.0099136	1626.8	976.5	2603.3	3.71035	5.29222	**154**
156	345.28	618.43	592.196	0.0097072	1634.7	962.6	2597.3	3.72259	5.27928	**156**
158	346.31	619.46	588.464	0.0095052	1642.6	948.5	2591.1	3.73404	5.26625	**158**
160	347.33	620.48	584.689	0.0093075	1650.5	934.4	2584.9	3.74710	5.25314	**160**
162	348.34	621.49	580.865	0.0091141	1658.5	920.0	2578.5	3.75938	5.23994	**162**
164	349.33	622.48	576.994	0.0089247	1666.5	905.6	2572.1	3.77170	5.22667	**164**
166	350.32	623.47	573.164	0.0087385	1674.5	891.0	2565.5	3.78424	5.21322	**166**
168	351.30	624.45	569.184	0.0085534	1683.0	875.6	2558.6	3.79737	5.19942	**168**
170	352.26	625.41	565.099	0.0083710	1691.7	859.9	2551.6	5.81070	5.18547	**170**
172	353.22	626.37	560.978	0.0081911	1700.4	844.0	2544.4	3.82400	5.17134	**172**
174	354.17	627.32	556.762	0.0080140	1709.0	828.1	2537.1	3.83721	5.15703	**174**
176	355.11	628.26	552.456	0.0078394	1717.6	811.9	2529.5	3.85036	5.14251	**176**
178	356.04	629.19	548.035	0.0076673	1726.2	795.6	2521.8	3.86345	5.12776	**178**
180	356.04	629.19	543.508	0.0074977	1734.8	779.1	2513.9	3.87654	5.11277	**180**
182	357.87	631.02	538.909	0.0073302	1743.4	762.4	2505.8	3.88964	5.09752	**182**
184	358.77	631.92	534.160	0.0071646	1752.1	745.3	2497.4	3.90279	5.08196	**184**
186	359.67	632.82	529.297	0.0070009	1760.9	727.9	2488.8	3.91602	5.06608	**186**
188	360.55	633.70	524.329	0.0068386	1769.7	710.1	2479.8	3.92937	5.04982	**188**
190	361.43	634.58	519.211	0.0066775	1778.7	691.9	2470.6	3.94288	5.03316	**190**
192	362.30	635.45	513.927	0.0065172	1787.7	673.4	2461.1	3.95660	5.01602	**192**
194	363.16	636.31	508.492	0.0063575	1797.0	654.1	2451.1	3.97058	4.99833	**194**
196	364.02	637.17	502.866	0.0061978	1806.5	634.2	2440.7	3.98490	4.98005	**196**
198	364.86	638.01	497.018	0.0060377	1616.3	613.5	2429.8	3.99963	4.96104	**198**
200	365.70	638.85	490.918	0.0058765	1826.5	591.8	2418.3	4.01487	4.94120	**200**
202	366.53	639.68	484.520	0.0057138	1837.0	569.2	2406.2	4.03075	4.92040	**202**
204	367.36	640.51	477.760	0.0055484	1848.1	545.2	2393.3	4.04743	4.89841	**204**
206	368.17	641.32	470.544	0.0053792	1859.9	519.4	2379.3	4.06512	4.87499	**206**
208	368.98	642.13	462.770	0.0052050	1872.5	491.7	2364.2	4.08412	4.84979	**208**
210	369.78	642.93	454.236	0.0050234	1886.2	461.4	2347.6	4.10483	4.82230	**210**
212	370.58	643.73	444.682	0.0048313	1901.5	427.4	2328.9	4.12788	4.79174	**212**
214	371.37	644.52	433.633	0.0046239	1919.0	388.4	2307.4	4.15432	4.75687	**214**
216	372.15	645.30	420.309	0.0043915	1939.9	341.6	2281.5	4.18603	4.71533	**216**
218	372.92	646.07	402.706	0.0041149	1967.2	280.8	2248.0	4.22755	4.66211	**218**
220	373.69	646.84	374.406	0.0037265	2011.0	184.4	2195.4	4.29451	4.57957	**220**
221.20	**374.15***	**647.30**	**315.457**	**0.00317**	**2107.4**	**0.0**	**2107.4**	**4.44286**	**4.44286**	**221.20**

*Critical point

TABLE: 1.8 Thermodynamic Properties of Steam (Superheat Region)

Temperature		Absolute Pressure (Saturation Temperature)									Temp.
t	T	0.01 bar (6.98°C/280.13K)			0.03 bar (24.10°C/297.25K)			0.05 bar (32.90°C/306.05K)			t
°C	K	v	i	s	v	i	s	v	i	s	°C
Satd. cond.		129.209	2514.4	8.9767	45.667	2545.6	8.5785	28.194	2561.6	8.3960	
40	313.15	144.472	2575.9	9.1842	48.124	2575.4	8.6760	28.854	2574.9	8.4390	40
60	333.15	153.713	2613.3	9.3001	51.211	2612.9	8.7922	30.711	2612.6	8.5555	60
80	353.15	162.951	2650.9	9.4096	54.296	2650.6	8.9019	32.565	2650.3	8.6655	80
100	373.15	172.187	2688.6	9.5136	57.378	2688.4	9.0060	34.417	2688.1	8.7698	100
120	393.15	181.421	2726.5	9.6125	60.460	2726.3	9.1051	36.267	2726.1	8.8690	120
140	413.15	190.655	2764.6	9.7070	63.540	2764.5	9.1997	38.117	2764.3	8.9636	140
160	433.15	199.888	2802.9	9.7975	66.619	2802.8	9.2902	39.966	2802.6	9.0542	160
180	453.15	209.120	2841.4	9.8843	69.698	2841.3	9.3771	41.814	2841.2	9.1412	180
200	473.15	218.352	2880.1	9.9679	72.777	2880.0	9.4607	43.661	2879.9	9.2248	200
220	493.15	227.584	2919.0	10.0484	75.855	2918.9	9.5412	45.509	2918.8	9.3054	220
240	513.15	236.815	2958.1	10.1262	78.932	2958.0	9.6190	47.356	2957.9	9.3832	240
260	533.15	246.046	2997.4	10.2014	82.010	2997.4	9.6943	49.203	2997.3	9.4584	260
280	553.15	255.277	3037.0	10.2743	85.088	3037.0	9.7672	51.050	3036.9	9.5313	280
300	573.15	264.508	3076.8	10.3450	88.165	3076.8	9.8379	52.896	3076.7	9.6021	300
320	593.15	273.739	3116.9	10.4137	91.242	3116.9	9.9066	54.743	3116.8	9.6708	320
340	613.15	282.969	3157.2	10.4805	94.320	3157.2	9.9735	56.590	3157.1	9.7377	340
360	633.15	292.200	3197.8	10.5457	97.397	3197.8	10.0386	58.436	3197.7	9.8028	360
380	653.15	301.431	3238.6	10.6091	100.474	3238.6	10.1021	60.283	3238.6	9.8663	380
400	673.15	310.661	3279.7	10.6711	103.551	3279.7	10.1641	62.129	3279.7	9.9283	400
420	693.15	319.892	3321.1	10.7317	106.628	3321.1	10.2246	63.975	3321.0	9.9888	420
440	713.15	329.122	3362.7	10.7909	109.705	3362.7	10.2838	65.822	3362.7	10.0480	440
460	733.15	338.353	3404.6	10.8488	112.782	3404.6	10.3418	67.668	3404.6	10.1060	460
480	753.15	347.583	3446.8	10.9056	115.859	3446.8	10.3985	69.514	3446.7	10.1627	480
500	773.15	356.813	3489.2	10.9612	118.936	3489.2	10.4541	71.360	3489.2	10.2184	500

Temperature		Absolute Pressure (Saturation Temperature)									Temp.
t	T	0.10 bar (45.83°C/318.98K)			0.14 bar (52.57°C/325.72K)			0.18 bar (57.83°C/330.98K)			t
°C	K	v	i	s	v	i	s	v	i	s	°C
Satd. cond.		14.675	2584.8	8.1511	10.694	2596.7	8.0334	8.445	2605.9	7.9460	
60	333.15	15.336	2611.6	8.2334	11.276	2629.9	8.1325	8.502	2610.1	7.9505	60
80	353.15	16.266	2649.5	8.3439	11.610	2648.9	8.1872	9.023	2648.3	8.0699	80
100	373.15	17.195	2687.5	8.4486	12.275	2687.0	8.2922	9.541	2686.5	8.1752	100
120	393.15	18.123	2725.6	8.5481	12.939	2725.2	8.3920	10.059	2724.8	8.2752	120
140	413.15	19.050	2763.9	8.6430	13.602	2763.6	8.4871	10.575	2763.2	8.3705	140
160	433.15	19.975	2802.3	8.7337	14.264	2802.0	8.5780	11.091	2801.8	8.4615	160
180	453.15	20.900	2840.9	8.8208	14.925	2840.7	8.6652	11.606	2840.4	8.5488	180
200	473.15	21.825	2879.6	8.9045	15.586	2879.5	8.7489	12.120	2879.3	8.6326	200
220	493.15	22.750	2918.6	8.9852	16.247	2918.4	8.8296	12.634	2918.3	8.7134	220
240	513.15	23.674	2957.8	9.0630	16.907	2957.6	8.9075	13.148	2957.5	8.7914	240
260	533.15	24.598	2997.2	9.1383	17.567	2997.0	8.9829	13.662	2996.9	8.8667	260
280	553.15	25.521	3036.8	9.2113	18.228	3036.7	9.0558	14.175	3036.6	8.9397	280
300	573.15	26.445	3076.6	9.2820	18.887	3076.5	9.1266	14.689	3076.4	9.0105	300
320	593.15	27.369	3116.7	9.3508	19.547	3116.6	9.1954	15.202	3116.5	9.0793	320
340	613.15	28.292	3157.0	9.4177	20.207	3157.0	9.2623	15.715	3156.9	9.1462	340
360	633.15	29.216	3197.6	9.4828	20.867	3197.6	9.3274	16.229	3197.5	9.2114	360
380	653.15	30.139	3238.5	9.5463	21.527	3238.4	9.3910	16.742	3238.3	9.2749	380
400	673.15	31.062	3279.6	9.6083	22.186	3279.5	9.4530	17.255	3279.5	9.3369	400
420	693.15	31.986	3321.0	9.6689	22.846	3320.9	9.5135	17.768	3320.8	9.3975	420
440	713.15	32.909	3362.6	9.7281	23.505	3362.5	9.5727	18.281	3362.5	9.4567	440
460	733.15	33.832	3404.5	9.7860	24.165	3404.4	9.6307	18.794	3404.4	9.5147	460
480	753.15	34.756	3446.7	9.8428	24.825	3446.6	9.6875	19.307	3446.6	9.5714	480
500	773.15	35.679	3489.1	9.8984	25.484	3489.1	9.7431	19.820	3489.0	9.6271	500
520	793.15	36.602	3531.9	9.9530	26.144	3531.7	9.7977	20.333	3531.8	9.6816	520

(Contd.)

TABLE: 1.8 Thermodynamic Properties of Steam (Superheat Region) [Contd.]

Temperature		Absolute Pressure (Saturation Temperature)									Temp.
t	*T*	0.22 bar (62.16°C/335.31K)			0.26 bar (65.87°C/339.02K)			0.30 bar (69.12°C/342.27K)			*t*
°C	K	*v*	*i*	*s*	*v*	*i*	*s*	*v*	*i*	*s*	°C
Satd.	**Cond.**	6.995	2613.5	7.8765	5.980	2619.9	7.8188	5.229	2625.4	7.7695	
80	353.15	7.376	2647.7	7.9759	6.237	2647.1	7.8974	5.401	2646.5	7.8300	**80**
100	373.15	7.802	2686.0	8.0816	6.598	2685.6	8.0034	5.714	2685.1	7.9363	**100**
120	393.15	8.226	2724.4	8.1818	6.957	2724.0	8.1039	6.027	2723.6	8.0370	**120**
140	413.15	8.649	2762.9	8.2772	7.316	2762.6	8.1995	6.338	2762.3	8.1329	**140**
160	433.15	9.071	2801.5	8.3684	7.673	2801.2	8.2908	6.648	2801.0	8.2243	**160**
180	453.15	9.493	2840.2	8.4558	8.031	2840.0	8.3783	6.958	2839.8	8.3119	**180**
200	473.15	9.914	2879.1	8.5397	8.387	2878.9	8.4623	7.268	2878.7	8.3960	**200**
220	493.15	10.335	2918.1	8.6206	8.744	2918.0	8.5432	7.577	2917.8	8.4769	**220**
240	513.15	10.756	2957.4	8.6985	9.100	2957.2	8.6212	7.885	2957.1	8.5550	**240**
260	533.15	11.176	2996.8	8.7739	9.456	2996.7	8.6967	8.194	2996.6	8.6305	**260**
280	553.15	11.597	3036.5	8.8470	9.812	3036.3	8.7697	8.502	3036.2	8.7035	**280**
300	573.15	12.017	3076.3	8.9178	10.167	3076.2	8.8406	8.811	3076.1	8.7744	**300**
320	593.15	12.437	3116.5	8.9866	10.523	3116.4	8.9094	9.119	3116.3	8.8432	**320**
340	613.15	12.857	3156.8	9.0535	10.878	3156.7	8.9763	9.427	3156.7	9.9102	**340**
360	633.15	13.277	3197.4	9.1187	11.234	3197.3	9.0415	9.735	3197.3	8.9754	**360**
380	653.15	13.697	3238.3	9.1822	11.589	3238.2	9.1050	10.043	3238.1	9.0389	**380**
400	673.15	14.117	3279.4	9.2442	11.945	3279.2	9.1671	10.351	3279.3	9.1010	**400**
420	693.15	14.537	3320.8	9.3048	12.300	3320.7	9.2276	10.659	3320.7	9.1615	**420**
440	713.15	14.957	3362.4	9.3640	12.655	3362.4	9.2869	10.967	3362.2	9.2208	**440**
460	733.15	15.377	3404.3	9.4220	13.010	3404.3	9.3449	11.275	3404.2	9.2788	**460**
480	753.15	15.796	3446.5	9.4788	13.366	3446.5	9.4016	11.583	3446.4	9.3355	**480**
500	773.15	16.216	3489.9	9.5344	13.721	3489.0	9.4573	11.891	3488.9	9.3912	**500**
520	793.15	16.636	3531.7	9.5890	14.076	3531.7	9.5118	12.199	3531.6	9.4458	**520**
540	813.15	17.056	3574.7	9.6425	14.431	3574.7	9.5654	12.507	3574.7	9.4993	**540**

Temperature		Absolute Pressure (Saturation Temperature)									Temp.
t	*T*	0.34 bar (72.03°C/345.18K)			0.38 bar (74.66°C/347.81K)			0.42 bar (77.06°C/350.21K)			*t*
°C	K	*v*	*i*	*s*	*v*	*i*	*s*	*v*	*i*	*s*	°C
Satd.	**Cond.**	4.650	2630.4	7.7266	4.190	2634.8	7.6884	3.815	2638.9	7.6542	
80	353.15	4.761	2645.8	7.7708	4.257	2645.2	7.7181	3.848	2644.6	7.6705	**80**
100	373.15	5.039	2684.6	7.8775	4.506	2684.1	7.8251	4.074	2683.6	7.7779	**100**
120	393.15	5.315	2723.2	7.9785	4.753	2722.8	7.9263	4.299	2722.4	7.8793	**120**
140	413.15	5.590	2761.9	8.0745	5.000	2761.6	8.0225	4.522	2761.3	7.9757	**140**
160	433.15	5.864	2800.7	8.1661	5.245	2800.4	8.1143	4.744	2800.1	8.0676	**160**
180	453.15	6.138	2839.5	8.2537	5.491	2839.3	8.2020	4.966	2839.1	8.1555	**180**
200	473.15	6.411	2878.5	8.3379	5.735	2878.3	8.2863	5.188	2878.1	8.2398	**200**
220	493.15	6.684	2917.6	8.4189	5.979	2917.5	8.3673	5.409	2917.3	8.3209	**220**
240	513.15	6.957	2956.9	8.4970	6.223	2956.8	8.4455	5.630	2956.7	8.3991	**240**
260	533.15	7.229	2996.4	8.5725	6.467	2996.3	8.5210	5.851	2996.2	8.4747	**260**
280	553.15	7.501	3036.1	8.6456	6.711	3036.0	8.5942	6.071	3035.9	8.5478	**280**
300	573.15	7.773	3076.0	8.7165	6.955	3076.0	8.6651	6.292	3075.9	8.6188	**300**
320	593.15	8.045	3116.2	8.7854	7.198	3116.1	8.7339	6.512	3116.0	8.6876	**320**
340	613.15	8.317	3156.6	8.8523	7.441	3156.5	8.8009	6.732	3156.4	8.7546	**340**
360	633.15	8.589	3197.2	8.9175	7.685	3197.1	8.8661	6.952	3197.1	8.8198	**360**
380	653.15	8.861	3238.1	8.9811	7.928	3238.0	8.9297	7.173	3238.0	8.8834	**380**
400	673.15	9.133	3279.2	9.0431	8.171	3279.2	8.9917	7.393	3279.1	8.9455	**400**
420	693.15	9.405	3320.6	9.1037	8.414	3320.6	9.0523	7.613	3320.5	9.0061	**420**
440	713.15	9.677	3362.3	9.1630	8.658	3362.2	9.1116	7.833	3362.2	9.0653	**440**
460	733.15	9.948	3404.2	9.2210	8.901	3404.1	9.1696	8.053	3404.1	9.1233	**460**
480	753.15	10.220	3446.4	9.2777	9.144	3446.3	9.2264	8.273	3446.3	9.1801	**480**
500	773.15	10.492	3488.9	9.3334	9.387	3488.8	9.2820	8.493	3488.8	9.2358	**500**
520	793.15	10.763	3531.6	9.3880	9.630	3531.6	9.3366	8.713	3531.5	9.2904	**520**
540	813.15	11.035	3574.6	9.4415	9.873	3574.6	9.3902	8.933	3574.5	9.3439	**540**

(Contd.)

TABLE: 1.8 Thermodynamic Properties of Steam (Superheat Region) [Contd.]

Temperature		Absolute Pressure (Saturation Temperature)									Temp.
t	*T*	0.46 bar (79.28°C/352.43K)			0.50 bar (81.35°C/354.50K)			0.55 bar (83.74°C/356.89K)			*t*
°C	K	*v*	*i*	*s*	*v*	*i*	*s*	*v*	*i*	*s*	°C
Satd.	**Cond.**	3.503	2642.6	7.6232	3.240	2646.0	7.5947	2.964	2649.9	7.5623	
100	373.15	3.718	2683.1	7.7348	3.418	2682.6	7.6953	3.105	2681.9	7.6499	**100**
120	393.15	3.923	2722.0	7.8365	3.607	2721.6	7.7972	3.277	2721.1	7.7522	**120**
140	413.15	4.127	2760.9	7.9331	3.796	2760.6	7.8940	3.449	2760.2	7.8492	**140**
160	433.15	4.331	2799.9	8.0251	3.983	2799.6	7.9861	3.619	2799.3	7.9415	**160**
180	453.15	4.533	2838.9	8.1131	4.170	2838.6	8.0742	3.789	2838.3	8.0298	**180**
200	473.15	4.736	2877.9	8.1975	4.356	2877.7	8.1587	3.959	2877.5	8.1143	**200**
220	493.15	4.938	2917.2	8.2787	4.542	2917.0	8.2399	4.128	2916.8	8.1956	**220**
240	513.15	5.140	2956.5	8.3569	4.728	2956.4	8.3182	4.297	2956.2	8.2740	**240**
260	533.15	5.341	2996.1	8.4325	4.913	2995.9	8.3939	4.466	2995.8	8.3497	**260**
280	553.15	5.543	3035.8	8.5057	5.099	3035.7	8.4671	4.634	3035.6	8.4229	**280**
300	573.15	5.744	3075.8	8.5766	5.284	3075.7	8.5380	4.803	3075.5	8.4939	**300**
320	593.15	5.945	3115.9	8.6455	5.469	3115.9	8.6070	4.971	3115.7	8.5628	**320**
340	613.15	6.146	3156.3	8.7125	5.654	3156.3	8.6740	5.140	3156.2	8.6299	**340**
360	633.15	6.347	3197.0	8.7778	5.839	3196.9	8.7392	5.308	3196.8	8.6951	**360**
380	653.15	6.548	3237.9	8.8414	6.024	3237.8	8.8028	5.476	3237.7	8.7587	**380**
400	673.15	6.749	3279.0	8.9034	6.209	3279.0	8.8649	5.644	3278.9	8.8208	**400**
420	693.15	6.950	3320.4	8.9640	6.394	3320.4	8.9255	5.812	3320.3	8.8814	**420**
440	713.15	7.151	3362.1	9.0233	6.579	3362.1	8.9848	5.981	3362.0	8.9407	**440**
460	733.15	7.352	3404.0	9.0813	6.764	3404.0	9.0428	6.149	3403.1	8.9987	**460**
480	753.15	7.553	3446.3	9.1381	6.949	3446.2	9.0996	6.317	3446.2	9.0555	**480**
500	773.15	7.754	3488.7	9.1938	7.133	3488.7	9.1552	6.485	3488.6	9.1112	**500**
520	793.15	7.955	3531.5	9.2483	7.318	3531.4	9.2098	6.653	3531.4	9.1658	**520**
540	813.15	8.156	3574.5	9.3019	7.503	3574.5	9.2634	6.821	3574.4	9.2194	**540**
560	833.15	8.357	3617.8	9.3545	7.688	3617.8	9.3160	6.989	3617.7	9.2720	**560**

Temperature		Absolute Pressure (Saturation Temperature)									Temp.
t	*T*	0.60 bar (85.95°C/359.10K)			0.70 bar (89.96°C/363.11K)			0.80 bar (93.51°C/366.66K)			*t*
°C	K	*v*	*i*	*s*	*v*	*i*	*s*	*v*	*i*	*s*	°C
Satd.	**Cond.**	2.732	2653.6	7.5327	2.365	2660.1	7.4804	2.087	2665.8	7.4352	
100	373.15	2.844	2681.3	7.6085	2.434	2680.0	7.5346	2.126	2678.8	7.4703	**100**
120	393.15	3.002	2720.6	7.7111	2.570	2719.6	7.6379	2.246	2718.6	7.5742	**120**
140	413.15	3.160	2759.8	7.8083	2.706	2759.0	7.7355	2.365	2758.1	7.6723	**140**
160	433.15	3.317	2798.9	7.9008	2.841	2798.2	7.8284	2.484	2797.5	7.7655	**160**
180	453.15	3.473	2838.1	7.9891	2.975	2837.5	7.9170	2.601	2836.9	7.8544	**180**
200	473.15	3.628	2877.3	8.0738	3.108	2876.8	8.0019	2.718	2876.3	7.9395	**200**
220	493.15	3.783	2916.6	8.1552	3.241	2916.2	8.0834	2.835	2915.8	8.0212	**220**
240	513.15	3.938	2956.0	8.2336	3.374	2955.7	8.1619	2.952	2955.3	8.0998	**240**
260	533.15	4.093	2995.6	8.3093	3.507	2995.3	8.2377	3.068	2995.0	8.1757	**260**
280	553.15	4.248	3035.4	8.3826	3.640	3035.2	8.3111	3.184	3034.9	8.2491	**280**
300	573.15	4.402	3075.4	8.4536	3.772	3075.2	8.3822	3.300	3075.0	8.3202	**300**
320	593.15	4.557	3115.6	8.5226	3.905	3115.4	8.4511	3.416	3115.2	8.3893	**320**
340	613.15	4.711	3156.1	8.5896	4.037	3155.9	8.5182	3.532	3155.7	8.4564	**340**
360	633.15	4.865	3196.7	8.6549	4.169	3196.6	8.5835	3.648	3196.4	8.5217	**360**
380	653.15	5.019	3237.7	8.7185	4.302	3237.5	8.6472	3.763	3237.3	8.5854	**380**
400	673.15	5.174	3278.8	8.7806	4.434	3278.7	8.7093	3.879	3278.5	8.6475	**400**
420	693.15	5.328	3320.2	8.8412	4.566	3320.1	8.7699	3.995	3320.0	8.7081	**420**
440	713.15	5.482	3361.9	8.9005	4.698	3361.8	8.8292	4.111	3361.7	8.7675	**440**
460	733.15	5.636	3403.9	8.9585	4.830	3403.8	8.8872	4.226	3403.6	8.8255	**460**
480	753.15	5.790	3446.1	9.0153	4.962	3446.0	8.9441	4.342	3445.9	8.8823	**480**
500	773.15	5.944	3488.6	9.0710	5.095	3488.5	8.9997	4.457	3488.4	8.9380	**500**
520	793.15	6.098	3531.3	9.1256	5.227	3531.2	9.0543	4.573	3531.1	8.9926	**520**
540	813.15	6.252	3574.4	9.1792	5.359	3574.3	9.1079	4.688	3574.2	9.0462	**540**
560	833.15	6.406	3617.7	9.2318	5.491	3617.6	9.1606	4.804	3617.5	9.0998	**560**

(Contd.)

TABLE: 1.8 Thermodynamic Properties of Steam (Superheat Region) [Contd.]

Temperature		Absolute Pressure (Saturation Temperature)									Temp.
t	*T*	0.90 bar (96.71°C/369.86K)			1.0 bar (99.63 °C/372.78K)			1.5 bar (111.37°C/384.52K)			*t*
°C	K	*v*	*i*	*s*	*v*	*i*	*s*	*v*	*i*	*s*	°C
Satd.	**Cond.**	1.869	2670.9	7.3954	1.694	2675.4	7.3598	1.159	2693.4	7.2234	
120	393.15	1.994	2717.5	7.5177	1.793	2716.5	7.4670	1.188	2711.2	7.2693	**120**
140	413.15	2.101	2757.3	7.6164	1.889	2756.4	7.5662	1.253	2752.2	7.3709	**140**
160	433.15	2.206	2796.9	7.7099	1.984	2796.2	7.6601	1.317	2792.7	7.4667	**160**
180	453.15	2.311	2836.4	7.7991	2.078	2835.8	7.7495	1.381	2832.9	7.5574	**180**
200	473.15	2.415	2875.8	7.8843	2.172	2875.4	7.8349	1.444	2872.9	7.6439	**200**
220	493.15	2.519	2915.4	7.9662	2.266	2915.0	7.9169	1.507	2912.9	7.7266	**220**
240	513.15	2.623	2955.0	8.0449	2.359	2954.6	7.9958	1.570	2952.9	7.8061	**240**
260	533.15	2.726	2994.7	8.1209	2.453	2994.4	8.0719	1.603	2992.9	7.8826	**260**
280	553.15	2.829	3034.6	8.1944	2.546	3034.4	8.1454	1.695	3033.0	7.9565	**280**
300	573.15	2.933	3074.7	8.2656	2.639	3074.5	8.2166	1.757	3073.3	8.0280	**300**
320	593.15	3.036	3115.0	8.3346	2.732	3114.8	8.2857	1.819	3113.7	8.0973	**320**
340	613.15	3.139	3155.5	8.4018	2.824	3155.3	8.3529	1.881	3154.3	8.1646	**340**
360	633.15	3.242	3196.2	8.4671	2.917	3196.0	8.4183	1.943	3195.1	8.2301	**360**
380	653.15	3.345	3237.2	8.5308	3.010	3237.0	8.4320	2.005	3236.2	8.2940	**380**
400	673.15	3.448	3278.4	8.5929	3.102	3278.2	8.5442	2.067	3277.5	8.3562	**400**
420	693.15	3.551	3319.8	8.6536	3.195	3319.7	8.6049	2.129	3319.0	8.4170	**420**
440	713.15	3.653	3361.5	8.7130	3.288	3361.4	8.6642	2.191	3360.7	8.4764	**440**
460	733.15	3.756	3403.5	8.7710	3.380	3403.4	8.7223	2.252	3402.8	8.5345	**460**
480	753.15	3.859	3445.7	8.8278	3.473	3445.6	8.7791	2.314	3445.0	8.5914	**480**
500	773.15	3.962	3488.3	8.8835	3.565	3488.1	8.8348	2.376	3487.6	8.6472	**500**
520	793.15	4.065	3531.0	8.9382	3.658	3530.9	8.8894	2.438	3530.4	8.7018	**520**
540	813.15	4.167	3574.1	8.9918	3.750	3574.0	8.9431	2.499	3573.5	8.7555	**540**
560	833.15	4.270	3617.4	9.0444	3.843	3617.3	8.9957	2.561	3616.9	8.8082	**560**
580	853.15	4.373	3661.0	9.0961	3.935	3660.9	9.0474	2.623	3660.5	8.8599	**580**

Temperature		Absolute Pressure (Saturation Temperature)									Temp.
t	*T*	2.0 bar (120.23°C/393.38K)			2.5 bar (127.43°C/400.58K)			3.0 bar (133.54°C/406.69K)			*t*
°C	K	*v*	*i*	*s*	*v*	*i*	*s*	*v*	*i*	*s*	°C
Satd.	**Cond.**	0.8854	2706.3	7.1268	0.7184	2716.4	7.0520	0.6056	2724.7	6.9909	
140	413.15	0.9349	2747.8	7.2298	0.7440	2743.3	7.1183	0.6167	2738.8	7.0254	**140**
160	433.15	0.9840	2789.1	7.3275	0.7840	2785.5	7.2179	0.6506	2781.8	7.1271	**160**
180	453.15	1.032	2830.0	7.4196	0.8232	2827.0	7.3115	0.6837	2824.0	7.2222	**180**
200	473.15	1.080	2870.5	7.5072	0.8620	2868.0	7.4001	0.7164	2865.5	7.3119	**200**
220	493.15	1.128	2910.8	7.5907	0.9004	2908.7	7.4845	0.7486	2906.6	7.3971	**220**
240	513.15	1.175	2951.1	7.6707	0.9385	2949.3	7.5651	0.7805	2947.5	7.4783	**240**
260	533.15	1.222	2991.4	7.7477	0.9763	2989.8	7.6425	0.8123	2988.2	7.5562	**260**
280	553.15	1.269	3031.7	7.8219	1.014	3030.3	7.7171	0.8438	3028.9	7.6311	**280**
300	573.15	1.316	3072.1	7.8937	1.052	3070.9	7.7891	0.8753	3069.7	7.7034	**300**
320	593.15	1.363	3112.6	7.9632	1.089	3111.5	7.8589	0.9066	3110.4	7.7734	**320**
340	613.15	1.410	3153.3	8.0307	1.127	3152.4	7.9266	0.9379	3151.4	7.8412	**340**
360	633.15	1.456	3194.2	8.0964	1.164	3193.3	7.9923	0.9691	3192.4	7.9072	**360**
380	653.15	1.503	3235.4	8.1603	1.201	3234.5	8.0564	1.000	3233.7	7.9713	**380**
400	673.15	1.549	3276.7	8.2226	1.238	3275.9	8.1188	1.031	3275.2	8.0338	**400**
420	693.15	1.596	3318.3	8.2835	1.276	3317.6	8.1797	1.062	3316.8	8.0949	**420**
440	713.15	1.642	3360.1	8.3429	1.313	3359.4	8.2393	1.094	3358.8	8.1545	**440**
460	733.15	1.688	3402.1	8.4011	1.350	3401.5	8.2975	1.125	3400.9	8.2128	**460**
480	753.15	1.735	3444.5	8.4581	1.387	3443.9	8.3545	1.156	3443.3	8.2698	**480**
500	773.15	1.781	3487.0	8.5139	1.424	3486.5	8.4104	1.187	3486.0	8.3257	**500**
520	793.15	1.828	3529.9	8.5686	1.462	3529.4	8.4651	1.218	3528.9	8.3805	**520**
540	813.15	1.874	3573.0	8.6223	1.499	3572.5	8.5188	1.248	3572.0	8.4343	**540**
560	833.15	1.920	3616.4	8.6750	1.536	3615.9	8.5716	1.279	3615.5	8.4870	**560**
580	853.15	1.967	3660.0	8.7268	1.573	3659.6	8.6234	1.310	3659.2	8.5389	**580**
600	873.15	2.013	3704.0	8.7776	1.610	3703.6	8.6743	1.341	3703.2	8.5898	**600**

(Contd.)

TABLE: 1.8 Thermodynamic Properties of Steam (Superheat Region) [Contd.]

Temperature		Absolute Pressure (Saturation Temperature)									Temp.
t	T	4.0 bar (143.62°C/416.77K)			0.5 bar (151.84°C/424.99K)			6.0 bar (158.84°C/431.99K)			t
°C	K	v	i	s	v	i	s	v	i	s	°C
Satd.	**Cond.**	0.4622	2737.6	6.8943	0.3747	2747.5	6.8192	0.3155	2755.5	6.7575	
160	433.15	0.4837	2774.2	6.9805	0.3835	2766.4	6.8631	0.3165	2758.2	6.7640	**160**
180	453.15	0.5093	2817.8	7.0788	0.4045	2811.4	6.9647	0.3346	2804.8	6.8691	**180**
200	473.15	0.5343	2860.4	7.1708	0.4250	2855.1	7.0592	0.3520	2849.7	6.9662	**200**
220	493.15	0.5589	2902.3	7.2576	0.4450	2898.0	7.1478	0.3690	2893.5	7.0567	**220**
240	513.15	0.5831	2943.9	7.3402	0.4647	2940.1	7.2317	0.3857	2936.4	7.1419	**240**
260	533.15	0.6072	2985.1	7.4190	0.4841	2981.9	7.3115	0.4021	2978.7	7.2228	**260**
280	553.15	0.6311	3026.2	7.4947	0.5034	3023.4	7.3879	0.4183	3020.6	7.3000	**280**
300	573.15	0.6549	3067.2	7.5675	0.5226	3064.8	7.4614	0.4344	3062.3	7.3740	**300**
320	593.15	0.6785	3108.3	7.6379	0.5416	3106.1	7.5322	0.4504	3103.9	7.4454	**320**
340	613.15	0.7021	3149.4	7.7061	0.5606	3147.4	7.6008	0.4663	3145.4	7.5143	**340**
360	633.15	0.7256	3190.6	7.7723	0.5795	3188.8	7.6673	0.4821	3187.0	7.5810	**360**
380	653.15	0.7491	3232.1	7.8367	0.5984	3230.4	7.7319	0.4979	3228.7	7.6459	**380**
400	673.15	0.7725	3273.6	7.8994	0.6172	3272.1	7.7948	0.5136	3270.6	7.7090	**400**
420	693.15	0.7959	3315.4	7.9606	0.6359	3314.0	7.8561	0.5293	3312.6	7.7705	**420**
440	713.15	0.8192	3357.4	8.0203	0.6547	3356.1	7.9160	0.5450	3354.8	7.8305	**440**
460	733.15	0.8426	3399.7	8.0787	0.6734	3398.4	7.9745	0.5606	3397.2	7.8891	**460**
480	753.15	0.8659	3442.1	8.1359	0.6921	3441.0	8.0318	0.5762	3439.8	7.9465	**480**
500	773.15	0.8892	3484.9	8.1919	0.7108	3483.8	8.0879	0.5918	3482.7	8.0027	**500**
520	793.15	0.9125	3527.8	8.2468	0.7294	3526.8	8.1428	0.6074	3525.8	8.0577	**520**
540	813.15	0.9357	3571.1	8.3006	0.7481	3570.1	8.1967	0.6230	3569.1	8.1117	**540**
560	833.15	0.9590	3614.6	8.3534	0.7667	3613.6	8.2496	0.6386	3612.7	8.1647	**560**
580	853.15	0.9822	3658.3	8.4053	0.7853	3657.4	8.3016	0.6541	3656.6	8.2167	**580**
600	873.15	1.005	3702.3	8.4563	0.8039	3701.5	8.3526	0.6696	3700.7	8.2678	**600**
620	893.15	1.029	3746.6	8.5065	0.8225	3745.8	8.4028	0.6851	3745.1	8.3180	**620**

Temperature		Absolute Pressure (Saturation Temperature)									Temp.
t	T	7.0 bar (164.96°C/438.11K)			8.0 bar (170.41°C/443.56K)			9.0 bar (175.36°C/448.51K)			t
°C	K	v	i	s	v	i	s	v	i	s	°C
Satd.	**Cond.**	0.2727	2762.0	6.7052	0.2403	2767.5	6.6596	0.2148	2772.1	6.6192	
180	453.15	0.2846	2798.0	6.7861	0.2471	2791.1	6.7122	0.2178	2783.9	6.6452	**180**
200	473.15	0.2999	2844.2	6.8859	0.2608	2838.6	6.8148	0.2303	2832.7	6.7508	**200**
220	493.15	0.3147	2888.9	6.9784	0.2740	2884.2	6.9094	0.2423	2879.5	6.8475	**220**
240	513.15	0.3292	2932.5	7.0651	0.2869	2928.6	6.9976	0.2539	2924.6	6.9373	**240**
260	533.15	0.3435	2975.4	7.1470	0.2995	2972.0	7.0806	0.2653	2968.7	7.0215	**260**
280	553.15	0.3575	3017.7	7.2250	0.3119	3014.9	7.1595	0.2764	3012.0	7.1012	**280**
300	573.15	0.3714	3059.8	7.2997	0.3241	3057.3	7.2348	0.2874	3054.7	7.1771	**300**
320	593.15	0.3852	3101.6	7.3715	0.3363	3099.4	7.3070	0.2983	3097.1	7.2499	**320**
340	613.15	0.3989	3143.4	7.4407	0.3483	3141.4	7.3767	0.3090	3139.4	7.3199	**340**
360	633.15	0.4125	3185.2	7.5078	0.3603	3183.4	7.4441	0.3197	3181.5	7.3876	**360**
380	653.15	0.4261	3227.1	7.5729	0.3723	3225.4	7.5094	0.3304	3223.7	7.4532	**380**
400	673.15	0.4396	3269.0	7.6362	0.3842	3267.5	7.5729	0.3410	3265.9	7.5169	**400**
420	693.15	0.4531	3311.2	7.6978	0.3960	3309.7	7.6347	0.3516	3308.3	7.5788	**420**
440	713.15	0.4666	3353.4	7.7580	0.4078	3352.1	7.6950	0.3621	3350.8	7.6393	**440**
460	733.15	0.4801	3395.9	7.8168	0.4196	3394.7	7.7539	0.3726	3393.4	7.6983	**460**
480	753.15	0.4935	3438.6	7.8742	0.4314	3437.5	7.8115	0.3831	3436.3	7.7560	**480**
500	773.15	0.5069	3481.6	7.9305	0.4432	3480.5	7.8678	0.3936	3479.4	7.8124	**500**
520	793.15	0.5203	3524.7	7.9856	0.4549	3523.7	7.9230	0.4041	3522.7	7.8677	**520**
540	813.15	0.5336	3568.1	8.0396	0.4666	3567.2	7.9771	0.4145	3566.2	7.9219	**540**
560	833.15	0.5470	3611.8	8.0927	0.4783	3610.9	8.0302	0.4249	3610.0	7.9750	**560**
580	853.15	0.5603	3655.7	8.1448	0.4900	3654.8	8.0824	0.4354	3654.0	8.0272	**580**
600	873.15	0.5737	3699.9	8.1959	0.5017	3699.1	8.1336	0.4458	3698.2	8.0785	**600**
620	893.15	0.5870	3744.3	8.2462	0.5134	3743.5	8.1839	0.4561	3742.7	8.1289	**620**
640	913.15	0.6003	3789.0	8.2957	0.5251	3788.2	8.2334	0.4665	3787.5	8.1785	**640**

(Contd.)

TABLE: 1.8 Thermodynamic Properties of Steam (Superheat Region) [Contd.]

Temperature		Absolute Pressure (Saturation Temperature)									Temp.
t	*T*	10.0 bar (179.88°C/453.03K)			12.0 bar (187.96°C/461.11K)			14.0 bar (195.04°C/468.19K)			*t*
°C	K	*v*	*i*	*s*	*v*	*i*	*s*	*v*	*i*	*s*	°C
Satd.	**Cond.**	0.1943	2776.2	6.5828	0.1632	2782.7	6.5194	0.1407	2787.8	6.4651	
200	473.15	0.2059	2826.8	6.6922	0.1692	2814.4	6.5872	0.1429	2801.4	6.4941	**200**
220	493.15	0.2169	2874.6	6.7911	0.1788	2864.5	6.6909	0.1515	2854.0	6.6030	**220**
240	513.15	0.2276	2920.6	6.8825	0.1879	2912.2	6.7858	0.1596	2903.6	6.7016	**240**
260	533.15	0.2379	2965.2	6.9680	0.1968	2958.2	6.8738	0.1674	2951.0	6.7922	**260**
280	553.15	0.2480	3009.0	7.0485	0.2054	3003.0	6.9562	0.1749	2996.9	6.8766	**280**
300	573.15	0.2580	3052.1	7.1251	0.2139	3046.9	7.0342	0.1823	3041.6	6.9561	**300**
320	593.15	0.2678	3094.9	7.1984	0.2222	3090.3	7.1085	0.1896	3085.6	7.0315	**320**
340	613.15	0.2776	3137.3	7.2689	0.2304	3133.2	7.1798	0.1967	3129.1	7.1036	**340**
360	633.15	0.2873	3179.7	7.3368	0.2386	3176.0	7.2484	0.2038	3172.3	7.1729	**360**
380	653.15	0.2969	3222.0	7.4027	0.2467	3218.7	7.3147	0.2108	3215.3	7.2398	**380**
400	673.15	0.3065	3264.4	7.4665	0.2547	3261.3	7.3790	0.2177	3258.2	7.3045	**400**
420	693.15	0.3160	3306.9	7.5287	0.2627	3304.0	7.4415	0.2246	3301.1	7.3673	**420**
440	713.15	0.3256	3349.5	7.5893	0.2707	3346.8	7.5024	0.2315	3344.1	7.4285	**440**
460	733.15	0.3350	3392.2	7.6484	0.2787	3389.7	7.5618	0.2384	3387.2	7.4881	**460**
480	753.15	0.3445	3435.1	7.7062	0.2866	3432.8	7.6198	0.2452	3430.5	7.5463	**480**
500	773.15	0.3540	3478.3	7.7627	0.2945	3476.1	7.6765	0.2452	3473.9	7.6032	**500**
520	793.15	0.3634	3521.6	7.8181	0.3024	3519.6	7.7320	0.2588	3517.5	7.6589	**520**
540	813.15	0.3728	3565.2	7.8724	0.3103	3563.3	7.7864	0.2656	3561.3	7.7135	**540**
560	833.15	0.3822	3609.0	7.9256	0.3181	3607.2	7.8398	0.2724	3605.4	7.7670	**560**
580	853.15	0.3916	3653.1	7.9779	0.3260	3651.4	7.8922	0.2791	3649.6	7.8195	**580**
600	873.15	0.4010	3697.4	8.0292	0.3338	3695.8	7.9436	0.2859	3694.1	7.8710	**600**
620	893.15	0.4104	3742.0	8.0796	0.3417	3740.4	7.9942	0.2926	3738.9	7.9217	**620**
640	913.15	0.4197	3786.8	8.1292	0.3495	3785.3	8.0439	0.2993	3783.8	7.9715	**640**
660	933.15	0.4291	3831.8	8.1780	0.3573	3830.4	8.0928	0.3060	3829.0	8.0205	**660**

Temperature		Absolute Pressure (Saturation Temperature)									Temp.
t	*T*	16.0 bar (201.37°C/474.52K)			18.0 bar (207.11°C/480.26K)			20.0 bar (212.37°C/485.52K)			*t*
°C	K	*v*	*i*	*s*	*v*	*i*	*s*	*v*	*i*	*s*	°C
Satd.	**Cond.**	0.1237	2791.7	6.4175	0.1103	2794.8	6.3751	0.09954	2797.2	6.3367	
220	493.15	0.1310	2843.1	6.5237	0.1150	2831.7	6.4509	0.1021	2819.9	6.3829	**220**
240	513.15	0.1383	2894.7	6.6263	0.1217	2885.4	6.5577	0.1084	2875.9	6.4943	**240**
260	533.15	0.1453	2943.6	6.7198	0.1282	2935.9	6.6543	0.1144	2928.1	6.5941	**260**
280	553.15	0.1521	2990.6	6.8063	0.1343	2984.1	6.7430	0.1200	2977.5	6.6852	**280**
300	573.15	0.1587	3036.2	6.8873	0.1402	3030.7	6.8257	0.1255	3025.0	6.7696	**300**
320	593.15	0.1651	3080.9	6.9639	0.1460	3076.1	6.9035	0.1308	3071.2	6.8487	**320**
340	613.15	0.1714	3124.9	7.0369	0.1517	3120.6	6.9774	0.1360	3116.3	6.9235	**340**
360	633.15	0.1777	3168.5	7.1069	0.1573	3164.7	7.0481	0.1411	3160.8	6.9950	**360**
380	653.15	0.1838	3211.8	7.1743	0.1629	3208.4	7.1160	0.1461	3204.9	7.0635	**380**
400	673.15	0.1900	3255.0	7.2394	0.1684	3251.9	7.1816	0.1511	3248.7	7.1295	**400**
420	693.15	0.1961	3298.2	7.3026	0.1738	3295.3	7.2452	0.1561	3292.4	7.1935	**420**
440	713.15	0.2021	3341.4	7.3641	0.1793	3338.7	7.3070	0.1610	3336.0	7.2555	**440**
460	733.15	0.2082	3384.7	7.4240	0.1847	3382.2	7.3671	0.1659	3379.7	7.3159	**460**
480	753.15	0.2142	3428.1	7.4824	0.1900	3425.8	7.4257	0.1707	3423.4	7.3748	**480**
500	773.15	0.2202	3471.7	7.5395	0.1954	3469.5	7.4830	0.1756	3467.3	7.4323	**500**
520	793.15	0.2261	3515.4	7.5954	0.2007	3513.4	7.5391	0.1804	3511.3	7.4885	**520**
540	813.15	0.2321	3559.4	7.6501	0.2060	3557.4	7.5939	0.1852	3555.5	7.5435	**540**
560	833.15	0.2380	3603.5	7.7037	0.2113	3601.7	7.6477	0.1900	3599.8	7.5974	**560**
580	853.15	0.2440	3647.9	7.7563	0.2166	3646.2	7.7004	0.1947	3644.4	7.6503	**580**
600	873.15	0.2499	3692.5	7.8080	0.2219	3690.9	7.7522	0.1995	3689.2	7.7022	**600**
620	893.15	0.2558	3737.3	7.8588	0.2272	3735.8	7.8031	0.2043	3734.2	7.7531	**620**
640	913.15	0.2617	3782.4	7.9086	0.2324	3780.9	7.8531	0.2090	3779.4	7.8032	**640**
660	933.15	0.2676	3827.7	7.9577	0.2377	3826.3	7.9022	0.2137	3824.9	7.8524	**660**
680	953.15	0.2734	3873.2	8.0060	0.2429	3871.9	7.9505	0.2185	3870.6	7.9009	**680**

(Contd.)

TABLE: 1.8 Thermodynamic Properties of Steam (Superheat Region) [Contd.]

Temperature		Absolute Pressure (Saturation Temperature)									Temp.
t	T	22.0 bar (217.24°C/490.39K)			24.0 bar (221.78°C/494.93K)			26.0 bar (226.04°C/499.19K)			t
°C	K	v	i	s	v	i	s	v	i	s	°C
Satd.	Cond.	0.09065	2799.1	6.3015	0.08320	2800.4	6.2690	0.07686	2801.4	6.2387	
240	513.15	0.09752	2866.0	6.4349	0.08839	2855.7	6.3788	0.08064	2845.2	6.3253	240
260	533.15	0.1031	2920.0	6.5382	0.09367	2911.6	6.4857	0.08567	2903.0	6.4360	260
280	553.15	0.1084	2970.8	6.6317	0.09863	2963.8	6.5818	0.09037	2956.7	6.5348	280
300	573.15	0.1134	3019.3	6.7179	0.1034	3013.4	6.6699	0.09483	3007.4	6.6249	300
320	593.15	0.1183	3066.2	6.7983	0.1079	3061.1	6.7517	0.09912	3056.0	6.7082	320
340	613.15	0.1231	3111.9	6.8742	0.1124	3107.5	6.8286	0.1033	3103.0	6.7862	340
360	633.15	0.1278	3156.9	6.9464	0.1167	3153.0	6.9016	0.1073	3149.0	6.8600	360
380	653.15	0.1324	3201.4	7.0155	0.1210	3197.8	6.9714	0.1113	3194.3	6.9304	380
400	673.15	0.1370	3245.5	7.0821	0.1252	3242.3	7.0384	0.1153	3239.0	6.9979	400
420	693.15	0.1415	3289.4	7.1464	0.1294	3286.5	7.1031	0.1191	3283.5	7.0630	420
440	713.15	0.1460	3333.3	7.2088	0.1335	3330.6	7.1658	0.1230	3327.8	7.1260	440
460	733.15	0.1505	3377.1	7.2694	0.1377	3374.6	7.2267	0.1268	3372.1	7.1872	460
480	753.15	0.1549	3421.1	7.3285	0.1418	3418.7	7.2860	0.1306	3416.3	7.2467	480
500	773.15	0.1593	3465.1	7.3862	0.1458	3462.8	7.3439	0.1344	3460.6	7.3048	500
520	793.15	0.1637	3509.2	7.4425	0.1499	3507.1	7.4004	0.1381	3505.1	7.3615	520
540	813.15	0.1681	3553.5	7.4977	0.1539	3551.6	7.4558	0.1419	3549.6	7.4170	540
560	833.15	0.1725	3598.0	7.5518	0.1579	3596.2	7.5099	0.1456	3594.3	7.4713	560
580	853.15	0.1768	3642.7	7.6047	0.1619	3640.9	7.5631	0.1493	3639.2	7.5246	580
600	873.15	0.1812	3687.6	7.6567	0.1659	3685.9	7.6152	0.1530	3684.3	7.5768	600
620	893.15	0.1855	3732.7	7.7078	0.1699	3731.8	7.6663	0.1567	3729.6	7.6281	620
640	913.15	0.1898	3778.0	7.7580	0.1739	3776.5	7.7166	0.1604	3775.0	7.6784	640
660	933.15	0.1942	3823.5	7.8073	0.1778	3822.1	7.7660	0.1640	3820.7	7.7279	660
680	953.15	0.1985	3869.2	7.8558	0.1818	3867.9	7.8146	0.1677	3866.6	7.7766	680
700	973.15	0.2028	3915.2	7.9035	0.1857	3914.0	7.8624	0.1714	3912.7	7.8245	700

Temperature		Absolute Pressure (Saturation Temperature)									Temp.
t	T	28.0 bar (230.05°C/503.20K)			30.0 bar (233.84°C/506.99K)			32.0 bar (237.45°C/510.60K)			t
°C	K	v	i	s	v	i	s	v	i	s	°C
Satd.	Cond.	0.07139	2802.0	6.2141	0.06663	2802.3	6.1837	0.06244	2802.3	6.1585	
240	513.15	0.07397	2834.2	6.2738	0.06816	2822.9	6.2241	0.06305	2811.2	6.1757	240
260	533.15	0.07880	2894.2	6.3886	0.07283	2885.1	6.3432	0.06759	2875.8	6.2993	260
280	553.15	0.08328	2949.5	6.4903	0.07712	2942.0	6.4479	0.07173	2934.4	6.4072	280
300	573.15	0.08751	3001.3	6.5824	0.08116	2995.1	6.5422	0.07559	2988.7	6.5037	300
320	593.15	0.09156	3050.8	6.6672	0.08500	3045.4	6.6285	0.07926	3040.0	6.5917	320
340	613.15	0.09548	3098.5	6.7464	0.08871	3093.9	6.7088	0.08279	3089.2	6.6733	340
360	633.15	0.09929	3145.0	6.8210	0.09232	3140.9	6.7844	0.08621	3136.8	6.7497	360
380	653.15	0.1030	3190.7	6.8921	0.09584	3187.0	6.8561	0.08955	3183.4	6.8221	380
400	673.15	0.1067	3235.8	6.9601	0.09931	3232.5	6.9246	0.09283	3229.2	6.8912	400
420	693.15	0.1104	3280.5	7.0256	0.1027	3277.5	6.9906	0.09606	3274.5	6.9576	420
440	713.15	0.1139	3325.1	7.0890	0.1061	3322.3	7.0543	0.09925	3319.5	7.0216	440
460	733.15	0.1175	3369.5	7.1504	0.1095	3367.0	7.1160	0.1024	3364.4	7.0836	460
480	753.15	0.1211	3413.9	7.2102	0.1128	3411.6	7.1760	0.1055	3409.2	7.1439	480
500	773.15	0.1246	3458.4	7.2685	0.1161	3456.2	7.2345	0.1086	3454.0	7.2026	500
520	793.15	0.1281	3503.0	7.3254	0.1194	3500.9	7.2916	0.1117	3498.8	7.2598	520
540	813.15	0.1316	3547.6	7.3810	0.1226	3545.7	7.3474	0.1148	3543.7	7.3158	540
560	833.15	0.1350	3592.5	7.4355	0.1259	3590.6	7.4020	0.1179	3588.8	7.3705	560
580	853.15	0.1385	3637.5	7.4888	0.1291	3635.7	7.4554	0.1209	3634.0	7.4241	580
600	873.15	0.1419	3682.6	7.5412	0.1323	3681.0	7.5079	0.1239	3679.3	7.4767	600
620	893.15	0.1454	3728.0	7.5925	0.1356	3726.4	7.5594	0.1270	3724.9	7.5283	620
640	913.15	0.1488	3773.6	7.6430	0.1388	3772.1	7.6099	0.1300	3770.6	7.5789	640
660	933.15	0.1522	3819.3	7.6926	0.1420	3817.9	7.6596	0.1330	3816.6	7.6287	660
680	953.15	0.1556	3865.3	7.7413	0.1451	3864.0	7.7084	0.1360	3862.7	7.6776	680
700	973.15	0.1590	3911.5	7.7893	0.1483	3910.3	7.7564	0.1390	3909.0	7.7257	700

(Contd.)

TABLE: 1.8 Thermodynamic Properties of Steam (Superheat Region) [Contd.]

Temperature		Absolute Pressure (Saturation Temperature)									Temp.
t	*T*	34.0 bar (240.88°C/514.03K)			36.0 bar (244.16°C/517.31K)			38.0 bar (247.31°C/520.46K)			*t*
°C	K	*v*	*i*	*s*	*v*	*i*	*s*	*v*	*i*	*s*	°C
Satd.	**Cond.**	0.05873	2802.1	6.1344	0.05541	2801.7	6.1115	0.05244	2801.1	6.0896	
260	533.15	0.06295	2866.2	6.2568	0.05880	2856.3	6.2155	0.05508	2846.1	6.1750	**260**
280	553.15	0.06695	2926.6	6.3681	0.06270	2918.6	6.3302	0.05888	2910.4	6.2935	**280**
300	573.15	0.07068	2982.2	6.4669	0.06630	2975.6	6.4315	0.06237	2968.9	6.3973	**300**
320	593.15	0.07419	3034.5	6.5566	0.06968	3028.9	6.5230	0.06564	3023.3	6.4906	**320**
340	613.15	0.07756	3084.4	6.6394	0.07291	3079.6	6.6070	0.06875	3074.8	6.5760	**340**
360	633.15	0.08082	3132.7	6.7168	0.07603	3128.4	6.6854	0.07174	3124.2	6.6553	**360**
380	653.15	0.08400	3179.7	6.7899	0.07906	3175.9	6.7592	0.07464	3172.2	6.7299	**380**
400	673.15	0.08711	3225.9	6.8595	0.08202	3222.5	6.8294	0.07747	3219.1	6.8007	**400**
420	693.15	0.09017	3271.5	6.9263	0.08494	3268.4	6.8967	0.08025	3265.4	6.8684	**420**
440	713.15	0.09319	3316.8	6.9907	0.08781	3314.0	6.9614	0.08300	3311.2	6.9336	**440**
460	733.15	0.09618	3367.8	7.0530	0.09065	3359.2	7.0241	0.08570	3356.6	6.9965	**460**
480	753.15	0.09915	3406.8	7.1135	0.09347	3404.4	7.0848	0.08838	3402.0	7.0575	**480**
500	773.15	0.1021	3451.7	7.1724	0.09626	3449.5	7.1439	0.09104	3447.2	7.1168	**500**
520	793.15	0.1050	3496.7	7.2299	0.09903	3494.6	7.2015	0.09367	3492.5	7.1746	**520**
540	813.15	0.1079	3541.7	7.2860	0.1018	3539.8	7.2578	0.09629	3537.8	7.2310	**540**
560	833.15	0.1108	3586.9	7.3408	0.1045	3585.1	7.3128	0.09890	3583.2	7.2862	**560**
580	853.15	0.1137	3632.2	7.3946	0.1072	3630.5	7.3667	0.1015	3628.7	7.3402	**580**
600	873.15	0.1165	3677.7	7.4473	0.1100	3676.1	7.4195	0.1041	3674.4	7.3931	**600**
620	893.15	0.1194	3723.3	7.4989	0.1127	3721.8	7.4712	0.1066	3720.2	7.4450	**620**
640	913.15	0.1222	3769.2	7.5497	0.1153	3767.7	7.5221	0.1092	3766.2	7.4959	**640**
660	933.15	0.1251	3815.2	7.5995	0.1180	3813.8	7.5720	0.1117	3812.4	7.5459	**660**
680	953.15	0.1279	3861.4	7.6485	0.1207	3860.1	7.6211	0.1143	3858.8	7.5951	**680**
700	973.15	0.1307	3907.8	7.6967	0.1234	3906.5	7.6693	0.1168	3905.3	7.6434	**700**
720	993.15	0.1335	3954.4	7.7441	0.1260	3953.2	7.7168	0.1193	3952.0	7.6909	**720**

Temperature		Absolute Pressure (Saturation Temperature)									Temp.
t	*T*	40.0 bar (250.33°C/523.48K)			42.0 bar (253.24°C/526.39K)			44.0 bar (256.05°C/529.20K)			*t*
°C	K	*v*	*i*	*s*	*v*	*i*	*s*	*v*	*i*	*s*	°C
Satd.	**Cond.**	0.04975	2800.3	6.0685	0.04731	2799.4	6.0482	0.04508	2798.3	6.0286	
260	533.15	0.05172	2835.6	6.1353	0.04865	2824.8	6.0962	0.04585	2813.6	6.0575	**260**
280	553.15	0.05544	2902.0	6.2576	0.05231	2893.5	6.2270	0.04946	2884.7	6.1884	**280**
300	573.15	0.05883	2962.0	6.3642	0.05562	2955.0	6.3320	0.05270	2947.8	6.3006	**300**
320	593.15	0.06200	3017.5	6.4593	0.05870	3011.6	6.4291	0.05569	3005.7	6.3998	**320**
340	613.15	0.06499	3069.8	6.5461	0.06160	3064.8	6.5173	0.05850	3059.7	6.4894	**340**
360	633.15	0.06787	3119.9	6.6265	0.06437	3115.5	6.5987	0.06119	3111.1	6.5719	**360**
380	653.15	0.07066	3168.4	6.7019	0.06706	3164.5	6.6749	0.06378	3160.6	6.6489	**380**
400	673.15	0.07338	3215.7	6.7733	0.06967	3212.3	6.7469	0.06629	3208.8	6.7216	**400**
420	693.15	0.07604	3262.3	6.8414	0.07222	3259.2	6.8156	0.06876	3256.0	6.7907	**420**
440	713.15	0.07866	3308.3	6.9069	0.07474	3305.5	6.8815	0.07117	3302.6	6.8570	**440**
460	733.15	0.08125	3354.0	6.9702	0.07722	3351.4	6.9450	0.07355	3348.8	6.9209	**460**
480	753.15	0.08381	3399.6	7.0314	0.07967	3397.1	7.0065	0.07590	3394.7	6.9826	**480**
500	773.15	0.08634	3445.0	7.0909	0.08209	3442.7	7.0662	0.07823	3440.5	7.0426	**500**
520	793.15	0.08886	3490.4	7.1489	0.08450	3488.3	7.1244	0.08054	3486.2	7.1010	**520**
540	813.15	0.09135	3535.8	7.2055	0.08689	3533.9	7.1812	0.08282	3531.9	7.1579	**540**
560	833.15	0.09384	3581.4	7.2608	0.08926	3579.5	7.2366	0.08510	3577.6	7.2135	**560**
580	853.15	0.09631	3627.0	7.3149	0.09162	3625.2	7.2909	0.08736	3623.5	7.2678	**580**
600	873.15	0.09876	3672.8	7.3680	0.09397	3671.1	7.3440	0.08960	3669.5	7.3211	**600**
620	893.15	0.1012	3718.7	7.4200	0.09630	3717.1	7.3961	0.09184	3715.6	7.3733	**620**
640	913.15	0.1036	3764.8	7.4710	0.09863	3763.3	7.4472	0.09407	3761.8	7.4245	**640**
660	933.15	0.1061	3811.0	7.5211	0.1009	3809.6	7.4974	0.09628	3808.2	7.4748	**660**
680	953.15	0.1085	3857.4	7.5703	0.1033	3856.1	7.5467	0.09849	3854.8	7.5242	**680**
700	973.15	0.1109	3904.1	7.6187	0.1056	3902.8	7.5952	0.1007	3901.6	7.5728	**700**
720	993.15	0.1133	3950.9	7.6663	0.1078	3949.7	7.6429	0.1029	3948.5	7.6205	**720**

(Contd.)

TABLE: 1.8 Thermodynamic Properties of Steam (Superheat Region) [Contd.]

Temperature		Absolute Pressure (Saturation Temperature)									Temp.
t	*T*	46.0 bar (258.75°C/531.90K)			48.0 bar (261.37°C/534.52K)			50.0 bar (263.91°C/537.06K)			*t*
°C	K	*v*	*i*	*s*	*v*	*i*	*s*	*v*	*i*	*s*	°C
Satd.	**Cond.**	0.04304	2797.0	6.0097	0.04116	2795.7	5.9913	0.03943	2794.2	5.9735	
280	553.15	0.04685	2875.6	6.1547	0.04444	2866.4	6.1214	0.04222	2856.9	6.0886	**280**
300	573.15	0.05003	2940.5	6.2700	0.04757	2933.1	6.2399	0.04530	2925.5	6.2105	**300**
320	593.15	0.05294	2999.6	6.3712	0.05042	2993.4	6.3434	0.04810	2987.2	6.3163	**320**
340	613.15	0.05568	3054.6	6.4624	0.05309	3049.3	6.4362	0.05070	3044.1	6.4106	**340**
360	633.15	0.05828	3106.7	6.5460	0.05561	3102.2	6.5209	0.05316	3097.6	6.4966	**360**
380	653.15	0.06079	3156.7	6.6239	0.05804	3152.8	6.5996	0.05551	3148.8	6.5762	**380**
400	673.15	0.06321	3205.3	6.6972	0.06039	3201.8	6.6736	0.05779	3198.3	6.6508	**400**
420	693.15	0.06559	3252.9	6.7668	0.06268	3249.7	6.7438	0.06001	3246.5	6.7215	**420**
440	713.15	0.06791	3299.8	6.8335	0.06493	3296.9	6.8108	0.06218	3294.9	6.7890	**440**
460	733.15	0.07020	3346.2	6.8977	0.06714	3343.5	6.8753	0.06431	3340.9	6.8538	**460**
480	753.15	0.07247	3392.3	6.9597	0.06931	3389.8	6.9376	0.06642	3387.4	6.9164	**480**
500	773.15	0.07470	3438.2	7.0199	0.07147	3435.9	6.9981	0.06849	3433.7	6.9770	**500**
520	793.15	0.07692	3484.1	7.0784	0.07360	3481.9	7.0568	0.07055	3479.8	7.0360	**520**
540	813.15	0.07912	3529.9	7.1355	0.07572	3527.9	7.1140	0.07259	3525.9	7.0934	**540**
560	833.15	0.08130	3575.8	7.1913	0.07782	3573.9	7.1699	0.07461	3572.0	7.1494	**560**
580	853.15	0.08347	3621.7	7.2458	0.07990	3620.0	7.2246	0.07662	3618.2	7.2042	**580**
600	873.15	0.08562	3667.8	7.2991	0.08197	3666.1	7.2781	0.07862	3664.5	7.2578	**600**
620	893.15	0.08777	3714.0	7.3515	0.08404	3712.4	7.3305	0.08060	3710.9	7.3103	**620**
640	913.15	0.08990	3760.3	7.4028	0.08609	3758.9	7.3819	0.08257	3757.4	7.3618	**640**
660	933.15	0.09203	3806.8	7.4531	0.08813	3805.5	7.4324	0.08454	3804.1	7.4124	**660**
680	953.15	0.09415	3853.5	7.5026	0.09016	3852.2	7.4819	0.08650	3850.9	7.4620	**680**
700	973.15	0.09626	3900.3	7.5513	0.09219	3899.1	7.5306	0.08845	3897.9	7.5108	**700**
720	993.15	0.09836	3947.4	7.5991	0.09421	3946.2	7.5785	0.09039	3945.0	7.5587	**720**
740	1013.15	0.1005	3994.5	7.6461	0.09622	3993.4	7.6256	0.09232	3992.3	7.6059	**740**

Temperature		Absolute Pressure (Saturation Temperature)									Temp.
t	*T*	52.0 bar (266.37°C/539.52K)			54.0 bar (268.76°C/541.91K)			56.0 bar (271.09°C/544.24K)			*t*
°C	K	*v*	*i*	*s*	*v*	*i*	*s*	*v*	*i*	*s*	°C
Satd.	**Cond.**	0.03782	2792.6	5.9561	0.03633	2790.8	5.9392	0.03495	2789.0	5.9227	
280	553.15	0.04016	2847.1	6.0560	0.03823	2837.0	6.0236	0.03644	2826.7	5.9914	**280**
300	573.15	0.04320	2917.8	6.1815	0.04125	2909.8	6.1530	0.03943	2901.7	6.1248	**300**
320	593.15	0.04595	2980.8	6.2897	0.04395	2974.3	6.2636	0.04210	2967.7	6.2380	**320**
340	613.15	0.04849	3038.7	6.3857	0.04644	3033.3	6.3614	0.04454	3027.7	6.3379	**340**
360	633.15	0.05089	3093.0	6.4729	0.04878	3088.3	6.4498	0.04683	3083.6	6.4272	**360**
380	653.15	0.05318	3144.8	6.5533	0.05102	3140.7	6.5312	0.04901	3136.6	6.5096	**380**
400	673.15	0.05539	3194.7	6.6287	0.05317	3191.1	6.4072	0.05110	3187.5	6.5863	**400**
420	693.15	0.05754	3243.3	6.6999	0.05525	3240.1	6.6790	0.05313	3236.9	6.6586	**420**
440	713.15	0.05964	3291.1	6.7678	0.05729	3288.2	6.7473	0.05511	3285.2	6.7274	**440**
460	733.15	0.06171	3338.2	6.8330	0.05929	3335.5	6.8128	0.05705	3332.8	6.7933	**460**
480	753.15	0.06374	3384.9	6.8958	0.06126	3382.5	6.8760	0.05896	3380.0	6.8567	**480**
500	773.15	0.06575	3431.4	6.9567	0.06320	3429.1	6.9371	0.06084	3426.8	6.9181	**500**
520	793.15	0.06773	3477.7	7.0158	0.06513	3475.6	6.9964	0.06270	3473.4	6.9776	**520**
540	813.15	0.06970	3523.9	7.0734	0.06703	3521.9	7.0542	0.06455	3519.9	7.0355	**540**
560	833.15	0.07165	3570.2	7.1296	0.06891	3568.3	7.1105	0.06637	3566.4	7.0920	**560**
580	853.15	0.07359	3616.5	7.1845	0.07079	3614.7	7.1655	0.06818	3612.9	7.1472	**580**
600	873.15	0.07552	3662.8	7.2382	0.07265	3661.2	7.2194	0.06998	3659.5	7.2011	**600**
620	893.15	0.07743	3709.3	7.2909	0.07449	3707.8	7.2721	0.07177	3706.2	7.2540	**620**
640	913.15	0.07933	3755.9	7.3425	0.07633	3754.5	7.3238	0.07355	3753.0	7.3058	**650**
660	933.15	0.08123	3802.7	7.3931	0.07816	3801.1	7.3745	0.07531	3799.9	7.3566	**660**
680	953.15	0.08311	3849.6	7.4428	0.07998	3848.3	7.4244	0.07707	3846.9	7.4065	**680**
700	973.15	0.08499	3896.6	7.4917	0.08179	3895.4	7.4733	0.07882	3894.1	7.4555	**700**
750	1023.15	0.08966	4015.0	7.6103	0.08629	4013.9	7.5920	0.08317	4012.8	7.5744	**750**
800	1073.15	0.09428	4134.4	7.7243	0.09076	4133.5	7.7061	0.08748	4132.5	7.6887	**800**

(Contd.)

TABLE: 1.8 Thermodynamic Properties of Steam (Superheat Region) [Contd.]

Temperature		Absolute Pressure (Saturation Temperature)									Temp.
t	T	58.0 bar (273.35°C/546.50K)			60.0 bar (275.55°C/548.70K)			64.0 bar (279.79°C/552.94K)			t
°C	K	v	i	s	v	i	s	v	i	s	°C
Satd.	**Cond.**	0.03365	2787.0	5.9066	0.03244	2785.0	5.8908	0.03023	2780.6	5.8601	
280	553.15	0.03476	2816.0	5.9592	0.03317	2804.9	5.9270	0.03026	2781.6	5.6619	**280**
300	573.15	0.03774	2893.5	6.0969	0.03614	2885.0	6.0692	0.03324	2867.5	6.0144	**300**
320	593.15	0.04036	2961.0	6.2128	0.03874	2954.2	6.1880	0.03580	2940.3	6.1393	**320**
340	613.15	0.04276	3022.2	6.3142	0.04111	3016.5	6.2913	0.03809	3004.9	6.2468	**340**
360	633.15	0.04501	3078.9	6.4052	0.04330	3074.0	6.3836	0.04021	3064.2	6.3418	**360**
380	653.15	0.04713	3132.4	6.4885	0.04539	3128.3	6.4680	0.04221	3119.8	6.4282	**380**
400	673.15	0.04918	3183.8	6.5660	0.04738	3180.1	6.5462	0.04412	3172.7	6.5079	**400**
420	693.15	0.05115	3233.6	6.6389	0.04931	3230.3	6.6196	0.04596	3223.7	6.5826	**420**
440	713.15	0.05308	3282.3	6.7081	0.05118	3279.3	6.6893	0.04774	3273.3	6.6532	**440**
460	733.15	0.05496	3330.2	6.7743	0.05302	3327.4	6.7559	0.04948	3322.0	6.7205	**460**
480	753.15	0.05682	3377.5	6.8381	0.05482	3375.0	6.8199	0.05119	3370.0	6.7851	**480**
500	773.15	0.05864	3424.5	6.8996	0.05659	3422.2	6.8818	0.05287	3417.6	6.8475	**500**
520	793.15	0.06045	3471.3	6.9594	0.05834	3469.1	6.9417	0.05453	3464.8	6.9078	**520**
540	813.15	0.06223	3517.9	7.0175	0.06008	3515.9	7.0000	0.05614	3511.9	6.9665	**540**
560	833.15	0.06400	3564.6	7.0741	0.06179	3562.7	7.0568	0.05779	3558.9	7.0236	**560**
580	853.15	0.06576	3611.2	7.1294	0.06349	3609.4	7.1122	0.05939	3605.9	7.0793	**580**
600	873.15	0.06750	3657.9	7.1835	0.06518	3656.2	7.1664	0.06099	3652.9	7.1337	**600**
620	893.15	0.06923	3704.6	7.2365	0.06686	3703.1	7.2195	0.06257	3699.9	7.1870	**620**
640	913.15	0.07095	3751.5	7.2884	0.06853	3750.0	7.2715	0.06414	3714.1	7.2392	**640**
660	933.15	0.07266	3798.5	7.3393	0.07019	3797.1	7.3225	0.06570	3794.3	7.2904	**660**
680	953.15	0.07436	3845.6	7.3892	0.07184	3844.3	7.3725	0.06726	3841.7	7.3406	**680**
700	973.15	0.07606	3892.9	7.4383	0.07348	3891.7	7.4217	0.06880	3889.2	7.3899	**700**
750	1023.15	0.08026	4011.7	7.5574	0.07755	4010.7	7.5409	0.07263	4008.5	7.5095	**750**
800	1073.15	0.08443	4131.6	7.6718	0.08159	4130.7	7.6554	0.07643	4128.8	7.6243	**800**

Temperature		Absolute Pressure (Saturation Temperature)									Temp.
t	T	68.0 bar (283.84°C/556.99K)			72.0 bar (287.70°C/560.85K)			76.0 bar (291.41°C/564.56K)			t
°C	K	v	i	s	v	i	s	v	i	s	°C
Satd.	**Cond.**	0.02827	2775.9	5.8306	0.02652	2770.9	5.8020	0.02495	2765.5	5.7742	
290	563.15	0.02923	2805.3	5.8830	0.02688	2782.5	5.8226	—	—	—	**290**
300	573.15	0.03065	2849.0	5.9599	0.02832	2829.5	5.9054	0.02620	2808.8	5.8503	**300**
320	593.15	0.03318	2925.8	6.0917	0.03084	2910.7	6.0447	0.02873	2895.0	5.9982	**320**
340	613.15	0.03542	2993.1	6.2033	0.03304	2980.8	6.1610	0.03090	2968.2	6.1196	**340**
360	633.15	0.03748	3054.2	6.3014	0.03504	3043.9	6.2623	0.03286	3033.4	6.2243	**360**
380	653.15	0.03941	3111.1	6.3899	0.03691	3102.3	6.3531	0.03467	3093.3	6.3174	**380**
400	673.15	0.04124	3165.1	6.4713	0.03868	3157.4	6.4362	0.03638	3149.6	6.4022	**400**
420	693.15	0.04300	3216.9	6.5472	0.04037	3210.1	6.5133	0.03801	3203.2	6.4807	**420**
440	713.15	0.04470	3267.2	6.6188	0.04200	3261.1	6.5859	0.03958	3254.9	6.5543	**440**
460	733.15	0.04636	3316.5	6.6869	0.04359	3310.9	6.6548	0.04111	3305.3	6.6240	**460**
480	753.15	0.04799	3365.0	6.7521	0.04514	3359.9	6.7207	0.04260	3354.7	6.6906	**480**
500	773.15	0.04959	3412.9	6.8150	0.04667	3408.2	6.7840	0.04406	3403.5	6.7545	**500**
520	793.15	0.05116	3460.5	6.8757	0.04817	3456.2	6.8452	0.04549	3451.8	6.8161	**520**
540	813.15	0.05272	3507.9	6.9347	0.04965	3503.8	6.9046	0.04690	3499.8	6.8759	**540**
560	833.15	0.05425	3555.1	6.9922	0.05111	3551.4	6.9623	0.04830	3547.6	6.9339	**560**
580	853.15	0.05577	3602.3	7.0481	0.05256	3598.8	7.0186	0.04968	3595.2	6.9905	**580**
600	873.15	0.05728	3649.6	7.1028	0.05399	3646.2	7.0735	0.05105	3642.9	7.0457	**600**
620	893.15	0.05878	3696.8	7.1563	0.05541	3693.7	7.1273	0.05240	3690.5	7.0996	**620**
640	913.15	0.06027	3744.1	7.2087	0.05682	3741.2	7.1799	0.05374	3738.2	7.1524	**640**
660	933.15	0.06174	3791.5	7.2601	0.05823	3788.7	7.2314	0.05508	3785.9	7.2041	**660**
680	953.15	0.06321	3839.0	7.3105	0.05962	3836.4	7.2820	0.05641	3833.8	7.2549	**680**
700	973.15	0.06467	3886.7	7.3599	0.06101	3884.2	7.3316	0.05772	3881.7	7.3046	**700**
750	1023.15	0.06829	4006.4	7.4799	0.06444	4004.2	7.4518	0.06099	4002.0	7.4252	**750**
800	1073.15	0.07188	4126.9	7.5949	0.06783	4125.1	7.5671	0.06421	4123.2	7.5408	**800**

(Contd.)

TABLE: 1.8 Thermodynamic Properties of Steam (Superheat Region) [Contd.]

Temperature		Absolute Pressure (Saturation Temperature)									Temp.
t	*T*	80.0 bar (294.97°C/568.12K)			84.0 bar (298.39°C/571.54K)			88.0 bar (301.70°C/574.85K)			*t*
°C	K	*v*	*i*	*s*	*v*	*i*	*s*	*v*	*i*	*s*	°C
Satd.	**Cond.**	0.02353	2759.9	5.7471	0.02223	2754.0	5.7207	0.02105	2747.8	5.6948	
320	593.15	0.02681	2878.7	5.9519	0.02506	2861.6	5.9056	0.02345	2843.6	5.8590	**320**
340	613.15	0.02896	2955.3	6.0790	0.02720	2941.9	6.0388	0.02559	2928.0	5.9990	**340**
360	633.15	0.03088	3022.7	6.1872	0.02909	3011.7	6.1509	0.02746	3000.4	6.1152	**360**
380	653.15	0.03265	3084.2	6.2828	0.03082	3074.8	6.2491	0.02915	3065.3	6.2162	**380**
400	673.15	0.03431	3141.6	6.3694	0.03243	3133.5	6.3376	0.03073	3125.3	6.3067	**400**
420	693.15	0.03589	3196.2	6.4493	0.03397	3189.1	6.4189	0.03222	3181.9	6.3894	**420**
440	713.15	0.03740	3248.7	6.5240	0.03543	3242.3	6.4947	0.03364	3235.9	6.4663	**440**
460	733.15	0.03887	3299.7	6.5945	0.03685	3293.9	6.5661	0.03501	3288.2	6.5386	**460**
480	753.15	0.04030	3349.6	6.6617	0.03823	3344.4	6.6340	0.03634	3339.1	6.6072	**480**
500	773.15	0.04170	3398.8	6.7262	0.03958	3394.0	6.6990	0.03764	3389.2	6.6728	**500**
520	793.15	0.04308	3447.4	6.7883	0.04090	3443.0	6.7615	0.03891	3438.6	6.7358	**520**
540	813.15	0.04443	3495.7	6.8484	0.04220	3491.6	6.8221	0.04016	3487.5	6.7968	**540**
560	833.15	0.04577	3543.8	6.9068	0.04348	3539.9	6.8808	0.04140	3536.1	6.8558	**560**
580	853.15	0.04709	3591.7	6.9636	0.04474	3588.1	6.9379	0.04261	3584.5	6.9132	**580**
600	873.15	0.04839	3639.5	7.0191	0.04600	3636.2	6.9936	0.04382	3632.8	6.9692	**600**
620	893.15	0.04969	3687.4	7.0732	0.04724	3684.2	7.0480	0.04501	3681.1	7.0238	**620**
640	913.15	0.05097	3735.2	7.1262	0.04846	3732.3	7.1012	0.04619	3729.3	7.0772	**640**
660	933.15	0.05225	3783.1	7.1781	0.04968	3780.3	7.1533	0.04735	3777.5	7.1295	**660**
680	953.15	0.05351	3831.1	7.2290	0.05089	3828.5	7.2044	0.04852	3825.9	7.1807	**680**
700	973.15	0.05477	3879.2	7.2791	0.05210	3876.7	7.2544	0.04967	3874.2	7.2310	**700**
720	993.15	0.05602	3827.4	7.3280	0.05329	3925.1	7.3036	0.05081	3922.7	7.2803	**720**
740	1013.15	0.05726	3975.7	7.3761	0.05448	3973.5	7.3519	0.05195	3971.3	7.3287	**740**
760	1033.15	0.05850	4024.1	7.4234	0.05566	4022.0	7.3993	0.05308	4019.9	7.3762	**760**
800	1073.15	0.06096	4121.3	7.5158	0.05801	4119.5	7.4918	0.05533	4117.6	7.4690	**800**

Temperature		Absolute Pressure (Saturation Temperature)									Temp.
t	*T*	92.0 bar (304.89°C/578.04K)			96.0 bar (307.97°C/581.12K)			100.0 bar (310.96°C/584.11K)			*t*
°C	K	*v*	*i*	*s*	*v*	*i*	*s*	*v*	*i*	*s*	°C
Satd.	**Cond.**	0.01996	2741.3	5.6694	0.01897	2734.7	5.6444	0.01804	2727.7	5.6198	
320	593.15	0.02195	2824.7	5.8118	0.02056	2804.7	5.7637	0.01926	2783.5	5.7145	**320**
340	613.15	0.02411	2913.7	5.9594	0.02274	2898.8	5.9199	0.02147	2883.4	5.8803	**340**
360	633.15	0.02596	2988.9	6.0801	0.02458	2977.0	6.0454	0.02331	2964.8	6.0110	**360**
380	653.15	0.02763	3055.7	6.1840	0.02622	3045.8	6.1524	0.02493	3035.7	6.1213	**380**
400	673.15	0.02917	3117.0	6.2765	0.02773	3108.5	6.2470	0.02641	3099.9	6.2182	**400**
420	693.15	0.03062	3174.6	6.3608	0.02915	3167.2	6.3329	0.02779	3159.7	6.3056	**420**
440	713.15	0.03200	3229.4	6.4388	0.03049	3222.9	6.4121	0.02911	3216.2	6.3861	**440**
460	733.15	0.03333	3282.4	6.5120	0.03179	3276.5	6.4863	0.03036	3270.5	6.4612	**460**
480	753.15	0.03462	3333.9	6.5814	0.03304	3328.5	6.5563	0.03158	3323.2	6.5321	**480**
500	773.15	0.03587	3384.4	6.6475	0.03425	3379.5	6.6231	0.03276	3374.6	6.5994	**500**
520	793.15	0.03710	3434.1	6.7111	0.03544	3429.6	6.6871	0.03391	3425.1	6.6640	**520**
540	813.15	0.03831	3483.4	6.7724	0.03661	3479.2	6.7489	0.03504	3475.1	6.7261	**540**
560	833.15	0.03950	3532.3	6.8318	0.03775	3528.4	6.8086	0.03615	3524.5	6.7863	**560**
580	853.15	0.04067	3580.7	6.8895	0.03888	3577.3	6.8667	0.03724	3573.7	6.8446	**580**
600	873.15	0.04182	3629.4	6.9457	0.04000	3626.1	6.9231	0.03832	3622.7	6.9013	**600**
620	893.15	0.04297	3677.9	7.0006	0.04110	3674.7	6.9782	0.03939	3671.6	6.9567	**620**
640	913.15	0.04410	3726.3	7.0542	0.04220	3723.3	7.0321	0.04044	3720.4	7.0107	**640**
660	933.15	0.04523	3774.7	7.1067	0.04328	3771.9	7.0847	0.04148	3769.1	7.0635	**660**
680	953.15	0.04634	3823.2	7.1581	0.04435	3820.6	7.1363	0.04252	3817.9	7.1153	**680**
700	973.15	0.04745	3871.7	7.2085	0.04542	3869.3	7.1868	0.04355	3866.8	7.1660	**700**
720	993.15	0.04855	3920.3	7.2579	0.04647	3918.0	7.2364	0.04457	3915.6	7.2157	**720**
740	1013.15	0.04964	3969.0	7.3064	0.04753	3966.8	7.2851	0.04558	3964.6	7.2645	**740**
760	1033.15	0.05073	4017.8	7.3541	0.04857	4015.7	7.3329	0.04658	4013.6	7.3124	**760**
800	1073.15	0.05288	4115.7	7.4471	0.05064	4113.9	7.4261	0.04858	4112.0	7.4058	**800**

(Contd.)

TABLE: 1.8 Thermodynamic Properties of Steam (Superheat Region) [Contd.]

Temperature		Absolute Pressure (Saturation Temperature)									Temp.
t	T	220.0 bar (373.69°C/646.84K)			250.0 bar			300.0 bar			t
°C	K	v	i	s	v	i	s	v	i	s	°C
Satd. Cond.		0.003727	2195.4	4.5796	(Supercritical Region)			(Supercritical Region)			
380	653.15	0.006111	2504.4	5.0559	0.002240	1941.0	4.1757	0.001874	1837.7	4.0021	380
400	673.15	0.008251	2738.8	5.4102	0.006014	2582.0	5.1455	0.002831	2161.8	4.4896	400
420	693.15	0.009588	2874.6	5.6091	0.007580	2774.1	5.4271	0.004921	2558.0	5.0706	420
440	713.15	0.01064	2977.5	5.7556	0.008696	2901.7	5.6087	0.006227	2754.0	5.3499	440
460	733.15	0.01155	3064.0	5.8753	0.009609	3002.3	5.7479	0.007189	2887.7	5.5349	460
480	753.15	0.01237	3141.0	5.9789	0.01041	3088.5	5.8640	0.007985	2993.9	5.6779	480
500	773.15	0.01312	3211.7	6.0716	0.01113	3165.9	5.9655	0.008681	3085.0	5.7972	500
520	793.15	0.01382	3278.0	6.1563	0.01180	3237.4	6.0568	0.009310	3166.6	5.9014	520
540	813.15	0.01449	3341.0	6.2347	0.01242	3304.7	6.1405	0.009890	3241.7	5.9949	540
560	833.15	0.01512	3401.6	6.3083	0.01301	3368.7	6.2183	0.01043	3312.1	6.0805	560
580	853.15	0.01574	3460.2	6.3779	0.01358	3430.2	6.2913	0.01095	3378.9	6.1597	580
600	873.15	0.01633	3517.4	6.4441	0.01413	3489.9	6.3604	0.01144	3443.0	6.2340	600
620	893.15	0.01690	3573.5	6.5076	0.01465	3548.1	6.4263	0.01191	3505.0	6.3042	620
640	913.15	0.01746	3628.7	6.5688	0.01517	3605.2	6.4895	0.01236	3565.3	6.3710	640
660	933.15	0.01801	3683.3	6.6279	0.01566	3661.4	6.5504	0.01280	3624.4	6.4350	660
680	953.15	0.01855	3737.4	6.6853	0.01615	3716.9	6.6093	0.01323	3682.4	6.4966	680
700	973.15	0.01907	3791.1	6.7410	0.01663	3771.9	6.6664	0.01365	3739.7	6.5560	700
720	993.15	0.01959	3844.5	6.7953	0.01710	3826.5	6.7219	0.01406	3796.3	6.6136	720
740	1013.15	0.02010	3897.6	6.8483	0.01756	3880.7	6.7760	0.01446	3852.4	6.6696	740
760	1033.15	0.02061	3950.5	6.9000	0.01802	3934.6	6.8287	0.01485	3908.1	6.7240	760
780	1053.15	0.02111	4003.3	6.9506	0.01846	3988.4	6.8802	0.01524	3963.5	6.7770	780
800	1073.15	0.02160	4055.9	7.0001	0.01891	4041.9	6.9306	0.01562	4018.5	6.8288	800

Temperature		Absolute Pressure									Temp.
t	T	350.0 bar			400.0 bar			450.0 bar			t
°C	K	v	i	s	v	i	s	v	i	s	°C
380	653.15	0.001754	1799.9	3.9304	0.001682	1776.4	3.8824	0.001630	1759.7	3.8430	380
400	673.15	0.002111	1993.1	4.2214	0.001909	1934.1	4.1190	0.001801	1900.6	4.0554	400
420	693.15	0.003082	2296.7	4.6656	0.002371	2145.7	4.4285	0.002088	2070.6	4.3042	420
440	713.15	0.004404	2577.2	5.0649	0.003200	2399.4	4.7893	0.002587	2277.0	4.5977	440
460	733.15	0.005430	2758.0	5.3151	0.004137	2617.1	5.0906	0.003266	2486.4	4.8874	460
480	753.15	0.006239	2890.4	5.4934	0.004941	2779.8	5.3097	0.003982	2667.5	5.1312	480
500	773.15	0.006925	2998.3	5.6349	0.005616	2906.8	5.4762	0.004625	2813.5	5.3226	500
520	793.15	0.007532	3091.8	5.7543	0.006205	3013.7	5.6128	0.005190	2933.8	5.4763	520
540	813.15	0.008083	3176.0	5.8592	0.006735	3108.0	5.7302	0.005698	3038.5	5.6066	540
560	833.15	0.008592	3253.5	5.9534	0.007219	3193.4	5.8340	0.006161	3132.2	5.7206	560
580	853.15	0.009069	3326.2	6.0396	0.007667	3272.4	5.9276	0.006587	3217.9	5.8222	580
600	873.15	0.009519	3395.1	6.1194	0.008088	3346.4	6.0135	0.006984	3297.4	5.9143	600
620	893.15	0.009949	3461.1	6.1942	0.008487	3416.7	6.0931	0.007358	3372.2	5.9990	620
640	913.15	0.01036	3524.9	6.2648	0.008868	3484.2	6.1678	0.007714	3443.4	6.0778	640
660	933.15	0.01076	3587.0	6.3321	0.009234	3549.4	6.2384	0.008055	3511.7	6.1519	660
680	953.15	0.01115	3647.7	6.3965	0.009588	3612.8	6.3056	0.008382	3577.9	6.2221	680
700	973.15	0.01152	3707.3	6.4584	0.009930	3674.8	6.3701	0.008699	3642.4	6.2890	700
720	993.15	0.01189	3766.1	6.5181	0.01026	3735.7	6.4320	0.009006	3705.5	6.3532	720
740	1013.15	0.01224	3824.1	6.5760	0.01059	3795.7	6.4918	0.009305	3767.5	6.4150	740
760	1033.15	0.01259	3881.6	6.6322	0.01091	3855.0	6.5498	0.009597	3828.5	6.4746	760
780	1053.15	0.01294	3938.5	6.6868	0.01122	3913.6	6.6060	0.009881	3888.8	6.5324	780
800	1073.15	0.01327	3995.1	6.7400	0.01152	3971.7	6.6606	0.01016	3948.4	6.5885	800

(Contd.)

TABLE: 1.8 Thermodynamic Properties of Steam (Superheat Region) [Contd.]

Temperature		Absolute Pressure									Temp.
t	T	500.0 bar			600.0 bar			700.0 bar			t
°C	K	v	i	s	v	i	s	v	i	s	°C
380	653.15	0.001589	1746.8	3.8110	0.001528	1728.4	3.7589	0.001482	1715.8	3.7167	**380**
400	673.15	0.001729	1877.7	4.0083	0.001632	1847.3	3.9383	0.001567	1827.8	3.8855	**400**
420	693.15	0.001938	2026.6	4.2262	0.001771	1975.0	4.1252	0.001673	1944.5	4.0564	**420**
440	713.15	0.002269	2199.7	4.4723	0.001962	2113.5	4.3221	0.001808	2066.6	4.2300	**440**
460	733.15	0.002747	2387.2	4.7316	0.002226	2263.2	4.5291	0.001982	2194.8	4.4073	**460**
480	753.15	0.003308	2564.9	4.9709	0.002565	2418.8	4.7385	0.002203	2329.3	4.5882	**480**
500	773.15	0.003882	2723.0	5.1782	0.002952	2570.6	4.9374	0.002467	2467.1	4.7688	**500**
520	793.15	0.004408	2854.9	5.3466	0.003358	2712.6	5.1189	0.002761	2602.9	4.9422	**520**
540	813.15	0.004888	2968.9	5.4886	0.003755	2838.3	5.2755	0.003068	2731.1	5.1020	**540**
560	833.15	0.005328	3070.7	5.6124	0.004135	2951.7	5.4132	0.003375	2848.0	5.2440	**560**
580	853.15	0.005734	3163.2	5.7221	0.004496	3055.8	5.5367	0.003677	2957.4	5.3739	**580**
600	873.15	0.006111	3248.3	5.8207	0.004835	3151.6	5.6477	0.003972	3060.4	5.4931	**600**
620	893.15	0.006465	3327.7	5.9106	0.005153	3240.2	5.7480	0.004254	3156.6	5.6021	**620**
640	913.15	0.006799	3402.7	5.9937	0.005452	3322.9	5.8396	0.004523	3246.3	5.7015	**640**
660	933.15	0.007118	3474.3	6.0712	0.005736	3400.9	5.9242	0.004779	3330.5	5.7926	**660**
680	953.15	0.007424	3543.3	6.1443	0.006008	3475.3	6.0031	0.005023	3410.1	5.8771	**680**
700	973.15	0.007720	3610.2	6.2138	0.006269	3547.0	6.0775	0.005257	3486.3	5.9562	**700**
720	993.15	0.008006	3675.4	6.2802	0.006521	3616.4	6.1481	0.005482	3559.7	6.0308	**720**
740	1013.15	0.008284	3739.3	6.3439	0.006766	3684.1	6.2156	0.005701	3630.9	6.1018	**740**
760	1033.15	0.008554	3802.2	6.4053	0.007003	3750.4	6.2804	0.005913	3700.2	6.1697	**760**
780	1053.15	0.008818	3864.1	6.4647	0.007235	3815.5	6.3428	0.006119	3768.4	6.2350	**780**
800	1073.15	0.009076	3925.3	6.5222	0.007460	3879.6	6.4031	0.006321	3835.3	6.2979	**800**

Temperature		Absolute Pressure									Temp.
t	T	800.0 bar			900.0 bar			1000.0 bar			t
°C	K	v	i	s	v	i	s	v	i	s	°C
380	653.15	0.001445	1707.0	3.6807	0.001415	1700.7	3.6492	0.001390	1696.3	3.6211	**380**
400	673.15	0.001518	1814.2	3.8425	0.001479	1804.6	3.8059	0.001446	1797.6	3.7738	**400**
420	693.15	0.001605	1924.1	4.0033	0.001553	1909.6	3.9597	0.001511	1899.0	3.9223	**420**
440	713.15	0.001710	2036.6	4.1633	0.001641	2015.7	4.1105	0.001587	2000.3	4.0664	**440**
460	733.15	0.001841	2152.5	4.3237	0.001745	2123.7	4.2599	0.001675	2102.7	4.2079	**460**
480	753.15	0.001999	2272.8	4.4855	0.001869	2234.9	4.4095	0.001777	2207.7	4.3492	**480**
500	773.15	0.002188	2397.4	4.6488	0.002013	2349.9	4.5602	0.001893	2316.1	4.4913	**500**
520	793.15	0.002405	2524.0	4.8104	0.002177	2467.6	4.7105	0.002024	2427.2	4.6331	**520**
540	813.15	0.002641	2648.2	4.9650	0.002361	2585.3	4.8570	0.002168	2538.7	4.7719	**540**
560	833.15	0.002886	2765.1	5.1072	0.002558	2699.6	4.9960	0.002326	2648.2	4.9050	**560**
580	853.15	0.003132	2874.9	5.2374	0.002760	2808.0	5.1246	0.002493	2454.5	5.0311	**580**
600	873.15	0.003379	2980.3	5.3595	0.002967	2913.5	5.2468	0.002668	2857.5	5.1505	**600**
620	893.15	0.003622	3080.4	5.4729	0.003172	3015.2	5.3620	0.002847	2960.7	5.2674	**620**
640	913.15	0.003859	3175.0	5.5777	0.003375	3111.8	5.4689	0.003020	3058.6	5.3758	**640**
660	933.15	0.004088	3264.2	5.6743	0.003576	3203.6	5.5684	0.003191	3150.8	5.4757	**660**
680	953.15	0.004308	3348.4	5.7636	0.003773	3291.0	5.6611	0.003364	3239.1	5.5693	**680**
700	973.15	0.004519	3428.7	5.8470	0.003964	3374.6	5.7479	0.003536	3324.4	5.6579	**700**
720	993.15	0.004723	3505.7	5.9253	0.004150	3454.8	5.8295	0.003705	3406.9	5.7418	**720**
740	1013.15	0.004921	3580.2	5.9996	0.004331	3532.3	5.9067	0.003871	3486.9	5.8216	**740**
760	1033.15	0.005113	3652.6	6.0704	0.004505	3607.5	5.9802	0.004032	3564.6	5.8976	**760**
780	1053.15	0.005299	3723.4	6.1382	0.004675	3680.7	6.0505	0.004189	3640.3	5.9701	**780**
800	1073.15	0.005480	3792.8	6.2034	0.004841	3752.4	6.1179	0.004341	3714.3	6.0397	**800**

Note: Data, contained in Tables 1.6, 1.7 & 1.8, are based on 1967 IFC Formulation for Industrial Use, satisfying the International Skeleton Tables (1963) and ITS - 68 temperature scale.

Properties of Refrigerants and *n*–Hexane

Table: 2 Physico–Chemical Constants of Selected Refrigerants

Refrigerant number[a]	Chemical name	Chemical formula	Molar mass[b]	Normal boiling point at 1.01325 bar, °C	Critical properties		
					p_c, bar	t_c, °C	δ_c, kg/m^3
R–12	Dichlorodifluoromethane	CCl_2F_2	120.9135	–29.79	41.18	111.80	578.0
R–22	Chlorodifluoromethane	$CHClF_2$	86.4684	–40.76	49.86	96.13	512.8
R–23	Trifluoromethane	CHF_3	70.0138	–82.02	48.32	26.14	526.5
R–134a	1,1,1,2–Tetrafluoroethane	CH_2FCF_3	102.0309	–26.09	40.59	101.06	512.0
R–717	Ammonia	NH_3	17.0305	–33.33	114.17	133.00	235.6
R–718	Water	H_2O	18.0153	100.00	221.20	374.15	315.5
R–728	Nitrogen	N_2	28.0134	–195.80	33.96	–146.96	313.3
R–744	Carbon dioxide	CO_2	44.0095	–78.46[c]	73.77	30.98	467.6

[a]ASHRAE standard designation of refrigerant
[b]Based on ^{12}C = 12, IUPAC, 1991.
[c]Sublimation temperature

TABLE: 3.1 Thermodynamic Properties of Saturated Dichlorodifluoromethane (Refrigerant R–12) up to 1 Atmosphere (Temperature Basis)

Temperature		Saturation pressure, p		Liquid density ρ_l	Specific volume of vapour, v''	Specific enthalpy			Specific entropy		Temp t
t	T					h	λ_v	H	s'	s''	
°C	K	bar	torr	kg/m^3	v'' m^3/kg	kJ/kg			kJ/(kg·K)		°C
–120	153.15	0.0014	1.050	1721.9	76.336	98.1	199.4	297.5	0.515	1.816	**–120**
–115	158.15	0.0025	1.875	1709.7	43.478	102.0	197.6	299.6	0.540	1.879	**–115**
–110	163.15	0.0044	3.300	1697.4	25.707	105.9	195.8	301.7	0.564	1.764	**–110**
–105	168.15	0.0073	5.475	1685.1	15.798	109.8	194.1	303.9	0.587	1.742	**–105**
–100	173.15	0.0119	8.926	1672.7	10.030	113.8	192.3	306.1	0.610	1.721	**–100**
–95	178.15	0.0186	13.95	1660.2	6.5617	117.7	190.6	308.3	0.633	1.703	**–95**
–90	183.15	0.0285	21.38	1647.6	4.4111	121.7	188.9	310.6	0.655	1.687	**–90**
–85	188.15	0.0424	31.80	1634.9	3.0414	125.7	187.2	312.9	0.677	1.672	**–85**
–80	193.15	0.0617	46.28	1622.1	2.1445	129.8	185.4	315.2	0.698	1.658	**–80**
–75	198.15	0.0877	65.78	1609.2	1.5442	133.8	183.8	317.6	0.719	1.646	**–75**
–70	203.15	0.1223	91.73	1596.2	1.1331	138.0	181.9	319.9	0.739	1.635	**–70**
–65	208.15	0.1675	125.6	1583.0	0.8460	142.1	180.2	322.3	0.759	1.625	**–65**
–60	213.15	0.2255	169.1	1569.7	0.6423	146.3	178.4	324.7	0.779	1.616	**–60**
–58	215.15	0.2528	189.6	1564.3	0.5774	148.0	177.6	325.6	0.787	1.613	**–58**
–56	217.15	0.2828	212.1	1559.0	0.5206	149.7	176.9	326.6	0.795	1.610	**–56**
–54	219.15	0.3156	236.7	1553.6	0.4701	151.4	176.1	327.5	0.803	1.607	**–54**
–52	221.15	0.3514	263.6	1548.1	0.4255	153.1	175.4	328.5	0.811	1.604	**–52**
–50	223.15	0.3903	292.7	1542.7	0.3860	154.8	174.7	329.5	0.818	1.601	**–50**
–48	225.15	0.4327	324.6	1537.2	0.3509	156.5	173.9	330.4	0.826	1.598	**–48**
–46	227.15	0.4786	359.0	1531.7	0.3195	158.3	173.1	331.4	0.834	1.596	**–46**
–44	229.15	0.5283	396.3	1526.1	0.2915	160.0	172.3	332.3	0.841	1.593	**–44**
–42	231.15	0.5821	436.6	1520.6	0.2664	161.8	171.5	333.3	0.849	1.591	**–42**
–40	233.15	0.6401	480.1	1515.0	0.2439	163.5	170.8	334.3	0.856	1.589	**–40**
–38	235.15	0.7026	527.0	1509.3	0.2237	165.3	169.9	335.2	0.864	1.587	**–38**
–36	237.15	0.7698	577.4	1503.7	0.2055	167.0	169.2	336.2	0.871	1.584	**–36**
–34	239.15	0.8420	631.6	1498.0	0.1891	168.8	168.3	337.1	0.879	1.583	**–34**
–32	241.15	0.9193	689.5	1492.3	0.1742	170.6	167.5	338.1	0.886	1.581	**–32**
–30	243.15	1.002	751.6	1486.5	0.1608	172.4	166.6	339	0.893	1.579	**–30**
–29.79*	**243.36**	**1.01325**	**760.0**	**1486.0**	**0.1590**	**173.0**	**165.9**	**338.9**	**0.896**	**1.578**	**–29.79**

*Normal boiling point

TABLE: 3.2 Thermodynamic Properties of Saturated Dichlorodifluoromethane (Refrigerant R–12) above 1 Atmosphere (Temperature Basis)

Temperature t °C	Temperature T K	Saturation pressure p bar	Liquid density ρ_l kg/m³	Specific volume of vapour v'' kg/m³	Specific enthalpy h kJ/kg	Specific enthalpy λ_v kJ/kg	Specific enthalpy H kJ/kg	Specific entropy s' kJ/(kg·K)	Specific entropy s'' kJ/(kg·K)	Temp. t °C
–29	244.15	1.05	1485	0.1543	173.55	165.86	339.41	0.8982	1.5776	**–29**
–28	245.15	1.09	1482	0.1484	174.44	165.45	339.89	0.9019	1.5767	**–28**
–27	246.15	1.14	1479	0.1428	175.34	165.02	340.36	0.9055	1.5759	**–27**
–26	247.15	1.19	1476	0.1374	176.23	164.60	340.83	0.9091	1.5751	**–26**
–25	248.15	1.23	1473	0.1322	177.12	164.18	341.30	0.9127	1.5743	**–25**
–24	249.15	1.29	1470	0.1273	178.02	163.76	341.78	0.9163	1.5735	**–24**
–23	250.15	1.34	1467	0.1226	178.92	163.33	342.25	0.9199	1.5728	**–23**
–22	251.15	1.39	1464	0.1182	179.81	162.91	342.72	0.9234	1.5720	**–22**
–21	252.15	1.45	1461	0.1139	180.71	162.48	343.19	0.9270	1.5713	**–21**
–20	253.15	1.51	1458	0.1098	181.62	162.03	343.65	0.9305	1.5706	**–20**
–19	254.15	1.57	1455	0.1059	182.52	161.60	344.12	0.9341	1.5699	**–19**
–18	255.15	1.63	1452	0.1021	183.42	161.17	344.59	0.9376	1.5693	**–18**
–17	256.15	1.69	1449	0.0986	184.33	160.72	345.05	0.9412	1.5686	**–17**
–16	257.15	1.76	1446	0.0951	185.24	160.28	345.52	0.9447	1.5680	**–16**
–15	258.15	1.82	1443	0.0918	186.15	159.83	345.98	0.9482	1.5673	**–15**
–14	259.15	1.89	1440	0.0887	187.06	159.38	346.44	0.9517	1.5667	**–14**
–13	260.15	1.96	1437	0.0857	187.97	158.94	346.91	0.9552	1.5661	**–13**
–12	261.15	2.04	1434	0.0828	188.89	158.48	347.37	0.9587	1.5655	**–12**
–11	262.15	2.11	1431	0.0800	189.81	158.02	347.83	0.9622	1.5650	**–11**
–10	263.15	2.19	1428	0.0774	190.72	157.57	348.29	0.9656	1.5644	**–10**
–9	264.15	2.27	1425	0.0748	191.64	157.10	348.74	0.9691	1.5638	**–9**
–8	265.15	2.35	1421	0.0724	192.56	156.64	349.20	0.9726	1.5633	**–8**
–7	266.15	2.43	1418	0.0700	193.49	156.17	349.66	0.9760	1.5628	**–7**
–6	267.15	2.52	1415	0.0678	194.41	155.70	350.11	0.9795	1.5623	**–6**
–5	268.15	2.61	1412	0.0656	195.34	155.22	350.56	0.9829	1.5618	**–5**
–4	269.15	2.7	1409	0.0635	196.27	154.74	351.01	0.9863	1.5613	**–4**
–3	270.15	2.79	1406	0.0615	197.20	154.26	351.46	0.9898	1.5608	**–3**
–2	271.15	2.88	1402	0.0596	198.13	153.78	351.91	0.9932	1.5603	**–2**
–1	272.15	2.98	1399	0.0577	199.06	153.30	352.36	0.9966	1.5599	**–1**
0	273.15	3.08	1396	0.0560	200.00	152.81	352.81	1.0000	1.5594	**0**
1	274.15	3.18	1393	0.0542	200.94	152.31	353.25	1.0034	1.5590	**1**
2	275.15	3.29	1390	0.0526	201.88	151.81	353.69	1.0068	1.5586	**2**
3	276.15	3.4	1386	0.0510	202.82	151.31	354.13	1.0102	1.5581	**3**
4	277.15	3.51	1383	0.0495	203.76	150.81	354.57	1.0136	1.5577	**4**
5	278.15	3.62	1380	0.0480	204.71	150.30	355.01	1.0169	1.5573	**5**
6	279.15	3.74	1377	0.0466	205.65	149.80	355.45	1.0203	1.5569	**6**
7	280.15	3.85	1373	0.0452	206.60	149.28	355.88	1.0237	1.5565	**7**
8	281.15	3.98	1370	0.0439	207.56	148.76	356.32	1.0270	1.5561	**8**
9	282.15	4.1	1367	0.0426	208.51	148.24	356.75	1.0304	1.5558	**9**
10	283.15	4.23	1363	0.0414	209.46	147.72	357.18	1.0337	1.5554	**10**
11	284.15	4.36	1360	0.0402	210.42	147.19	357.61	1.0371	1.5550	**11**
12	285.15	4.49	1356	0.0390	211.38	146.65	358.03	1.0404	1.5547	**12**
13	286.15	4.63	1353	0.0379	212.34	146.11	358.45	1.0437	1.5544	**13**
14	287.15	4.76	1350	0.0368	213.31	145.57	358.88	1.0471	1.5540	**14**
15	288.15	4.91	1346	0.0358	214.27	145.03	359.30	1.0504	1.5537	**15**

(Contd.)

TABLE: 3.2 Thermodynamic Properties of Saturated Dichlorodifluoromethane (Refrigerant R–12) above 1 Atmosphere (Temperature Basis) [Contd.]

Temperature		Saturation pressure	Liquid density	Specific volume of vapour v''	Specific enthalpy			Specific entropy		Temp.
t	T	p	ρ_l		h	λ_v	H	s'	s''	t
°C	K	bar	kg/m³	kg/m³	kJ/kg			kJ/(kg·K)		°C
16	289.15	5.05	1343	0.0348	215.24	144.47	359.71	1.0537	1.5533	**16**
17	290.15	5.2	1339	0.0338	216.21	143.92	360.13	1.0570	1.5530	**17**
18	291.15	5.35	1336	0.0329	217.18	143.36	360.54	1.0603	1.5527	**18**
19	292.15	5.51	1332	0.0320	218.16	142.79	360.95	1.0636	1.5524	**19**
20	293.15	5.66	1329	0.0311	219.14	142.22	361.36	1.0669	1.5521	**20**
21	294.15	5.83	1325	0.0303	220.12	141.65	361.77	1.0702	1.5518	**21**
22	295.15	5.99	1322	0.0295	221.10	141.07	362.17	1.0735	1.5515	**22**
23	296.15	6.16	1318	0.0287	222.08	140.49	362.57	1.0768	1.5512	**23**
24	297.15	6.33	1315	0.0279	223.07	139.90	362.97	1.0801	1.5509	**24**
25	298.15	6.51	1311	0.0272	224.06	139.31	363.37	1.0834	1.5506	**25**
26	299.15	6.68	1307	0.0264	225.05	138.71	363.76	1.0866	1.5503	**26**
27	300.15	6.87	1304	0.0257	226.05	138.10	364.15	1.0899	1.5500	**27**
28	301.15	7.05	1300	0.0251	227.04	137.50	364.54	1.0932	1.5498	**28**
29	302.15	7.24	1296	0.0244	228.04	136.89	364.93	1.0964	1.5495	**29**
30	303.15	7.44	1293	0.0238	229.04	136.27	365.31	1.0997	1.5492	**30**
31	304.15	7.63	1289	0.0232	230.05	135.64	365.69	1.1030	1.5489	**31**
32	305.15	7.84	1285	0.0226	231.06	135.01	366.07	1.1062	1.5487	**32**
33	306.15	8.04	1281	0.0220	232.07	134.37	366.44	1.1095	1.5484	**33**
34	307.15	8.25	1278	0.0214	233.08	133.73	366.81	1.1127	1.5481	**34**
35	308.15	8.46	1274	0.0209	234.10	133.08	367.18	1.1160	1.5478	**35**
36	309.15	8.68	1270	0.0203	235.12	132.42	367.54	1.1192	1.5476	**36**
37	310.15	8.9	1266	0.0198	236.14	131.76	367.90	1.1225	1.5473	**37**
38	311.15	9.13	1262	0.0193	237.16	131.10	368.26	1.1257	1.5470	**38**
39	312.15	9.35	1258	0.0189	238.19	130.42	368.61	1.1289	1.5468	**39**
40	313.15	9.59	1254	0.0184	239.22	129.74	368.96	1.1322	1.5465	**40**
41	314.15	9.83	1250	0.0179	240.25	129.06	369.31	1.1354	1.5462	**41**
42	315.15	10.07	1246	0.0175	241.29	128.36	369.65	1.1387	1.5459	**42**
43	316.15	10.31	1242	0.0170	242.33	127.66	369.99	1.1419	1.5457	**43**
44	317.15	10.57	1238	0.0166	243.38	126.95	370.33	1.1451	1.5454	**44**
45	318.15	10.82	1234	0.0162	244.42	126.24	370.66	1.1484	1.5451	**45**
46	319.15	11.08	1230	0.0158	245.47	125.51	370.98	1.1516	1.5448	**46**
47	320.15	11.35	1226	0.0154	246.53	124.77	371.30	1.1548	1.5446	**47**
48	321.15	11.61	1222	0.0151	247.59	124.03	371.62	1.1580	1.5443	**48**
49	322.15	11.89	1217	0.0147	248.65	123.29	371.94	1.1613	1.5440	**49**
50	323.15	12.17	1213	0.0143	249.71	122.53	372.24	1.1645	1.5437	**50**
51	324.15	12.45	1209	0.0140	250.78	121.77	372.55	1.1677	1.5434	**51**
52	325.15	12.74	1204	0.0137	251.85	121.00	372.85	1.1710	1.5431	**52**
53	326.15	13.03	1200	0.0133	252.93	120.21	373.14	1.1742	1.5428	**53**
54	327.15	13.33	1196	0.0130	254.01	119.42	373.43	1.1774	1.5425	**54**
55	328.15	13.63	1191	0.0127	255.10	118.62	373.72	1.1807	1.5421	**55**
56	329.15	13.94	1187	0.0124	256.18	117.81	373.99	1.1839	1.5418	**56**
57	330.15	14.25	1182	0.0121	257.28	116.99	374.27	1.1871	1.5415	**57**
58	331.15	14.57	1177	0.0118	258.38	116.15	374.53	1.1904	1.5411	**58**
59	332.15	14.89	1173	0.0115	259.48	115.31	374.79	1.1936	1.5408	**59**
60	333.15	15.22	1168	0.0112	260.58	114.47	375.05	1.1969	1.5404	**60**

(Contd.)

TABLE: 3.2 Thermodynamic Properties of Saturated Dichlorodifluoromethane (Refrigerant R–12) above 1 Atmosphere (Temperature Basis) [Contd.]

Temperature		Saturation pressure	Liquid density	Specific volume of vapour v''	Specific enthalpy			Specific entropy		Temp. t
t	T	p	ρ_l		h	λ_v	H	s'	s''	
°C	K	bar	kg/m³	kg/m³	kJ/kg			kJ/(kg·K)		°C
61	334.15	15.55	1163	0.0110	261.70	113.60	375.30	1.2001	1.5401	**61**
62	335.15	15.89	1159	0.0107	262.81	112.73	375.54	1.2033	1.5397	**62**
63	336.15	16.24	1154	0.0105	263.93	111.85	375.78	1.2066	1.5393	**63**
64	337.15	16.59	1149	0.0102	265.06	110.94	376.00	1.2099	1.5389	**64**
65	338.15	16.94	1144	0.0100	266.19	110.04	376.23	1.2131	1.5385	**65**
66	339.15	17.3	1139	0.0097	267.33	109.11	376.44	1.2164	1.5381	**66**
67	340.15	17.67	1134	0.0095	268.47	108.18	376.65	1.2197	1.5377	**67**
68	341.15	18.04	1129	0.0093	269.62	107.22	376.84	1.2229	1.5372	**68**
69	342.15	18.42	1124	0.0090	270.78	106.26	377.04	1.2262	1.5368	**69**
70	343.15	18.8	1118	0.0088	271.94	105.28	377.22	1.2295	1.5363	**70**
71	344.15	19.19	1113	0.0086	273.11	104.28	377.39	1.2328	1.5358	**71**
72	345.15	19.59	1108	0.0084	274.28	103.27	377.55	1.2361	1.5353	**72**
73	346.15	19.99	1102	0.0082	275.46	102.25	377.71	1.2394	1.5348	**73**
74	347.15	20.4	1097	0.0080	276.65	101.20	377.85	1.2427	1.5343	**74**
75	348.15	20.81	1091	0.0078	277.84	100.15	377.99	1.2461	1.5337	**75**
76	349.15	21.23	1085	0.0076	279.05	99.06	378.11	1.2494	1.5331	**76**
77	350.15	21.66	1079	0.0074	280.26	97.96	378.22	1.2527	1.5325	**77**
78	351.15	22.09	1074	0.0073	281.48	96.84	378.32	1.2561	1.5319	**78**
79	352.15	22.53	1068	0.0071	282.70	95.71	378.41	1.2595	1.5313	**79**
80	353.15	22.97	1061	0.0069	283.94	94.54	378.48	1.2629	1.5306	**80**
81	354.15	23.43	1055	0.0067	285.19	93.36	378.55	1.2663	1.5299	**81**
82	355.15	23.89	1049	0.0066	286.44	92.15	378.59	1.2697	1.5292	**82**
83	356.15	24.35	1042	0.0064	287.71	90.92	378.63	1.2731	1.5284	**83**
84	357.15	24.82	1036	0.0062	288.99	89.65	378.64	1.2766	1.5276	**84**
85	358.15	25.3	1029	0.0061	290.27	88.37	378.64	1.2801	1.5268	**85**
86	359.15	25.79	1022	0.0059	291.58	87.04	378.62	1.2835	1.5259	**86**
87	360.15	26.28	1015	0.0058	292.89	85.70	378.59	1.2871	1.5250	**87**
88	361.15	26.78	1008	0.0056	294.21	84.32	378.53	1.2906	1.5241	**88**
89	362.15	27.29	1001	0.0055	295.55	82.90	378.45	1.2942	1.5231	**89**
90	363.15	27.81	993	0.0053	296.91	81.44	378.35	1.2978	1.5220	**90**
91	364.15	28.33	985	0.0052	298.28	79.95	378.23	1.3014	1.5209	**91**
92	365.15	28.86	977	0.0050	299.67	78.41	378.08	1.3050	1.5198	**92**
93	366.15	29.4	969	0.0049	301.08	76.82	377.90	1.3087	1.5186	**93**
94	367.15	29.95	961	0.0048	302.50	75.19	377.69	1.3125	1.5173	**94**
95	368.15	30.5	952	0.0046	303.95	73.50	377.45	1.3163	1.5159	**95**
96	369.15	31.06	943	0.0045	305.42	71.75	377.17	1.3201	1.5144	**96**
97	370.15	31.63	934	0.0044	306.92	69.93	376.85	1.3240	1.5129	**97**
98	371.15	32.21	924	0.0042	308.44	68.04	376.48	1.3279	1.5112	**98**
99	372.15	32.8	914	0.0041	309.99	66.08	376.07	1.3319	1.5095	**99**
100	373.15	33.4	904	0.0040	311.58	64.02	375.60	1.3360	1.5076	**100**
101	374.15	34.0	893	0.0038	313.21	61.86	375.07	1.3402	1.5055	**101**
102	375.15	34.62	881	0.0037	314.88	59.59	374.47	1.3445	1.5033	**102**
103	376.15	35.24	869	0.0036	316.60	57.18	373.78	1.3488	1.5009	**103**
104	377.15	35.88	856	0.0034	318.38	54.61	372.99	1.3534	1.4982	**104**
105	378.15	36.53	842	0.0033	320.24	51.84	372.08	1.3581	1.4952	**105**

(Contd.)

TABLE: 3.2 Thermodynamic Properties of Saturated Dichlorodifluoromethane (Refrigerant R–12) above 1 Atmosphere (Temperature Basis) [Contd.]

Temperature		Saturation pressure	Liquid density	Specific volume of vapour v''	Specific enthalpy			Specific entropy		Temp.
t	T	p	ρ_l		h	λ_v	H	s'	s''	t
°C	K	bar	kg/m³	kg/m³	kJ/kg			kJ/(kg·K)		°C
106	379.15	37.18	827	0.0032	322.18	48.85	371.03	1.3630	1.4919	**106**
107	380.15	37.85	811	0.0030	324.25	45.53	369.78	1.3682	1.4880	**107**
108	381.15	38.53	792	0.0029	326.47	41.81	368.28	1.3739	1.4836	**108**
109	382.15	39.22	770	0.0027	328.94	37.48	366.42	1.3801	1.4782	**109**
110	383.15	39.92	743	0.0025	331.82	32.13	363.95	1.3874	1.4712	**110**
111.80*	**384.95**	**41.18**	**578**	**0.00173**	**346.80**	**0.00**	**346.80**	**1.4260**	**1.4260**	**111.80**

*Critical temperature

Reference State:

Specific enthalpy of saturated liquid at 0°C (273.15 K) = 200.00 kJ/kg

Specific entropy of saturated liquid at 0°C (273.15 K) = 1.0000 kJ/(kg·K)

TABLE: 3.3 Thermodynamic Properties of Dichlorodifluoromethane (Refrigerant R–12) (Superheat Region)

Temperature		Absolute Pressure (Saturation Temperature)									Temp.
t	*T*	0.1 bar (–73.03°C/200.12K)			0.3 bar (–54.92°C/218.23K)			0.5 bar (–45.12°C/228.03K)			*t*
°C	K	*v*	*i*	*s*	*v*	*i*	*s*	*v*	*i*	*s*	°C
Satd.	Cond.	1.3674	318.5	1.642	0.44603	327.1	1.608	0.30684	331.8	1.594	
–70	203.15	1.3879	320.0	1.649	–	–	–	–	–	–	–70
–60	213.15	1.4579	325.0	1.673	–	–	–	–	–	–	–60
–50	223.15	1.5272	330.1	1.697	0.50429	329.7	1.620	–	–	–	–50
–40	233.15	1.5969	335.4	1.720	0.52798	335.0	1.643	0.31417	334.5	1.607	–40
–30	243.15	1.6661	340.7	1.742	0.55127	340.4	1.666	0.32852	340.0	1.629	–30
–20	253.15	1.7349	346.2	1.764	0.57471	345.9	1.688	0.34270	345.6	1.652	–20
–10	263.15	1.8041	351.9	1.786	0.59809	351.6	1.710	0.35689	351.2	1.674	–10
0	273.15	1.8741	357.6	1.808	0.62150	357.3	1.731	0.37106	357.0	1.695	0
10	283.15	1.9429	363.4	1.829	0.64475	363.2	1.752	0.38506	362.9	1.717	10
20	293.15	2.0121	369.7	1.849	0.66800	369.1	1.773	0.39920	368.9	1.737	20
30	303.15	2.0812	375.4	1.869	0.69109	375.2	1.793	0.41322	375.0	1.758	30
40	313.15	2.1501	381.6	1.889	0.71429	381.3	1.813	0.42717	381.1	1.778	40
50	323.15	2.2188	387.8	1.909	0.73746	387.6	1.833	0.44111	387.4	1.797	50
60	333.15	2.2878	394.1	1.928	0.76046	393.9	1.852	0.45517	393.7	1.817	60
70	343.15	2.3568	400.5	1.947	0.78370	400.4	1.871	0.46904	400.2	1.836	70
80	353.15	2.4260	407.0	1.966	0.80710	406.9	1.890	0.48309	406.7	1.855	80
90	363.15	2.4950	413.6	1.984	0.82988	413.5	1.908	0.49702	413.3	1.873	90
100	373.15	2.5628	420.3	2.002	0.85324	420.1	1.926	0.51073	420.0	1.891	100
110	383.15	2.6323	427.0	2.020	0.87566	426.9	1.944	0.52466	426.7	1.909	110

Temperature		Absolute Pressure (Saturation Temperature)									Temp.
t	*T*	1.0 bar (–30.13°C/243.02K)			2.0 bar (–12.58°C/260.57K)			3.0 bar (–0.89°C/272.26K)			*t*
°C	K	*v*	*i*	*s*	*v*	*i*	*s*	*v*	*i*	*s*	°C
Satd.	Cond.	0.16051	338.3	1.576	0.083963	346.5	1.566	0.057274	351.7	1.557	
0	273.15	0.18318	356.3	1.646	0.089127	354.7	1.594	0.057670	353.1	1.562	0
10	283.15	0.19040	362.2	1.667	0.092937	360.8	1.616	0.060386	359.3	1.584	10
20	293.15	0.19759	368.2	1.688	0.096712	366.9	1.637	0.063012	365.6	1.606	20
30	303.15	0.20471	374.4	1.709	0.10050	373.2	1.658	0.065660	371.9	1.627	30
40	313.15	0.21191	380.6	1.729	0.10420	379.5	1.679	0.068213	378.3	1.648	40
50	323.15	0.21901	386.9	1.749	0.10781	385.8	1.699	0.070771	384.8	1.669	50
60	333.15	0.22609	393.3	1.768	0.11150	392.3	1.718	0.073260	391.3	1.688	60
70	343.15	0.23310	399.7	1.787	0.11510	398.9	1.738	0.075758	397.9	1.708	70
80	353.15	0.24021	406.3	1.806	0.11869	405.4	1.757	0.078247	404.5	1.727	80
90	363.15	0.24722	412.9	1.825	0.12229	312.1	1.775	0.080710	411.3	1.746	90
100	373.15	0.25426	419.6	1.843	0.12590	418.8	1.794	0.083160	418.1	1.764	100
110	383.15	0.26130	426.4	1.861	0.12950	425.6	1.812	0.085616	424.9	1.782	110
120	393.15	0.26831	433.2	1.878	0.13307	432.5	1.829	0.088028	431.8	1.800	120
130	403.15	0.27518	440.1	1.896	0.13663	439.5	1.847	0.090416	438.8	1.818	130
140	413.15	0.28217	447.1	1.913	0.14019	446.5	1.864	0.092851	445.8	1.835	140
150	423.15	0.28918	454.1	1.929	0.14370	453.5	1.881	0.095238	452.9	1.852	150
160	433.15	0.29621	461.2	1.946	0.14721	460.6	1.897	0.097561	460.1	1.869	160
170	443.15	0.30312	468.3	1.962	0.15081	467.8	1.914	0.10000	467.3	1.885	170
180	453.15	0.31017	475.5	1.978	0.15430	475.0	1.930	0.10240	474.5	1.901	180

(Contd.)

TABLE: 3.3 Thermodynamic Properties of Dichlorodifluoromethane (Refrigerant R–12) (Superheat Region) [Contd.]

Temperature		Absolute Pressure (Saturation Temperature)									Temp.
t	*T*	0.1 bar (8.14°C/281.29K)			5.0 bar (15.61°C/288.76K)			6.0 bar (22.04°C/295.19K)			*t*
°C	K	*v*	*i*	*s*	*v*	*i*	*s*	*v*	*i*	*s*	°C
Satd.	**Cond.**	0.043516	355.7	1.554	0.035088	358.9	1.551	0.029377	361.5	1.549	
10	283.15	0.044014	357.7	1.561	–	–	–	–	–	–	**10**
20	293.15	0.046125	364.1	1.583	0.035945	362.6	1.564	–	–	–	**20**
30	303.15	0.048193	370.6	1.605	0.037693	369.2	1.586	0.030647	367.7	1.570	**30**
40	313.15	0.050201	377.1	1.626	0.039386	375.8	1.608	0.032134	374.5	1.592	**40**
50	323.15	0.052192	383.6	1.646	0.041034	382.5	1.628	0.033568	381.3	1.613	**50**
60	333.15	0.054142	390.3	1.667	0.043271	389.2	1.649	0.034977	388.1	1.634	**60**
70	343.15	0.056085	396.9	1.686	0.044248	395.9	1.669	0.036350	394.9	1.654	**70**
80	353.15	0.058005	403.6	1.706	0.045830	402.7	1.688	0.037693	401.8	1.674	**80**
90	363.15	0.059880	410.4	1.724	0.047371	409.6	1.707	0.039032	408.7	1.693	**90**
100	373.15	0.061767	417.3	1.743	0.048900	416.5	1.726	0.040339	415.6	1.712	**100**
110	383.15	0.063613	424.2	1.761	0.050429	423.4	1.745	0.041632	422.6	1.731	**110**
120	393.15	0.065488	431.1	1.779	0.051948	430.4	1.763	0.042918	429.7	1.749	**120**
130	403.15	0.067295	438.1	1.797	0.053447	437.5	1.780	0.044189	436.8	1.767	**130**
140	413.15	0.069156	445.2	1.814	0.054945	444.6	1.798	0.045455	443.9	1.784	**140**
150	423.15	0.070972	452.3	1.831	0.056402	451.7	1.815	0.046707	451.1	1.804	**150**
160	433.15	0.072780	459.5	1.848	0.057870	458.9	1.832	0.047939	458.4	1.818	**160**
170	443.15	0.074571	466.7	1.864	0.059347	466.2	1.848	0.049188	465.6	1.835	**170**
180	453.15	0.076394	474.0	1.881	0.060827	473.5	1.865	0.050429	473.0	1.851	**180**
190	463.15	0.078186	481.3	1.897	0.062267	480.6	1.896	0.051653	480.3	1.867	**190**
200	473.15	0.080000	488.7	1.912	0.063694	488.2	1.896	0.052854	487.8	1.883	**200**

Temperature		Absolute Pressure (Saturation Temperature)									Temp.
t	*T*	7.0 bar (27.72°C/300.87K)			8.0 bar (32.83°C/305.98K)			9.0 bar (37.48°C/310.63K)			*t*
°C	K	*v*	*i*	*s*	*v*	*i*	*s*	*v*	*i*	*s*	°C
Satd.	**Cond.**	0.025233	363.8	1.548	0.022085	365.8	1.547	0.019604	367.5	1.545	
30	303.15	0.025582	366.1	1.555	–	–	–	–	–	–	**30**
40	313.15	0.026932	373.1	1.578	0.023026	371.6	1.565	0.019908	370.0	1.554	**40**
50	323.15	0.028217	380.0	1.600	0.024190	378.7	1.588	0.021030	377.3	1.576	**50**
60	333.15	0.029481	386.9	1.621	0.025342	385.7	1.609	0.022099	384.5	1.598	**60**
70	343.15	0.030703	393.9	1.641	0.026441	392.8	1.630	0.023132	391.6	1.619	**70**
80	353.15	0.031888	400.8	1.661	0.027518	399.8	1.650	0.024120	398.8	1.640	**80**
90	363.15	0.033058	407.8	1.681	0.028580	406.9	1.670	0.025082	405.9	1.660	**90**
100	373.15	0.034211	414.8	1.700	0.029612	414.0	1.689	0.026028	413.1	1.679	**100**
110	383.15	0.035348	421.9	1.719	0.030628	421.1	1.708	0.026947	420.3	1.698	**110**
120	393.15	0.036470	429.0	1.737	0.031626	428.2	1.726	0.027863	427.5	1.717	**120**
130	403.15	0.037580	436.1	1.755	0.032616	435.4	1.744	0.028760	434.7	1.735	**130**
140	413.15	0.038685	443.3	1.772	0.033602	442.6	1.762	0.029638	442.0	1.753	**140**
150	423.15	0.039777	450.5	1.790	0.034566	449.9	1.779	0.030516	449.3	1.770	**150**
160	433.15	0.040850	457.8	1.807	0.035524	457.2	1.797	0.031387	456.9	1.787	**160**
170	443.15	0.041929	465.1	1.823	0.036483	464.5	1.813	0.032237	464.0	1.804	**170**
180	453.15	0.042992	472.4	1.840	0.037425	471.9	1.830	0.033102	471.4	1.821	**180**
190	463.15	0.044053	479.8	1.856	0.038373	479.3	1.846	0.033944	478.8	1.837	**190**
200	473.15	0.045126	487.3	1.872	0.039293	486.8	1.862	0.034783	486.3	1.853	**200**

(Contd.)

TABLE: 3.3 Thermodynamic Properties of Dichlorodifluoromethane (Refrigerant R–12) (Superheat Region) [Contd.]

Temperature		Absolute Pressure (Saturation Temperature)									Temp.
t	*T*	10.0 bar (41.77°C/314.92K)			15.0 bar (59.42°C/332.57K)			20.0 bar (73.10°C/346.25K)			*t*
°C	K	*v*	*i*	*s*	*v*	*i*	*s*	*v*	*i*	*s*	°C
Satd.	**Cond.**	0.017603	369.1	1.544	0.011438	374.6	1.540	0.008218	377.6	1.534	
50	323.15	0.018491	375.8	1.566	–	–	–	–	–	–	**50**
60	333.15	0.019501	383.2	1.588	0.011480	375.5	1.543	–	–	–	**60**
70	343.15	0.020458	390.5	1.610	0.012320	383.8	1.567	–	–	–	**70**
80	353.15	0.021390	397.7	1.631	0.013091	391.8	1.590	0.008749	384.4	1.554	**80**
90	363.15	0.022282	405.0	1.651	0.013806	399.7	1.612	0.009443	393.3	1.579	**90**
100	373.15	0.023159	412.2	1.670	0.014480	407.4	1.633	0.010060	401.8	1.602	**100**
110	383.15	0.024010	419.4	1.690	0.015131	415.0	1.653	0.010630	410.1	1.624	**110**
120	393.15	0.024851	426.7	1.708	0.015760	422.6	1.673	0.011157	418.1	1.645	**120**
130	403.15	0.025667	434.0	1.727	0.016369	430.2	1.692	0.011684	426.1	1.665	**130**
140	413.15	0.026483	441.3	1.744	0.016961	437.8	1.710	0.012177	434.0	1.684	**140**
150	423.15	0.027278	448.6	1.762	0.017541	445.3	1.728	0.012653	441.8	1.703	**150**
160	433.15	0.028066	456.0	1.779	0.018113	452.9	1.746	0.013118	449.6	1.721	**160**
170	443.15	0.028852	463.4	1.796	0.018671	460.5	1.763	0.013570	457.4	1.739	**170**
180	453.15	0.029630	470.8	1.813	0.019231	468.1	1.780	0.014013	465.2	1.756	**180**
190	463.15	0.030404	478.3	1.829	0.019771	475.7	1.797	0.014451	473.0	1.773	**190**
200	473.15	0.031172	485.8	1.845	0.020309	483.3	1.813	0.014881	480.8	1.790	**200**

Temperature		Absolute Pressure (Saturation Temperature)									Temp.
t	*T*	25.0 bar (84.39°C/357.54K)			30.0 bar (94.07°C/367.22K)			35.0 bar (102.55°C/375.7K)			*t*
°C	K	*v*	*i*	*s*	*v*	*i*	*s*	*v*	*i*	*s*	°C
Satd.	**Cond.**	0.006196	378.6	1.527	0.004762	377.7	1.517	0.003628	374.2	1.502	
90	363.15	0.006627	385.0	1.545	–	–	–	–	–	–	**90**
100	373.15	0.007289	395.1	1.572	0.005260	385.9	1.539	–	–	–	**100**
110	383.15	0.007855	404.3	1.597	0.005910	397.3	1.569	0.004354	387.5	1.537	**110**
120	393.15	0.008368	413.1	1.619	0.006435	407.2	1.595	0.004980	400.1	1.570	**120**
130	403.15	0.008834	421.6	1.641	0.006897	416.5	1.619	0.005470	410.7	1.596	**130**
140	413.15	0.009276	429.9	1.661	0.007321	425.4	1.640	0.005889	420.4	1.620	**140**
150	423.15	0.009699	438.1	1.681	0.007710	434.0	1.661	0.006274	429.8	1.642	**150**
160	433.15	0.010105	446.2	1.700	0.008084	442.5	1.681	0.006627	438.6	1.663	**160**
170	443.15	0.010496	454.2	1.718	0.008439	450.8	1.700	0.006959	447.2	1.683	**170**
180	453.15	0.010877	462.2	1.736	0.008780	459.1	1.718	0.007273	455.8	1.702	**180**
190	463.15	0.011247	470.2	1.753	0.009107	467.2	1.736	0.007576	464.2	1.720	**190**
200	473.15	0.011610	478.1	1.770	0.009434	475.4	1.753	0.007868	472.5	1.738	**200**

(Contd.)

TABLE: 3.3 Thermodynamic Properties of Dichlorodifluoromethane (Refrigerant R–12) (Superheat Region) [Contd.]

Temperature		Absolute Pressure (Saturation Temperature)									Temp.
t	*T*	40.0 bar (110.13°C/383.28K)			45.0 bar			50.0 bar			*t*
°C	K	*v*	*i*	*s*	*v*	*i*	*s*	*v*	*i*	*s*	°C
Satd.	**Cond.**	0.002522	364.2	1.472	Non-Existent			Non-Existent			
120	393.15	0.003764	390.4	1.540	0.002521	373.2	1.492	0.001400	343.3	1.414	**120**
130	403.15	0.004340	403.7	1.575	0.003385	394.7	1.546	0.002510	381.9	1.511	**130**
140	413.15	0.004789	414.7	1.600	0.003896	408.1	1.579	0.003143	400.1	1.555	**140**
150	423.15	0.005173	424.8	1.624	0.004299	419.4	1.606	0.003582	413.3	1.587	**150**
160	433.15	0.005519	434.3	1.646	0.004649	429.7	1.630	0.003942	424.7	1.613	**160**
170	443.15	0.005841	443.4	1.667	0.004963	439.4	1.652	0.004255	435.1	1.637	**170**
180	453.15	0.006139	452.3	1.687	0.005252	448.7	1.673	0.004539	444.9	1.659	**180**
190	463.15	0.006423	461.0	1.706	0.005525	457.7	1.692	0.004803	454.3	1.679	**190**
200	473.15	0.006698	469.6	1.724	0.005784	466.6	1.711	0.005053	463.4	1.699	**200**

Adapted from V. V. Altunin, V. Z. Geller, E. A. Kremenevskaya, I. I. Perelshtein, E. K. Petrov and T. B. Selover (Jr.) – Editor, *Thermophysical Properties of Freons, Methane Series*, Part 2, Hemisphere Publishing Corporation, USA, 1987.

TABLE: 4.1 Thermodynamic Properties of Saturated Chlorodifluoromethane (Refrigerant R–22) up to 1 Atmosphere (Temperature Basis)

Temperature		Saturation pressure, p		Liquid density	Specific volume of	Specific enthalpy			Specific entropy		Temp.
t	T			ρ_l	vapour, v''	h	λ_v	H	s'	s''	t
°C	K	bar	torr	kg/m^3	m^3/kg	kJ/kg			kJ/(kg·K)		°C
–105	168.15	0.0122	9.15	1567.9	13.210	83.74	272.26	356.00	0.4667	2.0858	**–105**
–104	169.15	0.0135	10.13	1565.7	12.030	84.81	271.67	356.48	0.4730	2.0791	**–104**
–103	170.15	0.0149	11.18	1563.7	10.960	85.88	271.08	356.96	0.4792	2.0724	**–103**
–102	171.15	0.0164	12.30	1561.5	10.000	86.96	270.48	357.44	0.4855	2.0659	**–102**
–101	172.15	0.0181	13.58	1559.6	9.137	88.04	269.89	357.93	0.4917	2.0595	**–101**
–100	173.15	0.0199	14.93	1557.4	8.359	89.11	269.30	358.41	0.4979	2.0532	**–100**
–99	174.15	0.0218	16.35	1555.2	7.658	90.19	268.70	358.89	0.5041	2.0470	**–99**
–98	175.15	0.0239	17.93	1553.3	7.023	91.27	268.11	359.38	0.5103	2.0410	**–98**
–97	176.15	0.0262	19.65	1551.1	6.449	92.34	267.52	359.86	0.5164	2.0350	**–97**
–96	177.15	0.0286	21.45	1548.9	5.929	93.42	266.92	360.34	0.5225	2.0292	**–96**
–95	178.15	0.0313	23.48	1546.8	5.457	94.50	266.33	360.83	0.5285	2.0235	**–95**
–94	179.15	0.0341	25.58	1544.6	5.029	95.58	265.73	361.31	0.5345	2.0178	**–94**
–93	180.15	0.0372	27.90	1542.5	4.639	96.66	265.14	361.80	0.5405	2.0123	**–93**
–92	181.15	0.0405	30.38	1540.4	4.284	97.74	264.55	362.29	0.5465	2.0068	**–92**
–91	182.15	0.0440	33.00	1538.0	3.961	98.82	263.95	362.77	0.5524	2.0015	**–91**
–90	183.15	0.0478	35.85	1535.9	3.665	99.90	263.36	363.26	0.5583	1.9963	**–90**
–89	184.15	0.0519	38.93	1533.7	3.396	100.98	262.77	363.75	0.5642	1.9911	**–89**
–88	185.15	0.0563	42.23	1531.4	3.149	102.06	262.18	364.24	0.5700	1.9860	**–88**
–87	186.15	0.0609	45.68	1529.3	2.923	103.14	261.59	364.73	0.5758	1.9810	**–87**
–86	187.15	0.0659	49.43	1527.0	2.716	104.22	260.99	365.21	0.5816	1.9761	**–86**
–85	188.15	0.0712	53.40	1524.9	2.526	105.30	260.40	365.70	0.5873	1.9714	**–85**
–84	189.15	0.0768	57.60	1522.5	2.352	106.39	259.80	366.19	0.5931	1.9666	**–84**
–83	190.15	0.0829	62.18	1520.2	2.191	107.47	259.21	366.68	0.5988	1.9620	**–83**
–82	191.15	0.0893	66.98	1517.9	2.043	108.55	258.62	367.17	0.6044	1.9574	**–82**
–81	192.15	0.0961	72.08	1515.6	1.907	109.63	258.03	367.66	0.6100	1.9529	**–81**
–80	193.15	0.1034	77.56	1513.3	1.782	110.72	257.43	368.15	0.6157	1.9485	**–80**
–79	194.15	0.1111	83.33	1511.0	1.666	111.80	256.84	368.64	0.6213	1.9442	**–79**
–78	195.15	0.1193	89.48	1508.8	1.559	112.89	256.24	369.13	0.6269	1.9399	**–78**
–77	196.15	0.1280	96.01	1506.3	1.460	113.97	255.65	369.62	0.6324	1.9357	**–77**
–76	197.15	0.1372	102.91	1504.0	1.369	115.06	255.05	370.11	0.6379	1.9316	**–76**
–75	198.15	0.1469	110.18	1501.7	1.284	116.14	254.46	370.60	0.6434	1.9276	**–75**
–74	199.15	0.1572	117.91	1499.3	1.206	117.23	253.86	371.09	0.6489	1.9236	**–74**
–73	200.15	0.1681	126.09	1496.8	1.133	118.32	253.26	371.58	0.6543	1.9197	**–73**
–72	201.15	0.1796	134.71	1494.5	1.065	119.41	252.66	372.07	0.6597	1.9158	**–72**
–71	202.15	0.1917	143.79	1492.1	1.002	120.50	252.05	372.55	0.6651	1.9120	**–71**
–70	203.15	0.2045	153.39	1489.6	0.9435	121.58	251.46	373.04	0.6705	1.9083	**–70**
–69	204.15	0.2180	163.51	1487.2	0.8890	122.67	250.86	373.53	0.6758	1.9046	**–69**
–68	205.15	0.2322	174.16	1484.8	0.8382	123.76	250.26	374.02	0.6811	1.9010	**–68**
–66	207.15	0.2629	197.19	1479.9	0.7467	125.93	249.06	374.99	0.6917	1.8940	**–66**
–65	208.15	0.2794	209.57	1477.3	0.7055	127.02	248.46	375.48	0.6969	1.8906	**–65**

(Contd.)

TABLE: 4.1 Thermodynamic Properties of Saturated Chlorodifluoromethane (Refrigerant R–22) up to 1 Atmosphere (Temperature Basis) [Contd.]

Temperature		Saturation pressure, p		Liquid density	Specific volume of vapour, v''	Specific enthalpy			Specific entropy		Temp.
t	T			ρ_l		h	λ_v	H	s'	s''	t
°C	K	bar	torr	kg/m³	m³/kg	kJ/kg			kJ/(kg·K)		°C
–64	209.15	0.2968	222.62	1474.9	0.6670	128.11	247.85	375.96	0.7021	1.8887	**–64**
–63	210.15	0.3150	236.27	1472.3	0.6309	129.20	247.25	376.45	0.7073	1.8884	**–63**
–62	211.15	0.3341	250.60	1469.9	0.5972	130.30	246.64	376.94	0.7125	1.8806	**–62**
–61	212.15	0.3541	265.60	1467.4	0.5657	131.39	246.03	377.42	0.7177	1.8774	**–61**
–60	213.15	0.3752	281.42	1464.8	0.5361	132.48	245.42	377.90	0.7228	1.8742	**–60**
–59	214.15	0.3972	297.92	1462.2	0.5084	133.57	244.81	378.38	0.7279	1.8711	**–59**
–58	215.15	0.4202	315.18	1459.6	0.4824	134.66	244.21	378.87	0.7330	1.8680	**–58**
–57	216.15	0.4443	333.25	1457.1	0.4579	135.76	243.59	379.35	0.7380	1.8650	**–57**
–56	217.15	0.4695	352.15	1454.3	0.4350	136.86	242.97	379.83	0.7431	1.8620	**–56**
–55	218.15	0.4958	371.88	1451.8	0.4134	137.95	242.36	380.31	0.7481	1.8591	**–55**
–54	219.15	0.5234	392.58	1449.1	0.3932	139.04	241.75	380.79	0.7531	1.8562	**–54**
–53	220.15	0.5521	414.11	1446.5	0.3741	140.14	241.12	381.26	0.7581	1.8534	**–53**
–52	221.15	0.5821	436.61	1443.8	0.3561	141.24	240.50	381.74	0.7630	1.8506	**–52**
–51	222.15	0.6134	460.09	1441.1	0.3391	142.33	239.89	382.22	0.7680	1.8478	**–51**
–50	223.15	0.6459	484.46	1438.4	0.3232	143.43	239.26	382.69	0.7729	1.8451	**–50**
–49	224.15	0.6792	509.44	1435.8	0.3081	144.53	238.64	383.17	0.7778	1.8424	**–49**
–48	225.15	0.7152	536.44	1433.1	0.2939	145.63	238.01	383.64	0.7827	1.8398	**–48**
–47	226.15	0.7520	564.05	1430.2	0.2804	146.73	237.38	384.11	0.7875	1.8372	**–47**
–46	227.15	0.7903	592.77	1427.6	0.2677	147.83	236.75	384.58	0.7924	1.8347	**–46**
–45	228.15	0.8302	622.70	1424.7	0.2557	148.93	236.13	385.06	0.7972	1.8321	**–45**
–44	229.15	0.8715	653.68	1421.9	0.2444	150.03	235.49	385.52	0.8020	1.8297	**–44**
–43	230.15	0.9145	685.93	1419.0	0.2336	151.14	234.85	385.99	0.8068	1.8272	**–43**
–42	231.15	0.9592	719.46	1416.2	0.2235	152.24	234.21	386.45	0.8116	1.8248	**–42**
–41	232.15	1.006	754.56	1413.4	0.2138	153.35	233.57	386.92	0.8164	1.8225	**–41**
–40.76*	**232.39**	**1.01325**	**760.00**	**1409.1**	**0.2126**	**153.62**	**233.41**	**387.03**	**0.8189**	**1.8249**	**–40.76**

*Normal boiling point

TABLE: 4.2 Thermodynamic Properties of Saturated Chlorodifluoromethane (Refrigerant R–22) above 1 Atmosphere (Temperature Basis)

Temperature		Saturation pressure,	Liquid density	Specific volume of vapour, v''	Specific enthalpy			Specific entropy		Temp.
t	T	p	ρ_l		h	λ_v	H	s'	s''	t
°C	K	torr	kg/m³	m³/kg	kJ/kg			kJ/(kg·K)		°C
–40	233.15	1.054	1410.6	0.2047	154.46	232.92	387.38	0.8211	1.8220	**–40**
–39	234.15	1.104	1407.9	0.1960	155.57	232.27	387.84	0.8258	1.8178	**–39**
–38	235.15	1.155	1404.9	0.1878	156.68	231.62	388.30	0.8305	1.8155	**–38**
–37	236.15	1.209	1402.1	0.1800	157.78	230.98	388.76	0.8352	1.8133	**–37**
–36	237.15	1.265	1399.2	0.1726	158.90	230.31	389.21	0.8399	1.8111	**–36**
–35	238.15	1.322	1396.3	0.1655	160.02	229.65	389.67	0.8446	1.8089	**–35**
–34	239.15	1.382	1393.3	0.1588	161.13	228.99	390.12	0.8492	1.8068	**–34**
–33	240.15	1.443	1390.4	0.1525	162.25	228.33	390.58	0.8539	1.8047	**–33**
–32	241.15	1.507	1387.3	0.1464	163.37	227.66	391.03	0.8585	1.8026	**–32**
–31	242.15	1.573	1384.5	0.1406	164.48	226.99	391.47	0.8631	1.8005	**–31**
–30	243.15	1.641	1381.4	0.1351	165.60	226.32	391.92	0.8677	1.7985	**–30**
–29	244.15	1.712	1378.4	0.1299	166.72	225.65	392.37	0.8723	1.7965	**–29**
–28	245.15	1.784	1375.3	0.1249	167.84	224.97	392.81	0.8768	1.7945	**–28**
–27	246.15	1.860	1372.3	0.1202	168.96	224.29	393.25	0.8814	1.7925	**–27**
–26	247.15	1.937	1369.3	0.1156	170.09	223.60	393.69	0.8859	1.7906	**–26**
–25	248.15	2.017	1366.3	0.1113	171.22	222.91	394.13	0.8904	1.7887	**–25**
–24	249.15	2.100	1363.1	0.1072	172.35	222.21	394.56	0.8949	1.7868	**–24**
–23	250.15	2.185	1360.0	0.1032	173.48	221.52	395.00	0.8994	1.7850	**–23**
–22	251.15	2.273	1357.0	0.09945	174.61	220.82	395.43	0.9039	1.7832	**–22**
–21	252.15	2.363	1353.7	0.09585	175.75	220.11	395.86	0.9084	1.7814	**–21**
–20	253.15	2.456	1350.6	0.09241	176.88	219.40	396.28	0.9129	1.7796	**–20**
–19	254.15	2.552	1347.5	0.08911	178.02	218.69	396.71	0.9173	1.7778	**–19**
–18	255.15	2.651	1344.3	0.08596	179.16	217.97	397.13	0.9218	1.7761	**–18**
–17	256.15	2.753	1341.0	0.08294	180.30	217.25	397.55	0.9262	1.7744	**–17**
–16	257.15	2.858	1337.8	0.08005	181.45	216.52	397.97	0.9306	1.7727	**–16**
–15	258.15	2.966	1334.6	0.07729	182.59	215.79	398.38	0.9350	1.7710	**–15**
–14	259.15	3.076	1331.4	0.07464	183.73	215.06	398.79	0.9394	1.7693	**–14**
–13	260.15	3.190	1328.0	0.07210	184.88	214.32	399.20	0.9438	1.7677	**–13**
–12	261.15	3.308	1324.9	0.06966	186.03	213.58	399.61	0.9482	1.7660	**–12**
–11	262.15	3.428	1321.5	0.06732	187.19	212.83	400.02	0.9526	1.7644	**–11**
–10	263.15	3.552	1318.2	0.06508	188.34	212.08	400.42	0.9569	1.7629	**–10**
–9	264.15	3.679	1314.9	0.06293	189.50	211.32	400.82	0.9613	1.7613	**–9**
–8	265.15	3.809	1311.6	0.06086	190.65	210.56	401.21	0.9656	1.7597	**–8**
–7	266.15	3.943	1308.2	0.05888	191.82	209.79	401.61	0.9699	1.7583	**–7**
–6	267.15	4.081	1304.8	0.05697	192.98	209.02	402.00	0.9742	1.7567	**–6**
–5	268.15	4.222	1301.4	0.05514	194.15	208.24	402.39	0.9786	1.7552	**–5**
–4	269.15	4.367	1298.0	0.05338	195.31	207.46	402.77	0.9829	1.7537	**–4**
–3	270.15	4.515	1294.7	0.05169	196.48	206.67	403.15	0.9872	1.7522	**–3**
–2	271.15	4.667	1291.3	0.05006	197.65	205.88	403.53	0.9914	1.7507	**–2**
–1	272.15	4.823	1287.8	0.04849	198.83	205.08	403.91	0.9957	1.7493	**–1**
0	273.15	4.983	1284.4	0.04698	200.00	204.28	404.28	1.0000	1.7479	**0**
1	274.15	5.147	1280.9	0.04553	201.18	203.47	404.65	1.0042	1.7464	**1**
2	275.15	5.315	1277.5	0.04413	202.36	202.65	405.01	1.0085	1.7450	**2**
3	276.15	5.487	1274.0	0.04279	203.55	201.83	405.38	1.0127	1.7436	**3**
4	277.15	5.663	1270.5	0.04149	204.73	201.00	405.73	1.0170	1.7422	**4**
5	278.15	5.844	1266.9	0.04024	205.92	200.17	406.09	1.0212	1.7409	**5**

(Contd.)

TABLE: 4.2 Thermodynamic Properties of Saturated Chlorodifluoromethane (Refrigerant R–22) above 1 Atmosphere (Temperature Basis) [Contd.]

Temperature		Saturation pressure,	Liquid density	Specific volume of	Specific enthalpy			Specific entropy		Temp.
t	T	p	ρ_l	vapour, v''	h	λ_v	H	s'	s''	t
°C	K	torr	kg/m³	m³/kg	kJ/kg			kJ/(kg·K)		°C
6	279.15	6.028	1263.4	0.03903	207.11	199.33	406.44	1.0254	1.7395	**6**
7	280.15	6.217	1259.9	0.03787	208.30	198.49	406.79	1.0296	1.7382	**7**
8	281.15	6.411	1256.3	0.03675	209.49	197.64	407.13	1.0338	1.7368	**8**
9	282.15	6.608	1252.8	0.03567	210.69	196.78	407.47	1.0380	1.7355	**9**
10	283.15	6.811	1249.2	0.03462	211.89	195.92	407.81	1.0422	1.7341	**10**
11	284.15	7.018	1245.6	0.03362	213.09	195.05	408.14	1.0464	1.7328	**11**
12	285.15	7.229	1241.9	0.03264	214.30	194.17	408.47	1.0506	1.7315	**12**
13	286.15	7.445	1238.4	0.03170	215.51	193.28	408.79	1.0548	1.7302	**13**
14	287.15	7.667	1234.7	0.03079	216.72	192.39	409.11	1.0589	1.7289	**14**
15	288.15	7.892	1231.1	0.02992	217.94	191.49	409.43	1.0631	1.7276	**15**
16	289.15	8.123	1227.4	0.02907	219.16	190.58	409.74	1.0673	1.7264	**16**
17	290.15	8.359	1223.8	0.02825	220.38	189.67	410.05	1.0714	1.7251	**17**
18	291.15	8.600	1220.1	0.02745	221.60	188.75	410.35	1.0755	1.7238	**18**
19	292.15	8.846	1216.4	0.02669	222.83	187.82	410.65	1.0796	1.7226	**19**
20	293.15	9.097	1212.7	0.02594	224.05	186.89	410.94	1.0838	1.7213	**20**
21	294.15	9.353	1209.0	0.02523	225.28	185.95	411.23	1.0879	1.7200	**21**
22	295.15	9.615	1205.3	0.02453	226.51	185.00	411.51	1.0920	1.7188	**22**
23	296.15	9.882	1201.5	0.02386	227.74	184.04	411.78	1.0961	1.7175	**23**
24	297.15	10.154	1197.6	0.02321	228.99	183.07	412.06	1.1002	1.7163	**24**
25	298.15	10.432	1193.6	0.02257	230.23	182.09	412.32	1.1043	1.7151	**25**
26	299.15	10.716	1190.1	0.02196	231.48	181.10	412.58	1.1084	1.7138	**26**
27	300.15	11.006	1186.2	0.02137	232.74	180.10	412.84	1.1125	1.7126	**27**
28	301.15	11.300	1182.3	0.02080	233.99	179.10	413.09	1.1166	1.7113	**28**
29	302.15	11.601	1178.4	0.02024	235.24	178.09	413.33	1.1207	1.7101	**29**
30	303.15	11.908	1174.4	0.01970	236.51	177.06	413.57	1.1248	1.7089	**30**
31	304.15	12.221	1170.5	0.01917	237.78	176.02	413.80	1.1289	1.7076	**31**
32	305.15	12.539	1166.5	0.01867	239.05	174.98	414.03	1.1330	1.7064	**32**
33	306.15	12.864	1162.4	0.01817	240.32	173.93	414.25	1.1371	1.7051	**33**
34	307.15	13.196	1158.2	0.01769	241.60	172.86	414.46	1.1411	1.7039	**34**
35	308.15	13.532	1154.1	0.01723	242.88	171.78	414.66	1.1452	1.7027	**35**
36	309.15	13.876	1150.0	0.01678	244.17	170.69	414.86	1.1493	1.7015	**36**
37	310.15	14.226	1145.9	0.01634	245.46	169.59	415.05	1.1534	1.7001	**37**
38	311.15	14.582	1141.4	0.01591	246.76	168.48	415.24	1.1574	1.6989	**38**
39	312.15	14.945	1137.1	0.01550	248.05	167.36	415.41	1.1615	1.6976	**39**
40	313.15	15.315	1132.6	0.01510	249.36	166.22	415.58	1.1656	1.6964	**40**
41	314.15	15.691	1128.2	0.01470	250.67	165.07	415.74	1.1697	1.6951	**41**
42	315.15	16.074	1123.6	0.01432	251.98	163.91	415.89	1.1737	1.6938	**42**
43	316.15	16.464	1119.1	0.01395	253.31	162.73	416.04	1.1778	1.6925	**43**
44	317.15	16.862	1114.3	0.01359	254.63	161.54	416.17	1.1819	1.6912	**44**
45	318.15	17.266	1109.8	0.01324	255.97	160.33	416.30	1.1860	1.6899	**45**
46	319.15	17.677	1105.0	0.01290	257.31	159.11	416.42	1.1901	1.6886	**46**
47	320.15	18.096	1100.2	0.01257	258.65	157.87	416.52	1.1941	1.6872	**47**
48	321.15	18.521	1095.3	0.01225	260.01	156.61	416.62	1.1982	1.6859	**48**
49	322.15	18.964	1090.5	0.01193	261.38	155.33	416.71	1.2024	1.6845	**49**
50	323.15	19.395	1085.4	0.01163	262.75	154.03	416.78	1.2065	1.6832	**50**

(Contd.)

TABLE: 4.2 Thermodynamic Properties of Saturated Chlorodifluoromethane (Refrigerant R–22) above 1 Atmosphere (Temperature Basis) [Contd.]

Temperature		Saturation pressure,	Liquid density	Specific volume of vapour, v''	Specific enthalpy			Specific entropy		Temp.
t	T	p	ρ_l		h	λ_v	H	s'	s''	t
°C	K	torr	kg/m³	m³/kg	kJ/kg			kJ/(kg·K)		°C
51	324.15	19.843	1080.4	0.01133	264.13	152.72	416.85	1.2107	1.6818	**51**
52	325.15	20.299	1075.3	0.01104	265.52	151.38	416.90	1.2148	1.6804	**52**
53	326.15	20.763	1070.1	0.01076	266.93	150.02	416.95	1.2190	1.6790	**53**
54	327.15	21.235	1064.7	0.01048	268.33	148.65	416.98	1.2231	1.6775	**54**
55	328.15	21.714	1059.4	0.01021	269.74	147.26	417.00	1.2273	1.6761	**55**
56	329.15	22.202	1054.0	0.009946	271.17	146.84	418.01	1.2315	1.6746	**56**
57	330.15	22.698	1048.5	0.009690	272.59	144.40	416.99	1.2357	1.6731	**57**
58	331.15	23.202	1043.0	0.009439	274.03	142.94	416.97	1.2399	1.6715	**58**
59	332.15	23.715	1037.3	0.009195	275.48	141.45	416.93	1.2441	1.6699	**59**
60	333.15	24.236	1031.7	0.008956	276.94	139.94	416.88	1.2483	1.6684	**60**
61	334.15	24.65	1026.1	0.008723	278.41	138.40	416.81	1.2526	1.6668	**61**
62	335.15	25.30	1020.0	0.008495	279.90	136.83	416.73	1.2569	1.6652	**62**
63	336.15	25.85	1014.1	0.008272	281.39	135.23	416.62	1.2618	1.6635	**63**
64	337.15	26.41	1008.0	0.008054	282.91	133.59	416.50	1.2656	1.6618	**64**
65	338.15	26.97	1002.0	0.007841	284.44	131.92	416.36	1.2699	1.6600	**65**
66	339.15	27.55	996.02	0.007633	285.99	130.21	416.20	1.2743	1.6583	**66**
67	340.15	28.13	989.12	0.007429	287.56	128.46	416.02	1.2788	1.6564	**67**
68	341.15	28.72	983.28	0.007229	289.14	126.68	415.82	1.2833	1.6546	**68**
69	342.15	29.32	976.56	0.007033	290.73	124.86	415.59	1.2877	1.6526	**69**
70	343.15	29.94	969.93	0.006842	292.34	123.00	415.34	1.2922	1.6506	**70**
71	344.15	30.56	964.32	0.006654	293.96	121.10	415.06	1.2968	1.6486	**71**
72	345.15	31.19	956.02	0.006469	295.61	119.15	414.76	1.3014	1.6466	**72**
73	346.15	31.83	949.67	0.006288	297.27	117.15	414.42	1.3060	1.6444	**73**
74	347.15	32.48	941.62	0.006111	298.96	115.09	414.05	1.3107	1.6422	**74**
75	348.15	33.15	934.58	0.005936	300.68	112.97	413.65	1.3154	1.6399	**75**
76	349.15	33.82	926.78	0.005764	302.42	110.80	413.22	1.3201	1.6375	**76**
77	350.15	34.50	919.96	0.005595	304.17	108.57	412.74	1.3249	1.6350	**77**
78	351.15	35.20	911.58	0.005429	305.94	106.28	412.22	1.3298	1.6325	**78**
79	352.15	35.90	903.34	0.005265	307.74	103.92	411.66	1.3347	1.6298	**79**
80	353.15	36.62	894.45	0.005104	309.57	101.48	411.05	1.3396	1.6270	**80**
81	354.15	37.35	884.96	0.004944	311.44	98.94	410.38	1.3447	1.6241	**81**
82	355.15	38.08	876.42	0.004785	313.35	96.30	409.65	1.3499	1.6210	**82**
83	356.15	38.84	867.30	0.004629	315.31	93.54	408.85	1.3552	1.6178	**83**
84	357.15	39.60	856.90	0.004474	317.34	90.64	407.98	1.3606	1.6144	**84**
85	358.15	40.37	846.02	0.004319	319.45	87.59	407.04	1.3661	1.6107	**85**
86	359.15	41.16	829.19	0.004166	321.61	84.37	405.98	1.3720	1.6069	**86**
87	360.15	41.96	818.33	0.00401	323.8	81.0	404.8	1.378	1.603	**87**
88	361.15	42.77	806.45	0.00386	326.2	77.3	403.5	1.384	1.598	**88**
89	362.15	43.59	794.28	0.00370	328.7	73.4	402.1	1.390	1.593	**89**
90	363.15	44.43	782.47	0.00354	331.1	69.3	400.4	1.397	1.588	**90**
92	365.15	46.14	745.71	0.00321	337.0	59.4	396.4	1.413	1.575	**92**
94	367.15	47.91	690.61	0.00283	344.6	45.7	390.3	1.433	1.557	**94**
96.13*	**369.28**	**49.86**	**512.82**	**0.00195**	**367.3**	**0.0**	**367.3**	**1.493**	**1.493**	**96.13**

*Critical point

Reference State:

Specific enthalpy of saturated liquid at 0°C (273.15 K) = 200.00 kJ/kg

Specific entropy of saturated liquid at 0°C (273.15 K) = 1.0000 kJ/(kg·K)

TABLE: 4.3 Thermodynamic Properties of Chlorodifluoromethane (Refrigerant R–22) (Superheat Region)

Temperature		Absolute Pressure (Saturation Temperature)									Temp.
t	*T*	0.02 bar (–99.93°C/173.22K)			0.05 bar (–89.46°C/183.69K)			0.1 bar (–80.47°C/192.68K)			*t*
°C	K	*v*	*i*	*s*	*v*	*i*	*s*	*v*	*i*	*s*	°C
Satd.	Cond.	8.309	358.44	2.0528	3.517	363.53	1.9935	1.839	367.92	1.9506	
–90	183.15	8.789	363.44	2.0809	–	–	–	–	–	–	–90
–80	193.15	9.272	368.59	2.1083	3.701	368.43	2.0195	1.844	368.17	1.9518	–80
–70	203.15	9.754	373.87	2.1349	3.895	373.74	2.0463	1.942	373.51	1.9788	–70
–60	213.15	10.240	379.27	2.1608	4.089	379.16	2.0724	2.039	378.97	2.0050	–60
–50	223.15	10.720	384.80	2.1861	4.282	384.70	2.0978	2.137	384.54	2.0306	–50
–40	233.15	11.200	390.46	2.2109	4.475	390.37	2.1226	2.234	390.23	2.0555	–40
–30	243.15	11.680	396.23	2.2352	4.668	396.16	2.1469	2.331	396.03	2.0799	–30
–20	253.15	12.160	402.13	2.2590	4.861	402.07	2.1707	2.428	401.95	2.1037	–20
–10	263.15	12.640	408.15	2.2823	5.054	408.10	2.1941	2.524	407.99	2.1271	–10
0	273.15	13.130	414.30	2.3052	5.247	414.24	2.2170	2.621	414.15	2.1501	0
10	283.15	13.610	420.56	2.3277	5.440	420.51	2.2395	2.718	420.43	2.1727	10
20	293.15	14.090	426.94	2.3499	5.633	426.90	2.2617	2.814	426.82	2.1949	20
30	303.15	14.570	433.44	2.3717	5.825	433.40	2.2835	2.911	433.33	2.2167	30
40	313.15	15.050	440.06	2.3931	6.018	440.02	2.3050	3.007	439.95	2.2382	40
50	323.15	15.530	446.79	2.4143	6.210	446.75	2.3261	3.103	446.69	2.2549	50
60	333.15	16.010	453.64	2.4352	6.403	453.60	2.3470	3.200	453.54	2.2802	60
70	343.15	16.490	460.59	2.4557	6.595	460.55	2.3676	3.296	460.50	2.3008	70
80	353.15	16.970	467.65	2.4760	6.788	467.62	2.3879	3.393	467.57	2.3211	80
90	363.15	17.460	474.82	2.4960	6.980	474.79	2.4079	3.489	474.75	2.3412	90

Temperature		Absolute Pressure (Saturation Temperature)									Temp.
t	*T*	1.2 bar (–70.35°C/202.8K)			3.0 bar (–63.82°C/209.33K)			0.5 bar (–54.85°C/218.3K)			*t*
°C	K	*v*	*i*	*s*	*v*	*i*	*s*	*v*	*i*	*s*	°C
Satd.	Cond.	0.9633	372.87	1.9096	0.6602	376.05	1.8866	0.4103	380.38	1.8586	
–60	213.15	1.015	378.58	1.9371	0.6730	378.20	1.8967	–	–	–	–60
–50	223.15	1.064	384.21	1.9628	0.7063	383.87	1.9227	0.4202	383.19	1.8714	–50
–40	233.15	1.113	389.94	1.9879	0.7393	389.65	1.9481	0.4404	389.06	1.8971	–40
–30	243.15	1.162	395.78	2.0125	0.7722	395.52	1.9727	0.4605	395.00	1.9221	–30
–20	253.15	1.210	401.73	2.0365	0.8050	401.50	1.9968	0.4805	401.04	1.9464	–20
–10	263.15	1.259	407.79	2.0599	0.8376	407.59	2.0204	0.5003	407.18	1.9702	–10
0	273.15	1.308	413.97	2.0829	0.8702	413.79	2.0435	0.5200	413.42	1.9935	0
10	283.15	1.356	420.26	2.1056	0.9027	420.10	2.0662	0.5397	419.77	2.0163	10
20	293.15	1.405	426.67	2.1279	0.9351	426.52	2.0885	0.5593	426.22	2.0387	20
30	303.15	1.453	433.19	2.1498	0.9675	433.06	2.1104	0.5789	432.78	2.0607	30
40	313.15	1.502	439.83	2.1713	0.9999	439.70	2.1320	0.5984	439.45	2.0823	40
50	323.15	1.550	446.58	2.1925	1.032	446.45	2.1533	0.6179	446.23	2.1036	50
60	333.15	1.598	453.43	2.2134	1.064	453.32	2.1742	0.6374	453.11	2.1246	60
70	343.15	1.647	460.40	2.2340	1.097	460.30	2.1948	0.6569	460.10	2.1453	70
80	353.15	1.695	467.48	2.2543	1.129	467.38	2.2151	0.6763	467.19	2.1657	80
90	363.15	1.743	474.66	2.2743	1.161	474.57	2.2352	0.6957	474.39	2.1858	90
100	373.15	1.791	481.95	2.2942	1.193	481.87	2.2550	0.7151	481.70	2.2056	100
110	383.15	1.840	489.34	2.3138	1.226	489.27	2.2746	0.7345	489.11	2.2251	110
120	393.15	1.889	496.84	2.3331	1.258	496.77	2.2939	0.7538	496.62	2.2445	120

(Contd.)

TABLE: 4.3 Thermodynamic Properties of Chlorodifluoromethane (Refrigerant R–22) (Superheat Region) [Contd.]

Temperature		Absolute Pressure (Saturation Temperature)									Temp.
t	*T*	0.7 bar (–48.43°C/224.32K)			1.0 bar (–41.12°C/183.69K)			1.5 bar (–32.11°C/241.04K)			*t*
°C	K	*v*	*i*	*s*	*v*	*i*	*s*	*v*	*i*	*s*	°C
Satd. Cond.		0.2998	383.44	1.8409	0.2149	386.86	1.8227	0.1470	390.98	1.8028	
–40	233.15	0.3123	388.46	1.8629	0.2161	387.54	1.8257	–	–	–	**–40**
–30	243.15	0.3269	394.48	1.8881	0.2266	393.68	1.8514	0.1486	392.32	1.8083	**–30**
–20	253.15	0.3413	400.58	1.9127	0.2370	399.88	1.8764	0.1558	398.68	1.8339	**–20**
–10	263.15	0.3557	406.77	1.9367	0.2472	406.14	1.9007	0.1628	405.08	1.8588	**–10**
0	273.15	0.3699	413.05	1.9602	0.2574	412.49	1.9243	0.1698	411.54	1.8829	**0**
10	283.15	0.3841	419.43	1.9831	0.2674	418.93	1.9475	0.1767	418.07	1.9063	**10**
20	293.15	0.3983	425.92	2.0056	0.2775	425.46	1.9702	0.1835	424.68	7.9292	**20**
30	303.15	0.4123	432.50	2.0277	0.2874	432.08	1.9924	0.1902	431.38	1.9517	**30**
40	313.15	0.4264	439.19	2.0493	0.2973	438.81	2.0142	0.1969	438.16	1.9738	**40**
50	323.15	0.4404	445.98	2.0707	0.3072	445.63	2.0356	0.2037	445.04	1.9954	**50**
60	333.15	0.4544	452.88	2.0918	0.3171	452.55	2.0568	0.2103	452.01	2.0166	**60**
70	343.15	0.4683	459.89	2.1125	0.3269	459.69	2.0776	0.2169	459.07	2.0375	**70**
80	353.15	0.4823	467.00	2.1329	0.3367	466.72	2.0980	0.2236	466.24	2.0581	**80**
90	363.15	0.4962	474.22	2.1531	0.3465	473.95	2.1182	0.2302	473.50	2.0784	**90**
100	373.15	0.5101	481.53	2.1729	0.3563	481.28	2.1382	0.2367	480.86	2.0984	**100**
110	383.15	0.5240	488.95	2.1925	0.3661	488.72	2.1578	0.2433	488.32	2.1181	**110**
120	393.15	0.5378	496.47	2.2119	0.3758	496.25	2.1772	0.2499	495.88	2.1376	**120**
130	403.15	0.5517	504.09	2.2311	0.3856	503.88	2.1964	0.2564	503.53	2.1568	**130**
140	413.15	0.5656	511.81	2.2500	0.3953	511.61	2.2153	0.2629	511.27	2.1758	**140**

Temperature		Absolute Pressure (Saturation Temperature)									Temp.
t	*T*	2.0 bar (–24.00°C/249.15K)			2.5 bar (–19.54°C/253.61K)			3.0 bar (–14.69°C/258.46K)			*t*
°C	K	*v*	*i*	*s*	*v*	*i*	*s*	*v*	*i*	*s*	°C
Satd. Cond.		0.1129	394.83	1.7923	0.09088	396.48	1.7788	0.076440	398.51	1.7705	
–20	253.15	0.1151	397.44	1.8028	–	–	–	–	–	–	**–20**
–10	263.15	0.1206	403.99	1.8281	0.09521	402.87	1.8035	0.07827	401.72	1.7828	**–10**
0	273.15	0.1259	410.57	1.8526	0.09964	409.58	1.8285	0.08207	408.57	1.8083	**0**
10	283.15	0.1312	417.20	1.8765	0.1040	416.31	1.8528	0.08579	415.41	1.8329	**10**
20	293.15	0.1364	423.89	1.8997	0.1083	423.09	1.8763	0.08943	422.28	1.8567	**20**
30	303.15	0.1416	430.65	1.9224	0.1125	429.93	1.8992	0.09301	429.20	1.8799	**30**
40	313.15	0.1468	437.50	1.9446	0.1166	436.64	1.9216	0.09655	436.17	1.9026	**40**
50	323.15	0.1519	444.43	1.9664	0.1208	443.82	1.9436	0.1000	443.21	1.9247	**50**
60	333.15	0.1569	451.45	1.9878	0.1249	450.89	1.9651	0.1035	450.32	1.9464	**60**
70	343.15	0.1620	458.56	2.0088	0.1289	458.04	1.9863	0.1069	457.51	1.9676	**70**
80	353.15	0.1670	465.76	2.0295	0.1330	465.27	2.0070	0.1104	464.78	1.9885	**80**
90	363.15	0.1720	473.05	2.0498	0.1370	472.60	2.0275	0.1138	472.14	2.0091	**90**
100	373.15	0.1769	480.44	2.0699	0.1411	480.02	2.0476	0.1171	479.59	2.0293	**100**
110	383.15	0.1819	487.92	2.0897	0.1451	487.52	2.0675	0.1205	487.12	2.0492	**110**
120	393.15	0.1868	495.50	2.1092	0.1490	495.12	2.0871	0.1238	494.75	2.0689	**120**
130	403.15	0.1918	503.17	2.1285	0.1530	502.82	2.1064	0.1272	502.46	2.0883	**130**
140	413.15	0.1967	510.94	2.1475	0.1570	510.60	2.1255	0.1305	510.26	2.1074	**140**
150	423.15	0.2016	518.79	2.1663	0.1609	518.47	2.1443	0.1338	518.15	2.1262	**150**
160	433.15	0.2065	526.73	2.1849	0.1649	526.43	2.1629	0.1371	526.13	2.1449	**160**

(Contd.)

TABLE: 4.3 Thermodynamic Properties of Chlorodifluoromethane (Refrigerant R–22) (Superheat Region) [Contd.]

Temperature		Absolute Pressure (Saturation Temperature)									Temp.
t	T	3.5 bar (–10.41°C/262.74K)			4.0 bar (–6.58°C/266.57K)			5.0 bar (0.10°C/273.25K)			t
°C	K	v	i	s	v	i	s	v	i	s	°C
Satd.	**Cond.**	0.06600	400.25	1.7635	0.05808	401.77	1.7576	0.04683	404.32	1.7481	
0	273.15	0.06951	407.52	1.7907	0.06007	406.46	1.7749	–	–	–	**0**
10	283.15	0.07278	414.49	1.8157	0.06301	413.54	1.8004	0.04929	411.59	1.7739	**10**
20	293.15	0.07597	421.45	1.8399	0.06587	420.61	1.8250	0.05170	418.87	1.7992	**20**
30	303.15	0.07911	428.45	1.8633	0.06867	427.69	1.8487	0.05404	426.13	1.8235	**30**
40	313.15	0.08219	435.49	1.8862	0.07142	434.80	1.8717	0.05632	433.39	1.8470	**40**
50	323.15	0.08523	442.58	1.9085	0.07412	441.95	1.8942	0.05856	440.67	1.8699	**50**
60	333.15	0.08824	449.75	1.9303	0.07678	449.17	1.9162	0.06075	447.99	1.8922	**60**
70	343.15	0.09122	456.98	1.9517	0.07943	456.44	1.9378	0.06291	455.36	1.9140	**70**
90	363.15	0.09712	471.68	1.9934	0.08464	471.22	1.9796	0.06717	470.29	1.9563	**90**
100	373.15	0.1000	479.16	2.0137	0.08722	478.73	2.0000	0.06927	477.86	1.9769	**100**
110	383.15	0.1029	486.72	2.0337	0.08978	486.31	2.0201	0.07135	485.51	1.9971	**110**
120	393.15	0.1058	494.37	2.0534	0.09233	493.98	2.0398	0.07342	493.23	2.0170	**120**
130	403.15	0.1087	502.10	2.0728	0.09487	501.74	2.0593	0.07548	501.02	2.0366	**130**
140	413.15	0.1116	509.92	2.0919	0.09740	509.58	2.0785	0.07753	508.90	2.0559	**140**
150	423.15	0.1144	517.83	2.1108	0.09991	517.51	2.0975	0.07957	516.86	2.0749	**150**
160	433.15	0.1173	525.82	2.1295	0.1024	525.52	2.1162	0.08160	524.90	2.0937	**160**
170	443.15	0.1201	533.90	2.1480	0.1049	533.61	2.1346	0.08362	533.02	2.1122	**170**
180	453.15	0.1230	542.06	2.1662	0.1074	541.79	2.1529	0.08564	541.22	2.1305	**180**

Temperature		Absolute Pressure (Saturation Temperature)									Temp.
t	T	6.0 bar (5.85°C/279.0K)			7.0 bar (10.92°C/284.07K)			8.0 bar (15.47°C/288.62K)			t
°C	K	v	i	s	v	i	s	v	i	s	°C
Satd.	**Cond.**	0.03921	406.39	1.7397	0.03370	408.11	1.7329	0.02951	409.58	1.7270	
10	283.15	0.04012	409.55	1.7510	–	–	–	–	–	–	**10**
20	293.15	0.04223	417.08	1.7771	0.03543	415.19	1.7575	0.03031	413.22	1.7396	**20**
30	303.15	0.04426	424.53	1.8021	0.03726	422.86	1.7832	0.03198	421.13	1.7661	**30**
40	313.15	0.04623	431.94	1.8262	0.03902	430.45	1.8078	0.03359	428.90	1.7914	**40**
50	323.15	0.04816	439.36	1.8495	0.04072	438.01	1.8316	0.03513	436.63	1.8156	**50**
60	333.15	0.05004	446.79	1.8721	0.04238	445.56	1.8546	0.03663	444.32	1.8390	**60**
70	343.15	0.05189	454.26	1.8942	0.04401	453.13	1.8770	0.03810	451.99	1.8617	**70**
80	353.15	0.05372	461.77	1.9158	0.04561	460.73	1.8988	0.03953	459.68	1.8838	**80**
90	363.15	0.05552	469.34	1.9369	0.04719	468.38	1.9202	0.04094	467.40	1.9054	**90**
100	373.15	0.05730	476.97	1.9576	0.04875	476.08	1.9411	0.04233	475.17	1.9265	**100**
110	383.15	0.06082	492.44	1.9780	0.05029	483.84	1.9616	0.04370	482.99	1.9472	**110**
120	393.15	0.06082	492.44	1.9980	0.05181	491.66	1.9818	0.04505	490.87	1.9675	**120**
130	403.15	0.06256	500.29	2.0177	0.05332	499.55	2.0016	0.04639	498.81	1.9874	**130**
140	413.15	0.06428	508.21	2.0371	0.05482	507.51	2.0211	0.04772	506.81	2.0070	**140**
150	423.15	0.06600	516.20	2.0563	0.05631	515.55	2.0403	0.04904	514.88	2.0263	**150**
160	433.15	0.06771	524.28	2.0751	0.05779	523.66	2.0593	0.05036	523.03	2.0453	**160**
170	443.15	0.06942	532.44	2.0937	0.05927	531.84	2.0780	0.05166	531.25	2.0641	**170**
180	453.15	0.07111	540.67	2.1121	0.06074	540.11	2.0964	0.05296	539.54	2.0826	**180**
190	463.15	0.07280	548.98	2.1303	0.06220	548.45	2.1146	0.05425	547.90	2.1009	**190**

(Contd.)

TABLE: 4.3 Thermodynamic Properties of Chlorodifluoromethane (Refrigerant R–22) (Superheat Region) [Contd.]

Temperature		Absolute Pressure (Saturation Temperature)									Temp.
t	T	9.0 bar (19.62°C/292.77K)			10.0 bar (23.44°C/296.59K)			12.0 bar (30.30°C/303.45K)			t
°C	K	v	i	s	v	i	s	v	i	s	°C
Satd.	**Cond.**	0.26230	410.83	1.7218	0.02357	411.90	1.7170	0.01954	413.64	1.7085	
30	303.15	0.02786	419.33	1.7503	0.02455	417.44	1.7355	–	–	–	**30**
40	313.15	0.02935	427.32	1.7762	0.02595	425.67	1.7622	0.02080	422.17	1.7362	**40**
50	323.15	0.03078	435.20	1.8010	0.02728	433.74	1.7875	0.02201	430.66	1.7629	**50**
60	333.15	0.03216	443.02	1.8249	0.02856	441.71	1.8118	0.02315	438.97	1.7882	**60**
70	343.15	0.03350	450.82	1.8480	0.02980	449.62	1.8352	0.02424	447.15	1.8124	**70**
80	353.15	0.03480	458.61	1.8703	0.03100	457.52	1.8579	0.02530	455.27	1.8357	**80**
90	363.15	0.03607	466.42	1.8921	0.03218	465.41	1.8800	0.02633	463.36	1.8583	**90**
100	373.15	0.03733	474.26	1.9134	0.03333	473.33	1.9015	0.02733	471.43	1.8802	**100**
110	383.15	0.03857	482.14	1.9342	0.03447	481.27	1.9225	0.02831	479.52	1.9016	**110**
120	393.15	0.03979	490.07	1.9546	0.03559	489.26	1.9431	0.02927	487.63	1.9225	**120**
130	403.15	0.04100	498.06	1.9747	0.03669	497.30	1.9633	0.03022	493.77	1.9430	**130**
140	413.15	0.04220	506.11	1.9945	0.03778	505.40	1.9831	0.03115	503.96	1.9630	**140**
150	423.15	0.04339	514.22	2.0139	0.03887	513.55	2.0026	0.03207	512.20	1.9827	**150**
160	433.15	0.04457	522.40	2.0330	0.03994	521.77	2.0218	0.03298	520.50	2.0021	**160**
170	443.15	0.04574	530.65	2.0518	0.04100	530.05	2.0407	0.03389	528.85	2.0211	**170**
180	453.15	0.04690	538.97	2.0704	0.04206	538.40	2.0593	0.03479	537.26	2.0399	**180**
190	463.15	0.04806	547.37	2.0887	0.04311	546.82	2.0777	0.03568	548.73	2.0584	**190**
200	473.15	0.04921	555.83	2.1068	0.04416	555.31	2.0958	0.03657	554.27	2.0767	**200**
210	483.15	0.05036	564.36	2.1246	0.04520	563.87	2.1137	0.03745	562.87	2.0947	**210**

Temperature		Absolute Pressure (Saturation Temperature)									Temp.
t	T	14.0 bar (36.37°C/309.52K)			16.0 bar (41.81°C/314.96K)			18.0 bar (46.77°C/319.92)			t
°C	K	v	i	s	v	i	s	v	i	s	°C
Satd.	**Cond.**	0.01662	414.94	1.7010	0.01440	415.87	1.6940	0.01265	416.50	1.6875	
50	323.15	0.01820	427.36	1.7404	0.01530	423.78	1.7189	–	–	–	**50**
60	333.15	0.01926	436.07	1.7668	0.01631	432.97	1.7469	0.01398	429.65	1.7278	**60**
70	343.15	0.02026	444.57	1.7919	0.01725	441.84	1.7731	0.01488	438.96	1.7553	**70**
80	353.15	0.02121	452.94	1.8159	0.01813	450.50	1.7980	0.01572	447.95	1.7812	**80**
90	363.15	0.02213	461.23	1.8391	0.01898	459.03	1.8218	0.01652	456.75	1.8058	**90**
100	373.15	0.02303	469.48	1.8615	0.01980	467.47	1.8447	0.01728	465.41	1.8293	**100**
110	383.15	0.02390	477.72	1.8833	0.02059	475.87	1.8669	0.01801	473.98	1.8520	**110**
120	393.15	0.02475	485.96	1.9046	0.02136	484.25	1.8885	0.01872	482.50	1.8739	**120**
130	403.15	0.02559	494.22	1.9253	0.02212	492.63	1.9096	0.01941	491.01	1.8952	**130**
140	413.15	0.02642	502.51	1.9456	0.02286	501.02	1.9301	0.02009	499.52	1.9161	**140**
150	423.15	0.02723	510.83	1.9655	0.02358	509.44	1.9503	0.02075	508.04	1.9365	**150**
160	433.15	0.02803	519.20	1.9851	0.02430	517.90	1.9700	0.02140	516.58	1.9565	**160**
170	443.15	0.02882	527.62	2.0043	0.02501	526.40	1.9894	0.02204	525.15	1.9760	**170**
180	453.15	0.02960	536.10	2.0232	0.02571	534.94	2.0085	0.02268	533.76	1.9952	**180**
190	463.15	0.03038	544.64	2.0418	0.02640	543.53	2.0272	0.02331	542.41	2.0141	**190**
200	473.15	0.03115	553.23	2.0602	0.02709	552.18	2.0457	0.02393	551.11	2.0327	**200**
210	483.15	0.03192	561.88	2.0783	0.02777	560.88	2.0639	0.02454	559.87	2.0510	**210**
220	493.15	0.03268	570.59	2.0961	0.02845	569.64	2.0818	0.02515	568.67	2.0691	**220**
230	503.15	0.03344	579.36	2.1137	0.02912	578.45	2.0995	0.02576	577.53	2.0869	**230**

(Contd.)

TABLE: 4.3 Thermodynamic Properties of Chlorodifluoromethane (Refrigerant R–22) (Superheat Region) [Contd.]

Temperature		Absolute Pressure (Saturation Temperature)									Temp.
t	*T*	20.0 bar (51.35°C/324.5K)			22.0 bar (55.59°C/328.74K)			24.0 bar (59.55°C/332.7K)			*t*
°C	K	*v*	*i*	*s*	*v*	*i*	*s*	*v*	*i*	*s*	°C
Satd.	**Cond.**	0.01123	416.67	1.6813	0.01005	417.00	1.6752	0.009063	416.90	1.6691	
60	333.15	0.01208	426.03	1.7091	0.01049	422.01	1.6903	0.009108	417.46	1.6708	**60**
70	343.15	0.01297	435.89	1.7383	0.01138	432.58	1.7216	0.01002	428.98	1.7048	**70**
80	353.15	0.01378	445.28	1.7653	0.01217	442.44	1.7500	0.01082	439.42	1.7349	**80**
90	363.15	0.01454	454.37	1.7907	0.01290	451.88	1.7763	0.01153	449.27	1.7624	**90**
100	373.15	0.01526	463.26	1.8148	0.01359	461.05	1.8011	0.01219	458.74	1.7881	**100**
110	383.15	0.01594	472.03	1.8380	0.01424	470.03	1.8249	0.01281	467.96	1.8125	**110**
120	393.15	0.01660	480.72	1.8604	0.01486	478.89	1.8478	0.01341	477.02	1.8358	**120**
130	403.15	0.01724	489.37	1.8821	0.01546	487.68	1.8699	0.01398	485.97	1.8583	**130**
140	413.15	0.01787	497.99	1.9033	0.01605	496.43	1.8913	0.01454	494.85	1.8801	**140**
150	423.15	0.01848	506.61	1.9239	0.01662	505.16	1.9122	0.01508	503.69	1.9012	**150**
160	433.15	0.01908	515.24	1.9440	0.01718	513.89	1.9326	0.01560	512.52	1.9218	**160**
170	443.15	0.01967	523.89	1.9638	0.01778	522.63	1.9525	0.01611	521.34	1.9420	**170**
180	453.15	0.02025	532.57	1.9832	0.01827	531.38	1.9720	0.01662	530.17	1.9620	**180**
190	463.15	0.02083	541.29	2.0022	0.01880	540.16	1.9912	0.01712	539.02	1.9810	**190**
200	473.15	0.02140	550.05	2.0209	0.01933	548.98	2.0100	0.01761	547.90	2.0000	**200**
210	483.15	0.02196	558.86	2.0393	0.01985	557.84	2.0285	0.01809	556.81	2.0186	**210**
220	493.15	0.02252	567.71	2.0574	0.02036	566.74	2.0468	0.01857	565.76	2.0369	**220**
230	503.15	0.02307	576.61	2.0753	0.02088	575.69	2.0648	0.01904	574.76	2.0550	**230**
240	513.15	0.02362	585.57	2.0929	0.02138	584.68	2.0825	0.01951	583.80	2.0728	**240**

Temperature		Absolute Pressure (Saturation Temperature)									Temp.
t	*T*	27.0 bar (65.05°C/338.2K)			30.0 bar (70.10°C/343.25K)			35.0 bar (77.72°C/350.87K)			*t*
°C	K	*v*	*i*	*s*	*v*	*i*	*s*	*v*	*i*	*s*	°C
Satd.	**Cond.**	0.007831	416.36	1.6600	0.006822	415.31	1.6505	0.005476	412.38	1.6332	
70	343.15	0.008304	422.82	1.6789	–	–	–	–	–	–	**70**
80	353.15	0.009122	434.48	1.7124	0.007715	428.84	1.6893	0.005727	416.61	1.6452	**80**
90	363.15	0.009831	445.09	1.7421	0.008443	440.51	1.7219	0.006579	431.59	1.6871	**90**
100	373.15	0.01047	455.10	1.7693	0.009080	451.21	1.7510	0.007245	443.98	1.7207	**100**
110	383.15	0.01106	464.73	1.7948	0.009659	461.33	1.7778	0.007820	455.18	1.7504	**110**
120	393.15	0.01163	474.11	1.8189	0.01020	471.08	1.8029	0.008338	465.70	1.7775	**120**
130	403.15	0.01217	483.82	1.8420	0.01071	480.58	1.8268	0.008818	475.79	1.8028	**130**
140	413.15	0.01268	492.41	1.8643	0.01119	489.92	1.8497	0.009270	485.59	1.8268	**140**
150	423.15	0.01318	501.44	1.8859	0.01166	499.15	1.8717	0.009699	495.19	1.8498	**150**
160	433.15	0.01366	510.43	1.9069	0.01211	508.30	1.8931	0.01011	504.65	1.8719	**160**
170	443.15	0.01413	519.39	1.9273	0.01255	517.40	1.9139	0.01051	514.02	1.8933	**170**
180	453.15	0.01459	528.34	1.9473	0.01298	526.48	1.9341	0.01090	523.33	1.9141	**180**
190	463.15	0.01505	537.29	1.9669	0.01340	535.55	1.9539	0.01127	532.60	1.9343	**190**
200	473.15	0.01549	546.27	1.9861	0.01381	544.63	1.9733	0.01164	541.85	1.9541	**200**
210	483.15	0.01593	555.27	2.0049	0.01421	553.72	1.9923	0.01200	551.10	1.9734	**210**
220	493.15	0.01637	564.30	2.0234	0.01461	562.83	2.0110	0.01236	560.35	1.9923	**220**
230	503.15	0.01679	573.37	2.0416	0.01501	571.96	2.0294	0.01270	569.61	2.0109	**230**
240	513.15	0.01722	582.47	2.0595	0.01540	581.13	2.0474	0.01305	578.89	2.0292	**240**
250	523.15	0.01764	591.61	2.0771	0.01578	590.33	2.0651	0.01339	588.20	2.0472	**250**

(Contd.)

TABLE: 4.3 Thermodynamic Properties of Chlorodifluoromethane (Refrigerant R–22) (Superheat Region) [Contd.]

Temperature		Absolute Pressure (Saturation Temperature)									Temp.
t	*T*	40.0 bar (84.52°C/357.67K)			45.0 bar (90.67°C/363.82K)			50.0 bar (Supercritical)			*t*
°C	K	*v*	*i*	*s*	*v*	*i*	*s*	*v*	*i*	*s*	°C
Satd.	**Cond.**	0.004394	407.50	1.6112	0.003433	399.21	1.5841	**Non-Existent**			
90	363.15	0.005024	419.70	1.6464	–	–	–	–	–	–	**90**
100	373.15	0.005800	435.41	1.6891	0.004572	424.47	1.6528	0.003365	407.51	1.6020	**100**
110	383.15	0.006403	448.28	1.7231	0.005254	440.28	1.6947	0.004275	430.58	1.6631	**110**
120	393.15	0.006921	459.84	1.7529	0.005794	453.36	1.7284	0.004864	446.06	1.7030	**120**
130	403.15	0.007387	470.67	1.7801	0.006260	465.15	1.7580	0.005342	459.16	1.7360	**130**
140	413.15	0.007818	481.02	1.8055	0.006680	476.19	1.7850	0.005760	471.05	1.7651	**140**
150	423.15	0.008223	491.06	1.8295	0.007433	486.74	1.8103	0.006139	482.22	1.7918	**150**
160	433.15	0.008607	500.88	1.8524	0.007433	496.97	1.8342	0.006491	492.92	1.8168	**160**
170	443.15	0.008975	510.55	1.8745	0.007780	506.97	1.8570	0.006823	503.29	1.8405	**170**
180	453.15	0.009331	520.11	1.8958	0.008113	516.81	1.8789	0.007138	513.43	1.8631	**180**
190	463.15	0.009676	529.60	1.9165	0.008434	526.53	1.9001	0.007440	523.41	1.8849	**190**
200	473.15	0.01001	539.04	1.9367	0.008745	536.17	1.9207	0.007731	533.27	1.9060	**200**
210	483.15	0.01034	548.45	1.9564	0.009047	545.76	1.9408	0.008014	543.05	1.9264	**210**
220	493.15	0.01066	557.85	1.9756	0.009343	555.31	1.9604	0.008290	552.77	1.9463	**220**
230	503.15	0.01098	567.24	1.9945	0.009633	564.85	1.9795	0.008559	562.45	1.9657	**230**
240	513.15	0.01129	576.64	2.0130	0.009917	574.38	1.9983	0.008822	572.10	1.9847	**240**
250	523.15	0.01159	586.05	2.0312	0.01020	563.90	2.0167	0.009080	581.74	2.0034	**250**

Adapted from A. V. Kletskii, *Thermophysical Properties of Freon–22,* Translated from Russian by the Israel Program for Scientific Translations, 1971 and Published by the U. S. Department of Commerce, USA.

TABLE: 5.1 Thermodynamic Properties of Saturated Trifluoromethane (Refrigerant R–23) up to 1 Atmosphere (Temperature Basis)

Temperature		Saturation pressure, p		Liquid density	Specific volume of vapour v''	Specific enthalpy			Specific entropy		Temp
t	T			ρ_l		h	λ_v	H	s'	s''	t
°C	K	bar	torr	kg/m³	m³/kg	kJ/kg			kJ/(kg·K)		°C
–155.13[a]	**118.02**	**0.0006**	**0.45**	**1701.9**	**241.55**	**–3.67**	**292.88**	**289.21**	**–0.07052**	**2.4111**	**–155.13**
–150	123.15	0.0014	1.05	1685.3	105.93	2.54	289.19	291.73	–0.01896	2.3293	**–150**
–145	128.15	0.0030	2.25	1668.7	51.151	8.55	285.64	294.19	0.02884	2.2578	**–145**
–140	133.15	0.0060	4.50	1652.0	26.330	14.53	282.12	296.65	0.07463	2.1934	**–140**
–135	138.15	0.0114	8.55	1635.0	14.347	20.51	278.61	299.11	0.1187	2.1353	**–135**
–130	143.15	0.0206	15.45	1617.9	8.2196	26.48	274.78	301.26	0.1612	2.0828	**–130**
–125	148.15	0.0355	26.63	1600.7	4.9239	32.46	271.55	304.01	0.2022	2.0352	**–125**
–120	153.15	0.0588	44.10	1583.3	3.0696	38.45	268.00	306.45	0.2420	1.9919	**–120**
–115	158.15	0.0938	70.36	1565.8	1.9826	44.45	264.42	308.87	0.2805	1.9525	**–115**
–110	163.15	0.1447	108.53	1548.2	1.3219	50.47	260.80	311.27	0.3180	1.9165	**–110**
–109	164.15	0.1573	118.00	1544.7	1.2302	51.59	260.14	311.73	0.3241	1.9099	**–109**
–108	165.15	0.1704	127.79	1541.1	1.1452	52.81	259.36	312.17	0.3309	1.9036	**–108**
–107	166.15	0.1844	138.31	1537.5	1.0664	54.03	258.58	312.62	0.3377	1.8973	**–107**
–106	167.15	0.1995	149.64	1533.9	0.99357	55.26	257.80	313.06	0.3446	1.8911	**–106**
–105	168.15	0.2167	162.54	1530.3	0.90678	56.50	257.14	313.64	0.3544	1.8836	**–105**
–104	169.15	0.2332	174.90	1526.6	0.84552	57.71	256.38	314.09	0.3611	1.8776	**–104**
–103	170.15	0.2519	188.95	1523.0	0.78871	58.93	255.60	314.53	0.3679	1.8716	**–103**
–102	171.15	0.2720	204.05	1519.4	0.73601	60.16	254.82	314.98	0.3748	1.8657	**–102**
–101	172.15	0.2936	220.25	1515.7	0.68711	61.38	254.04	315.42	0.3816	1.8598	**–101**
–100	173.15	0.3157	236.79	1512.3	0.63808	62.56	253.42	315.98	0.3898	1.8534	**–100**
–99	174.15	0.3386	253.94	1508.5	0.60002	63.83	252.60	316.43	0.3967	1.8480	**–99**
–98	175.15	0.3629	272.21	1504.9	0.56082	65.05	251.82	316.87	0.4035	1.8423	**–98**
–97	176.15	0.3889	291.68	1501.3	0.52438	66.28	251.04	317.32	0.4103	1.8367	**–97**
–96	177.15	0.4165	312.41	1497.6	0.49050	67.50	250.26	317.77	0.4172	1.8312	**–96**
–95	178.15	0.4488	336.63	1494.1	0.45941	68.64	249.63	318.27	0.4244	1.8256	**–95**
–94	179.15	0.4773	358.00	1490.4	0.42997	69.88	248.83	318.71	0.4309	1.8201	**–94**
–93	180.15	0.5106	383.01	1486.8	0.40264	71.10	248.05	319.15	0.4377	1.8148	**–93**
–92	181.15	0.5461	409.62	1483.1	0.37718	72.33	247.27	319.60	0.4446	1.8095	**–92**
–91	182.15	0.5838	437.91	1479.5	0.35345	73.55	246.49	320.05	0.4514	1.8043	**–91**
–90	183.15	0.6239	467.96	1475.7	0.33767	74.76	245.75	320.51	0.4582	1.8000	**–90**
–89	184.15	0.6654	499.10	1472.0	0.31902	76.00	244.94	320.94	0.4649	1.7950	**–89**
–88	185.15	0.7074	530.60	1468.2	0.30152	77.22	244.16	321.39	0.4718	1.7900	**–88**
–87	186.15	0.7518	563.90	1464.4	0.28506	78.45	243.38	321.83	0.4786	1.7850	**–87**
–86	187.15	0.7987	599.09	1460.6	0.26959	79.67	242.61	322.28	0.4855	1.7800	**–86**
–85	188.15	0.8501	637.63	1456.9	0.25282	80.91	241.78	322.69	0.4912	1.7763	**–85**
–84	189.15	0.9007	675.55	1453.0	0.23927	82.12	240.98	323.10	0.4974	1.7717	**–84**
–83	190.15	0.9560	717.03	1449.2	0.22648	83.34	240.20	323.54	0.5037	1.7670	**–83**
–82.018[b]	**191.132**	**1.01325**	**760.00**	**1445.4**	**0.21446**	**84.60**	**239.36**	**323.96**	**0.5106**	**1.7630**	**–82.018**

[a]Triple point
[b]Normal boiling point

TABLE: 5.2 Thermodynamic Properties of Saturated Trifluoromethane (Refrigerant R–23) above 1 Atmosphere (Temperature Basis)

Temperature		Saturation pressure,	Liquid density	Specific volume of	Specific enthalpy			Specific entropy		Temp
t	T	p	ρ_l	vapour v''	h	λ_v	H	s'	s''	t
°C	K	bar	kg/m³	m³/kg	kJ/kg			kJ/(kg·K)		°C
–82	191.15	1.0133	1445.3	0.21443	84.62	239.40	324.02	0.5106	1.7627	**–82**
–81	192.15	1.0749	1441.6	0.20309	85.79	238.62	324.42	0.5169	1.7580	**–81**
–80	193.15	1.1370	1437.9	0.19247	87.10	237.71	324.81	0.5236	1.7543	**–80**
–79	194.15	1.2100	1434.0	0.18241	88.32	236.94	325.25	0.5300	1.7498	**–79**
–78	195.15	1.2764	1430.2	0.17291	89.54	236.16	325.70	0.5363	1.7458	**–78**
–77	196.15	1.3460	1426.4	0.16394	90.77	235.38	326.15	0.5426	1.7418	**–77**
–76	197.15	1.4191	1422.6	0.15548	91.99	234.60	326.59	0.5489	1.7378	**–76**
–75	198.15	1.4955	1418.6	0.14875	93.35	233.50	326.84	0.5554	1.7338	**–75**
–74	199.15	1.5760	1414.6	0.14184	94.64	232.66	327.30	0.5616	1.7300	**–74**
–73	200.15	1.6602	1410.5	0.13528	95.96	231.79	327.75	0.5679	1.7262	**–73**
–72	201.15	1.7485	1406.4	0.12906	97.27	230.92	328.19	0.5743	1.7224	**–72**
–71	202.15	1.8410	1402.3	0.12315	98.59	230.05	328.64	0.5806	1.7186	**–71**
–70	203.15	1.9370	1398.9	0.11652	99.64	229.15	328.79	0.5866	1.7146	**–70**
–69	204.15	2.0221	1394.2	0.11124	100.95	228.12	329.08	0.5930	1.7112	**–69**
–68	205.15	2.1274	1390.1	0.10622	102.27	227.12	329.38	0.5993	1.7076	**–68**
–67	206.15	2.2376	1386.0	0.10145	103.58	226.11	329.69	0.6056	1.7039	**–67**
–66	207.15	2.3531	1381.9	0.096910	104.90	225.10	330.00	0.6119	1.7004	**–66**
–65	208.15	2.4737	1378.7	0.092387	106.00	224.65	330.65	0.6174	1.6966	**–65**
–64	209.15	2.5887	1373.8	0.088497	107.32	223.61	330.92	0.6234	1.6933	**–64**
–63	210.15	2.7086	1369.7	0.084597	108.63	222.60	331.23	0.6293	1.6898	**–63**
–62	211.15	2.8335	1365.6	0.080885	109.94	221.60	331.54	0.6353	1.6864	**–62**
–61	212.15	2.9635	1361.5	0.077352	111.26	220.59	331.85	0.6412	1.6830	**–61**
–60	213.15	3.1188	1358.1	0.074058	112.43	219.97	332.40	0.6476	1.6796	**–60**
–59	214.15	3.2396	1353.3	0.070916	113.74	218.93	332.67	0.6535	1.6762	**–59**
–58	215.15	3.3862	1349.3	0.067861	115.05	217.93	332.98	0.6595	1.6729	**–58**
–57	216.15	3.5386	1345.2	0.064951	116.37	216.93	333.29	0.6654	1.6696	**–57**
–56	217.15	3.6971	1341.1	0.062179	117.68	215.92	333.60	0.6714	1.6664	**–56**
–55	218.15	3.8859	1332.9	0.059945	118.93	215.09	334.02	0.6775	1.6635	**–55**
–54	219.15	4.0334	1332.9	0.057595	120.30	214.02	334.33	0.6835	1.6604	**–54**
–53	220.15	4.2116	1328.9	0.055344	121.62	213.02	334.64	0.6895	1.6572	**–53**
–52	221.15	4.3967	1324.8	0.053191	122.93	212.02	334.94	0.6954	1.6541	**–52**
–51	222.15	4.5892	1319.5	0.051130	124.24	211.01	335.25	0.7014	1.6509	**–51**
–50	223.15	4.7893	1315.3	0.048948	125.53	209.99	335.52	0.7071	1.6482	**–50**
–49	224.15	4.9803	1310.4	0.047070	126.84	209.03	335.87	0.7131	1.6452	**–49**
–48	225.15	5.1783	1305.8	0.045271	128.15	208.03	336.18	0.7190	1.6421	**–48**
–47	226.15	5.3832	1301.2	0.043547	129.47	207.02	336.49	0.7250	1.6391	**–47**
–46	227.15	5.5953	1296.7	0.041897	130.78	206.02	336.80	0.7309	1.6361	**–46**
–45	228.15	5.8439	1292.9	0.040274	132.22	204.64	336.86	0.7364	1.6334	**–45**
–44	229.15	6.0697	1287.5	0.038763	133.53	203.57	337.11	0.7428	1.6302	**–44**
–43	230.15	6.3056	1283.0	0.037312	134.85	202.50	337.35	0.7488	1.6273	**–43**
–42	231.15	6.5496	1278.4	0.035922	136.16	201.43	337.59	0.7547	1.6244	**–42**
–41	232.15	6.8020	1273.8	0.034589	137.47	200.34	337.81	0.7607	1.6215	**–41**
–40	233.15	7.0653	1269.7	0.033364	139.02	199.02	338.04	0.7655	1.6191	**–40**

(Contd.)

TABLE: 5.2 Thermodynamic Properties of Saturated Trifluoromethane (Refrigerant R–23) above 1 Atmosphere (Temperature Basis) [Contd.]

Temperature		Saturation pressure,	Liquid density	Specific volume of	Specific enthalpy			Specific entropy		Temp
t	T	p	ρ_l	vapour v''	h	λ_v	H	s'	s''	t
°C	K	bar	kg/m³	m³/kg	kJ/kg			kJ/(kg·K)		°C
–39	234.15	7.3007	1264.7	0.032153	140.26	197.98	338.24	0.7712	1.6160	**–39**
–38	235.15	7.5783	1260.1	0.030975	141.58	196.87	338.44	0.7770	1.6132	**–38**
–37	236.15	7.8653	1255.6	0.029845	142.89	195.75	338.64	0.7828	1.6105	**–37**
–36	237.15	8.1618	1251.0	0.028760	144.20	194.62	338.83	0.7886	1.6077	**–36**
–35	238.15	8.4697	1245.6	0.027802	145.95	193.09	339.04	0.7945	1.6052	**–35**
–34	239.15	8.7647	1239.9	0.026831	147.40	191.77	339.17	0.8002	1.6023	**–34**
–33	240.15	9.0687	1234.3	0.025892	148.92	190.42	339.34	0.8060	1.5996	**–33**
–32	241.15	9.3820	1228.7	0.024989	150.43	189.06	339.49	0.8118	1.5969	**–32**
–31	242.15	9.7047	1223.2	0.024122	151.95	187.69	339.64	0.8176	1.5943	**–31**
–30	243.15	10.0739	1220.5	0.023283	153.03	186.79	339.82	0.8233	1.5915	**–30**
–29	244.15	10.3795	1214.7	0.022485	154.51	185.43	339.95	0.8292	1.5891	**–29**
–28	245.15	10.7320	1209.2	0.021714	156.03	184.04	340.07	0.8349	1.5865	**–28**
–27	246.15	11.0951	1203.6	0.020972	157.54	182.64	340.18	0.8407	1.5839	**–27**
–26	247.15	11.4688	1198.1	0.020258	159.05	181.24	340.29	0.8465	1.5814	**–26**
–25	248.15	11.8955	1194.2	0.019578	160.26	180.10	340.36	0.8521	1.5779	**–25**
–24	249.15	12.2497	1187.0	0.018919	161.75	178.72	340.48	0.8580	1.5754	**–24**
–23	250.15	12.6573	1181.4	0.018283	163.27	177.30	340.56	0.8638	1.5729	**–23**
–22	251.15	13.0768	1175.8	0.017671	164.78	175.86	340.64	0.8696	1.5705	**–22**
–21	252.15	13.5084	1170.3	0.017081	166.29	174.41	340.70	0.8753	1.5681	**–21**
–20	253.15	13.9529	1166.6	0.016514	167.68	172.94	340.62	0.8810	1.5642	**–20**
–19	254.15	14.4192	1159.2	0.015967	169.31	171.33	340.64	0.8868	1.5615	**–19**
–18	255.15	14.8621	1153.6	0.015440	170.83	169.82	340.65	0.8927	1.5584	**–18**
–17	256.15	15.3167	1148.1	0.014933	172.34	168.30	340.64	0.8986	1.5554	**–17**
–16	257.15	15.7834	1142.5	0.014444	173.85	166.76	340.61	0.9044	1.5524	**–16**
–15	258.15	16.2654	1137.2	0.013959	175.32	165.23	340.55	0.9101	1.5502	**–15**
–14	259.15	16.7541	1137.6	0.013478	176.87	163.63	340.50	0.9160	1.5472	**–14**
–13	260.15	17.2586	1130.1	0.013014	178.38	162.04	340.42	0.9219	1.5443	**–13**
–12	261.15	17.7762	1122.6	0.012568	179.90	160.43	340.32	0.9277	1.5414	**–12**
–11	262.15	18.3073	1115.2	0.012138	181.41	158.80	340.21	0.9336	1.5385	**–11**
–10	263.15	18.8535	1105.9	0.011810	183.21	156.88	340.09	0.9395	1.5357	**–10**
–9	264.15	19.3868	1100.2	0.011410	184.72	155.21	339.93	0.9454	1.5328	**–9**
–8	265.15	19.9594	1092.8	0.011025	186.24	153.53	339.77	0.9512	1.5299	**–8**
–7	266.15	20.5466	1085.3	0.010654	187.75	151.83	339.58	0.9571	1.5271	**–7**
–6	267.15	21.1488	1077.9	0.010297	189.26	150.12	339.38	0.9630	1.5243	**–6**
–5	268.15	21.7393	1072.1	0.009988	191.42	147.74	339.16	0.9694	1.5204	**–5**
–4	269.15	22.3420	1062.9	0.009656	193.29	145.64	338.93	0.9752	1.5177	**–4**
–3	270.15	22.9598	1055.5	0.009336	195.20	143.48	338.67	0.9811	1.5150	**–3**
–2	271.15	23.5923	1048.0	0.009027	197.10	141.28	338.38	0.9870	1.5126	**–2**
–1	272.15	24.2397	1040.6	0.008730	199.01	139.04	338.05	0.9928	1.5085	**–1**
0	273.15	24.9469	1035.1	0.008427	200.00	137.64	337.64	1.0000	1.5039	**0**
1	274.15	25.5809	1025.6	0.008153	201.91	135.38	337.29	1.0066	1.5000	**1**
2	275.15	26.2751	1018.2	0.007888	203.81	133.04	336.85	1.0132	1.4960	**2**
3	276.15	26.9856	1010.7	0.007632	205.72	130.66	336.37	1.0198	1.4920	**3**
4	277.15	27.7127	1003.2	0.007385	207.62	128.24	335.86	1.0263	1.4880	**4**
5	278.15	28.5030	993.9	0.007075	209.08	126.28	335.36	1.0317	1.4857	**5**

(Contd.)

TABLE: 5.2 Thermodynamic Properties of Saturated Trifluoromethane (Refrigerant R–23) above 1 Atmosphere (Temperature Basis) [Contd.]

Temperature		Saturation pressure,	Liquid density	Specific volume of	Specific enthalpy			Specific entropy		Temp
t	T	p	ρ_l	vapour v''	h	λ_v	H	s'	s''	t
°C	K	bar	kg/m³	m³/kg	kJ/kg			kJ/(kg·K)		°C
6	279.15	29.2424	985.1	0.006807	210.97	123.76	334.73	1.0376	1.4816	**6**
7	280.15	30.0217	974.7	0.006548	212.88	121.24	334.11	1.0442	1.4777	**7**
8	281.15	30.8188	964.4	0.006299	214.78	118.67	333.45	1.0508	1.4739	**8**
9	282.15	31.6342	954.0	0.006060	216.68	116.08	332.76	1.0573	1.4701	**9**
10	283.15	32.4384	946.8	0.005887	218.84	113.17	332.01	1.0650	1.4647	**10**
11	284.15	33.2610	933.3	0.005676	220.77	110.49	331.26	1.0714	1.4611	**11**
12	285.15	34.1101	923.0	0.005463	222.68	107.78	330.46	1.0780	1.4573	**12**
13	286.15	34.9779	912.6	0.005259	224.58	105.04	329.62	1.0846	1.4536	**13**
14	287.15	35.8646	902.2	0.005063	226.48	102.26	328.74	1.0911	1.4499	**14**
15	288.15	36.7907	890.4	0.004817	229.64	97.42	327.06	1.1011	1.4392	**15**
16	289.15	37.6961	880.9	0.004658	231.71	94.02	325.73	1.1111	1.4292	**16**
17	290.15	38.6417	868.6	0.004479	233.98	90.56	324.54	1.1214	1.4192	**17**
18	291.15	39.6076	853.8	0.004268	236.52	86.43	322.96	1.1318	1.4092	**18**
19	292.15	40.5942	836.5	0.004035	239.34	81.64	320.99	1.1423	1.3992	**19**
20	293.15	41.6103	816.4	0.003790	242.36	76.81	319.17	1.1430	1.4050	**20**
21	294.15	42.6311	794.5	0.003542	245.80	70.08	315.88	1.1583	1.3792	**21**
22	295.15	43.6821	769.7	0.003298	249.44	63.30	312.74	1.1689	1.3692	**22**
23	296.15	44.7553	742.4	0.003061	253.36	55.85	309.21	1.1796	1.3592	**23**
24	297.15	45.8511	712.6	0.002836	257.55	47.75	305.30	1.1903	1.3492	**24**
25	298.15	46.9869	680.0	0.002625	261.94	39.61	301.55	1.2067	1.3396	**25**
26	299.15	48.1122	645.5	0.002428	266.76	29.54	296.30	1.2176	1.3292	**26**
26.143*	**299.293**	**48.32**	**526.5**	**0.001899**	**280.97**	**0.00**	**280.97**	**1.2697**	**1.2697**	**26.143**

*Critical point

Reference State:

Specific enthalpy of saturated liquid at 0°C (273.15 K) = 200.00 kJ/kg

Specific entropy of saturated liquid at 0°C (273.15 K) = 1.0000 kJ/(kg·K)

TABLE: 5.3 Thermodynamic Properties of Trifluoromethane (Refrigerant R–23) (Superheat Region)

Temperature		Absolute Pressure (Saturation Temperature)									Temp.
t	*T*	1.0 bar (–82.245°C/190.905K)			2.0 bar (–69.361°C/203.789K)			5.0 bar (–48.938°C/224.212K)			*t*
°C	K	*v*	*i*	*s*	*v*	*i*	*s*	*v*	*i*	*s*	°C
Satd.	**Cond.**	0.21713	323.87	1.7640	0.11304	329.04	1.7123	0.046933	335.82	1.6450	
–80	193.15	0.22021	325.41	1.7720	–	–	–	–	–	–	**–80**
–70	203.15	0.23359	332.13	1.8059	–	–	–	–	–	–	**–70**
–60	213.15	0.24656	338.70	1.8375	0.11976	335.89	1.7452	–	–	–	**–60**
–50	223.15	0.25928	345.25	1.8676	0.12663	342.95	1.7775	–	–	–	**–50**
–40	233.15	0.27182	351.84	1.8964	0.13328	349.89	1.8079	0.049865	343.36	1.6780	**–40**
–30	243.15	0.28424	358.49	1.9244	0.13979	356.81	1.8370	0.052930	351.33	1.7114	**–30**
–20	253.15	0.29656	365.24	1.9515	0.14620	363.76	1.8650	0.055847	359.03	1.7425	**–20**
–10	263.15	0.30882	372.09	1.9781	0.15253	370.78	1.8922	0.058665	366.63	1.7719	**–10**
0	273.15	0.32103	379.07	2.0041	0.15881	377.89	1.9187	0.061410	374.18	1.8001	**0**
10	283.15	0.33319	386.17	2.0297	0.16504	385.10	1.9446	0.064098	381.76	1.8273	**10**
20	293.15	0.34532	393.40	2.0548	0.17123	392.42	1.9701	0.066747	389.38	1.8538	**20**
30	303.15	0.35741	400.77	2.0795	0.17740	399.86	1.9950	0.069363	397.08	1.8796	**30**
40	313.15	0.36949	408.27	2.1038	0.18354	407.43	2.0196	0.071948	404.87	1.9049	**40**
50	323.15	0.38155	415.91	2.1278	0.18966	415.13	2.0438	0.074516	412.75	1.9297	**50**
60	333.15	0.39359	423.68	2.1515	0.19576	422.96	2.0676	0.077059	420.74	1.9540	**60**
70	343.15	0.40561	431.60	2.1749	0.20185	430.92	2.0912	0.079586	428.85	1.9780	**70**
80	353.15	0.41762	439.64	2.1980	0.20793	439.01	2.1144	0.082102	437.07	2.0016	**80**
90	363.15	0.42963	447.83	2.2209	0.21399	447.23	2.1374	0.084610	445.41	2.0249	**90**
100	373.15	0.44162	456.15	2.2435	0.22005	455.58	2.1600	0.087100	453.87	2.0479	**100**
150	423.15	0.50145	499.70	2.3529	0.25020	499.26	2.2698	0.099443	497.96	2.1587	**150**
200	473.15	0.56117	546.33	2.4570	0.28022	545.99	2.3741	0.111657	544.96	2.2636	**200**

Temperature		Absolute Pressure (Saturation Temperature)									Temp.
t	*T*	10.0 bar (–30.217°C/242.933K)			20.0 bar (–7.949°C/265.201K)			50.0 bar			*t*
°C	K	*v*	*i*	*s*	*v*	*i*	*s*	*v*	*i*	*s*	°C
Satd.	**Cond.**	0.023460	339.79	1.5921	0.011027	339.77	1.5295	**Non-Existent**			
–30	243.15	0.023505	340.01	1.5930	–	–	–	–	–	–	**–30**
–20	253.15	0.025434	349.82	1.6326	–	–	–	–	–	–	**–20**
–10	263.15	0.027177	358.83	1.6675	–	–	–	–	–	–	**–10**
0	273.15	0.028804	367.41	1.6995	0.012088	350.14	1.5681	–	–	–	**0**
10	283.15	0.030354	375.76	1.7295	0.013217	361.69	1.6085	–	–	–	**10**
20	293.15	0.031846	384.00	1.7581	0.014217	371.62	1.6440	–	–	–	**20**
30	303.15	0.033296	392.20	1.7856	0.015138	381.28	1.6765	0.003056	317.37	1.3894	**30**
40	313.15	0.034711	400.40	1.8122	0.016004	390.63	1.7068	0.004265	347.50	1.4874	**40**
50	323.15	0.036100	408.64	1.8381	0.016831	399.79	1.7356	0.004975	364.87	1.5420	**50**
60	333.15	0.037467	416.94	1.8634	0.017626	408.84	1.7632	0.005542	379.00	1.5851	**60**
70	343.15	0.038817	425.31	1.8882	0.018398	417.86	1.7898	0.006035	391.60	1.6224	**70**
80	353.15	0.040149	433.77	1.9125	0.019150	426.87	1.8157	0.006481	403.35	1.6561	**80**
90	363.15	0.041470	442.31	1.9368	0.019886	435.90	1.8409	0.006896	414.56	1.6874	**90**
100	373.15	0.042779	450.96	1.9598	0.020608	444.97	1.8656	0.007286	425.43	1.7170	**100**
150	423.15	0.049196	495.76	2.0724	0.024074	491.32	1.9821	0.009021	477.64	1.8483	**150**
200	473.15	0.055472	543.24	2.1784	0.027385	539.81	2.0904	0.010560	529.49	1.9641	**200**

TABLE: 6.1 Thermodynamic Properties of Saturated 1,1,1,2,-Tetrafluoroethane (R–134a) up to 1 Atmosphere (Temperature Basis)

Temperature		Saturation pressure, p		Liquid density	Specific volume of vapour v''	Specific enthalpy			Specific entropy		Temp
t	T			ρ_l		h	λ_v	H	s'	s''	t
°C	K	bar	torr	kg/m³	m³/kg	kJ/kg			kJ/(kg·K)		°C
–103.3*	**169.85**	**0.0039**	**2.93**	**1591.1**	**35.4988**	**71.45**	**263.49**	**334.94**	**0.4126**	**1.9639**	**–103.3**
–103	170.15	0.0041	3.09	1590.3	34.3001	71.82	263.33	335.15	0.4147	1.9619	**–103**
–102	171.15	0.0046	3.42	1587.7	30.9905	73.00	262.73	335.73	0.4214	1.9569	**–102**
–101	172.15	0.0051	3.80	1585.0	28.0168	74.19	262.13	336.32	0.4281	1.9501	**–101**
–100	173.15	0.0056	4.21	1582.4	25.3433	75.37	261.53	336.90	0.4348	1.9450	**–100**
–99	174.15	0.0062	4.66	1579.7	22.9381	76.56	260.93	337.49	0.4415	1.9401	**–99**
–98	175.15	0.0069	5.16	1577.0	20.7730	77.74	260.33	338.08	0.4482	1.9351	**–98**
–97	176.15	0.0076	5.71	1574.4	18.8229	78.93	259.73	338.66	0.4549	1.9302	**–97**
–96	177.15	0.0084	6.32	1571.7	17.0654	80.11	259.14	339.25	0.4616	1.9252	**–96**
–95	178.15	0.0093	6.98	1569.1	15.4805	81.30	258.54	339.83	0.4683	1.9203	**–95**
–94	179.15	0.0103	7.71	1566.4	14.0505	82.48	257.94	340.42	0.4750	1.9153	**–94**
–93	180.15	0.0113	8.51	1563.7	12.7968	83.67	257.34	341.00	0.4817	1.9104	**–93**
–92	181.15	0.0125	9.39	1561.1	11.6271	84.85	256.74	341.59	0.4884	1.9054	**–92**
–91	182.15	0.0138	10.36	1558.4	10.6644	86.03	256.11	342.15	0.4951	1.9014	**–91**
–90	183.15	0.0152	11.42	1555.8	9.7852	87.23	255.53	342.75	0.5018	1.8972	**–90**
–89	184.15	0.0168	12.58	1553.1	8.9826	88.42	254.94	343.36	0.5085	1.8931	**–89**
–88	185.15	0.0183	13.75	1550.4	8.2497	89.61	254.35	343.96	0.5152	1.8889	**–88**
–87	186.15	0.0200	14.98	1547.8	7.5801	90.80	253.76	344.57	0.5213	1.8847	**–87**
–86	187.15	0.0218	16.32	1545.1	6.9680	92.00	253.18	345.17	0.5275	1.8805	**–86**
–85	188.15	0.0237	17.77	1542.5	6.4082	93.19	252.59	345.78	0.5337	1.8763	**–85**
–84	189.15	0.0258	19.34	1539.8	5.8959	94.38	252.00	346.38	0.5399	1.8722	**–84**
–83	190.15	0.0280	21.04	1537.1	5.4271	95.57	251.41	346.99	0.5461	1.8688	**–83**
–82	191.15	0.0305	22.87	1534.5	4.9976	96.77	250.83	347.59	0.5523	1.8653	**–82**
–81	192.15	0.0331	24.86	1531.7	4.6042	97.96	250.24	348.20	0.5585	1.8617	**–81**
–80	193.15	0.0369	27.68	1529.0	4.2435	99.15	249.65	348.80	0.5647	1.8582	**–80**
–79	194.15	0.0391	29.32	1526.3	3.9497	100.34	249.07	349.42	0.5710	1.8547	**–79**
–78	195.15	0.0424	31.83	1523.5	3.6690	101.55	248.49	350.04	0.5772	1.8511	**–78**
–77	196.15	0.0460	34.53	1520.8	3.4006	102.75	247.91	350.66	0.5834	1.8476	**–77**
–76	197.15	0.0499	37.45	1518.1	3.1612	103.96	247.32	351.28	0.5896	1.8441	**–76**
–75	198.15	0.0541	40.60	1515.3	2.9399	105.16	246.74	351.90	0.5958	1.8412	**–75**
–74	199.15	0.0587	44.00	1512.6	2.7350	106.37	246.16	352.52	0.6020	1.8382	**–74**
–73	200.15	0.0635	47.66	1509.9	2.5453	107.57	245.57	353.14	0.6082	1.8353	**–73**
–72	201.15	0.0689	51.70	1507.1	2.3696	108.78	244.99	353.76	0.6144	1.8323	**–72**
–71	202.15	0.0739	55.44	1504.4	2.2068	109.98	244.40	354.38	0.6201	1.8293	**–71**
–70	203.15	0.0792	59.43	1501.7	2.0559	111.19	243.82	355.01	0.6259	1.8263	**–70**
–69	204.15	0.0849	63.70	1498.9	1.9212	112.39	243.24	355.63	0.6318	1.8234	**–69**
–68	205.15	0.0910	68.24	1496.2	1.7911	113.60	242.65	356.25	0.6376	1.8204	**–68**
–67	206.15	0.0974	73.08	1493.5	1.6891	114.82	242.07	356.89	0.6434	1.8170	**–67**
–66	207.15	0.104	78.24	1490.8	1.5847	116.04	241.48	357.52	0.6492	1.8145	**–66**
–65	208.15	0.112	83.74	1488.0	1.4793	117.26	240.89	358.15	0.6550	1.8126	**–65**
–64	209.15	0.119	89.59	1485.3	1.3887	118.48	240.30	358.78	0.6609	1.8101	**–64**
–63	210.15	0.128	95.82	1482.6	1.3040	119.70	239.71	359.41	0.6667	1.8077	**–63**
–62	211.15	0.137	102.46	1479.8	1.2249	120.92	239.12	360.04	0.6725	1.8052	**–62**
–61	212.15	0.146	109.51	1477.1	1.1509	122.14	238.53	360.67	0.6783	1.8027	**–61**
–60	213.15	0.159	119.26	1474.4	1.0817	123.36	237.94	361.30	0.6841	1.8002	**–60**

(Contd.)

TABLE: 6.1 Thermodynamic Properties of Saturated 1,1,1,2,-Tetrafluoroethane (R–134a) up to 1 Atmosphere (Temperature Basis) [Contd.]

Temperature		Saturation pressure, p		Liquid density	Specific volume of vapour v''	Specific enthalpy			Specific entropy		Temp
t	T			ρ_l		h	λ_v	H	s'	s''	t
°C	K	bar	torr	kg/m³	m³/kg	kJ/kg			kJ/(kg·K)		°C
–59	214.15	0.167	125.00	1471.8	1.0170	124.58	237.35	361.94	0.6900	1.7986	**–59**
–58	215.15	0.178	133.49	1469.0	0.95639	125.80	236.76	362.57	0.6958	1.7965	**–58**
–57	216.15	0.190	142.51	1466.2	0.90327	127.02	236.17	363.20	0.7016	1.7944	**–57**
–56	217.15	0.203	152.09	1463.4	0.85321	128.24	235.58	363.83	0.7074	1.7923	**–56**
–55	218.15	0.216	162.27	1460.5	0.80613	129.47	235.01	364.48	0.7133	1.7902	**–55**
–54	219.15	0.231	173.08	1457.7	0.76185	130.71	234.40	365.11	0.7188	1.7882	**–54**
–53	220.15	0.246	184.55	1454.9	0.72019	131.95	233.80	365.75	0.7243	1.7861	**–53**
–52	221.15	0.263	197.63	1452.1	0.68098	133.19	233.19	366.38	0.7298	1.7840	**–52**
–51	222.15	0.279	209.15	1449.2	0.64407	134.43	232.59	367.02	0.7353	1.7823	**–51**
–50	223.15	0.295	221.29	1446.4	0.60931	135.67	231.99	367.65	0.7408	1.7806	**–50**
–49	224.15	0.312	234.07	1443.6	0.57657	136.90	231.38	368.29	0.7463	1.7789	**–49**
–48	225.15	0.330	247.53	1440.8	0.54572	138.14	230.78	368.92	0.7519	1.7772	**–48**
–47	226.15	0.349	261.70	1437.9	0.51665	139.38	230.17	369.56	0.7574	1.7755	**–47**
–46	227.15	0.369	276.61	1435.1	0.48925	140.62	229.57	370.19	0.7629	1.7738	**–46**
–45	228.15	0.390	292.30	1432.3	0.46729	141.86	228.96	370.83	0.7684	1.7721	**–45**
–44	229.15	0.412	308.80	1429.4	0.44439	143.10	228.36	371.46	0.7739	1.7704	**–44**
–43	230.15	0.435	326.16	1426.6	0.42178	144.37	227.73	372.10	0.7794	1.7688	**–43**
–42	231.15	0.459	344.41	1423.8	0.40129	145.63	227.10	372.73	0.7849	1.7673	**–42**
–41	232.15	0.485	363.60	1421.0	0.38187	146.89	226.48	373.36	0.7904	1.7659	**–41**
–40	233.15	0.512	383.77	1418.1	0.36347	148.15	225.85	374.00	0.7960	1.7644	**–40**
–39	234.15	0.540	404.96	1414.9	0.34603	149.40	225.23	374.63	0.8013	1.7630	**–39**
–38	235.15	0.570	427.23	1412.0	0.32950	150.66	224.60	375.26	0.8068	1.7615	**–38**
–37	236.15	0.601	450.61	1409.0	0.31382	151.92	223.97	375.90	0.8124	1.7601	**–37**
–36	237.15	0.628	471.24	1406.0	0.29895	153.18	223.35	376.53	0.8179	1.7586	**–36**
–35	238.15	0.660	494.75	1403.1	0.28484	154.44	222.72	377.16	0.8223	1.7577	**–35**
–34	239.15	0.692	519.33	1400.1	0.27145	155.70	222.10	377.79	0.8278	1.7566	**–34**
–33	240.15	0.727	545.01	1397.1	0.25874	156.96	221.47	378.43	0.8328	1.7554	**–33**
–32	241.15	0.762	571.85	1394.2	0.24668	158.22	220.85	379.06	0.8380	1.7543	**–32**
–31	242.15	0.800	599.89	1391.2	0.23622	159.49	220.21	379.69	0.8431	1.7531	**–31**
–30	243.15	0.844	633.05	1388.2	0.22598	160.77	219.55	380.32	0.8483	1.7519	**–30**
–29	244.15	0.880	659.78	1385.3	0.21622	162.05	218.89	380.94	0.8535	1.7508	**–29**
–28	245.15	0.922	691.73	1382.3	0.20692	163.33	218.23	381.57	0.8587	1.7496	**–28**
–27	246.15	0.969	726.57	1379.3	0.19806	164.61	217.58	382.19	0.8645	1.7485	**–27**
–26.09*	**247.06**	**1.01325**	**760.00**	**1376.6**	**0.19035**	**165.78**	**216.98**	**382.76**	**0.8691**	**1.7474**	**–26.09**

*Normal boiling point

TABLE: 6.2 Thermodynamic Properties of Saturated 1,1,1,2–Tetrafluoroethane (Refrigerant R–134a) above 1 Atmosphere (Temperature Basis)

Temperature		Saturation pressure,	Liquid density	Specific volume of	Specific enthalpy			Specific entropy		Temp
t	T	p	ρ_l	vapour v''	h	λ_v	H	s'	s''	t
°C	K	bar	kg/m³	m³/kg	kJ/kg			kJ/(kg·K)		°C
–26	247.15	1.017	1376.4	0.18960	165.90	216.92	382.82	0.8696	1.7471	**–26**
–25	248.15	1.064	1373.4	0.18155	167.18	216.26	383.44	0.8747	1.7461	**–25**
–24	249.15	1.113	1370.4	0.17386	168.46	215.61	384.07	0.8798	1.7451	**–24**
–23	250.15	1.164	1367.4	0.16662	169.74	214.95	384.69	0.8849	1.7442	**–23**
–22	251.15	1.214	1364.5	0.15962	171.02	214.29	385.32	0.8900	1.7432	**–22**
–21	252.15	1.269	1361.5	0.15350	172.31	213.64	385.94	0.8951	1.7423	**–21**
–20	253.15	1.327	1358.5	0.14745	173.59	212.98	386.57	0.9002	1.7413	**–20**
–19	254.15	1.387	1355.4	0.14165	174.91	212.27	387.17	0.9053	1.7404	**–19**
–18	255.15	1.444	1352.[illegible]	0.13611	176.22	211.57	387.78	0.9104	1.7394	**–18**
–17	256.15	1.509	1349.0	0.13080	177.52	210.87	388.39	0.9155	1.7386	**–17**
–16	257.15	1.576	1345.9	0.12572	178.83	210.17	389.00	0.9206	1.7378	**–16**
–15	258.15	1.640	1342.7	0.12085	180.14	209.47	389.61	0.9257	1.7371	**–15**
–14	259.15	1.713	1339.6	0.11619	181.45	208.78	390.22	0.9308	1.7363	**–14**
–13	260.15	1.781	1336.4	0.11173	182.76	208.08	390.83	0.9356	1.7355	**–13**
–12	261.15	1.859	1333.3	0.10746	184.06	207.38	391.44	0.9406	1.7347	**–12**
–11	262.15	1.929	1330.1	0.10336	185.37	206.68	392.05	0.9456	1.7339	**–11**
–10	263.15	2.013	1327.0	0.09943	186.68	205.98	392.66	0.9506	1.7331	**–10**
–9	264.15	2.101	1323.8	0.09584	187.99	205.29	393.27	0.9556	1.7323	**–9**
–8	265.15	2.172	1320.6	0.09239	189.30	204.59	393.88	0.9606	1.7316	**–8**
–7	266.15	2.257	1317.5	0.08908	190.64	203.82	394.46	0.9656	1.7311	**–7**
–6	267.15	2.344	1314.3	0.08591	191.98	203.07	395.05	0.9705	1.7304	**–6**
–5	268.15	2.434	1311.2	0.08285	193.31	202.32	395.64	0.9755	1.7298	**–5**
–4	269.15	2.527	1308.0	0.07992	194.65	201.57	396.23	0.9805	1.7292	**–4**
–3	270.15	2.624	1304.9	0.07710	195.99	200.82	396.81	0.9852	1.7286	**–3**
–2	271.15	2.723	1301.7	0.07439	197.33	200.07	397.40	0.9901	1.7280	**–2**
–1	272.15	2.824	1298.1	0.07178	198.67	199.32	397.99	0.9950	1.7271	**–1**
0	273.15	2.931	1294.7	0.06927	200.00	198.57	398.58	1.0000	1.7266	**0**
1	274.15	3.036	1291.2	0.06694	201.34	197.82	399.17	1.0048	1.7261	**1**
2	275.15	3.150	1287.8	0.06462	202.68	197.07	399.75	1.0097	1.7256	**2**
3	276.15	3.259	1284.4	0.06250	204.02	196.32	400.34	1.0146	1.7251	**3**
4	277.15	3.374	1281.0	0.06043	205.36	195.57	400.93	1.0195	1.7246	**4**
5	278.15	3.496	1277.5	0.05838	206.73	194.79	401.52	1.0244	1.7241	**5**
6	279.15	3.618	1274.1	0.05646	208.10	193.97	402.07	1.0292	1.7236	**6**
7	280.15	3.743	1270.7	0.05461	209.48	193.16	402.63	1.0340	1.7231	**7**
8	281.15	3.873	1267.2	0.05283	210.85	192.34	403.19	1.0388	1.7226	**8**
9	282.15	4.013	1263.8	0.05111	212.22	191.53	403.75	1.0436	1.7225	**9**
10	283.15	4.148	1260.4	0.04945	213.59	190.71	404.31	1.0484	1.7221	**10**
11	284.15	4.287	1256.9	0.04785	214.97	189.90	404.87	1.0532	1.7216	**11**
12	285.15	4.430	1253.5	0.04631	216.34	189.09	405.43	1.0580	1.7212	**12**
13	286.15	4.577	1250.1	0.04483	217.71	188.27	405.99	1.0628	1.7208	**13**
14	287.15	4.729	1246.6	0.04339	219.09	187.46	406.55	1.0676	1.7204	**14**
15	288.15	4.885	1243.2	0.04208	220.46	186.64	407.10	1.0724	1.7199	**15**
16	289.15	5.045	1239.8	0.04074	221.83	185.83	407.66	1.0772	1.7195	**16**
17	290.15	5.206	1236.4	0.03955	223.23	184.95	408.19	1.0820	1.7191	**17**
18	291.15	5.376	1232.9	0.03833	224.64	184.06	408.70	1.0868	1.7187	**18**
19	292.15	5.541	1229.5	0.03716	226.06	183.17	409.22	1.0914	1.7181	**19**
20	293.15	5.713	1226.1	0.03603	227.47	182.27	409.74	1.0961	1.7177	**20**

(Contd.)

TABLE: 6.2 Thermodynamic Properties of Saturated 1,1,1,2–Tetrafluoroethane (Refrigerant R–134a) above 1 Atmosphere (Temperature Basis) [Contd.]

Temperature		Saturation pressure,	Liquid density	Specific volume of	Specific enthalpy			Specific entropy		Temp
t	T	p	ρ_l	vapour v''	h	λ_v	H	s'	s''	t
°C	K	bar	kg/m^3	m^3/kg	kJ/kg			kJ/(kg·K)		°C
21	294.15	5.895	1221.8	0.03494	228.88	181.38	410.26	1.1009	1.7173	21
22	295.15	6.076	1217.9	0.03388	230.30	180.48	410.78	1.1056	1.7169	22
23	296.15	6.266	1214.1	0.03286	231.71	179.59	411.30	1.1104	1.7165	23
24	297.15	6.458	1210.2	0.03187	233.12	178.70	411.82	1.1151	1.7164	24
25	298.15	6.654	1206.4	0.03092	234.54	177.80	412.34	1.1198	1.7163	25
26	299.15	6.856	1202.5	0.02999	235.95	176.91	412.86	1.1246	1.7159	26
27	300.15	7.060	1198.6	0.02910	237.36	176.01	413.37	1.1293	1.7154	27
28	301.15	7.272	1194.8	0.02824	238.78	175.12	413.89	1.1341	1.7150	28
29	302.15	7.482	1190.9	0.02744	240.25	174.11	414.36	1.1386	1.7147	29
30	303.15	7.697	1187.0	0.02665	241.71	173.09	414.80	1.1433	1.7144	30
31	304.15	7.924	1183.2	0.02588	243.17	172.06	415.24	1.1480	1.7141	31
32	305.15	8.151	1179.3	0.02513	244.63	171.04	415.67	1.1527	1.7138	32
33	306.15	8.390	1175.4	0.02441	246.10	170.02	416.11	1.1574	1.7135	33
34	307.15	8.628	1171.6	0.02371	247.56	168.99	416.55	1.1621	1.7132	34
35	308.15	8.871	1167.7	0.02303	249.02	167.97	416.99	1.1668	1.7129	35
36	309.15	9.122	1163.9	0.02238	250.49	166.94	417.43	1.1715	1.7126	36
37	310.15	9.372	1160.0	0.02174	251.95	165.92	417.87	1.1762	1.7123	37
38	311.15	9.634	1156.1	0.02113	253.41	164.89	418.31	1.1808	1.7120	38
39	312.15	9.896	1151.9	0.02054	254.88	163.87	418.74	1.1855	1.7114	39
40	313.15	10.171	1147.4	0.01996	256.34	162.84	419.18	1.1902	1.7110	40
41	314.15	10.443	1142.8	0.01940	257.85	162.02	419.87	1.1952	1.7106	41
42	315.15	10.719	1138.2	0.01886	259.38	160.89	420.27	1.1999	1.7102	42
43	316.15	11.018	1133.6	0.01834	260.91	159.75	420.66	1.2046	1.7098	43
44	317.15	11.307	1129.0	0.01783	262.43	158.62	421.06	1.2093	1.7094	44
45	318.15	11.604	1124.5	0.01735	263.96	157.49	421.45	1.2140	1.7092	45
46	319.15	11.906	1119.9	0.01687	265.48	156.36	421.85	1.2187	1.7088	46
47	320.15	12.212	1115.3	0.01641	267.01	155.23	422.24	1.2234	1.7084	47
48	321.15	12.529	1110.7	0.01596	268.54	154.10	422.64	1.2281	1.7080	48
49	322.15	12.853	1106.1	0.01552	270.06	152.97	423.03	1.2328	1.7076	49
50	323.15	13.184	1101.6	0.01509	271.59	151.84	423.43	1.2375	1.7071	50
51	324.15	13.527	1097.0	0.01468	273.12	150.71	423.83	1.2422	1.7067	51
52	325.15	13.873	1092.4	0.01427	274.64	149.58	424.22	1.2469	1.7062	52
53	326.15	14.156	1087.8	0.01389	276.28	148.33	424.62	1.2515	1.7057	53
54	327.15	14.501	1083.2	0.01351	277.88	147.13	425.01	1.2563	1.7052	54
55	328.15	14.898	1078.7	0.01314	279.49	145.66	425.14	1.2610	1.7047	55
56	329.15	15.259	1074.1	0.01278	281.09	144.34	425.43	1.2658	1.7042	56
57	330.15	15.627	1069.5	0.01243	282.69	143.03	425.72	1.2705	1.7038	57
58	331.15	16.004	1064.9	0.01209	284.29	141.72	426.01	1.2753	1.7033	58
59	332.15	16.388	1058.1	0.01176	285.89	140.41	426.30	1.2800	1.7030	59
60	333.15	16.780	1052.4	0.01145	287.50	139.10	426.59	1.2848	1.7024	60
61	334.15	17.210	1047.6	0.01113	289.10	137.79	426.88	1.2896	1.7018	61
62	335.15	17.620	1041.9	0.01083	290.70	136.48	427.18	1.2943	1.7012	62
63	336.15	18.038	1036.2	0.01054	292.39	135.04	427.43	1.2991	1.7006	63
64	337.15	18.464	1030.4	0.01026	294.08	133.52	427.60	1.3038	1.7000	64
65	338.15	18.884	1025.7	0.00997	295.78	132.00	427.78	1.3089	1.6994	65

(Contd.)

TABLE: 6.2 Thermodynamic Properties of Saturated 1,1,1,2–Tetrafluoroethane (Refrigerant R–134a) above 1 Atmosphere (Temperature Basis) [Contd.]

Temperature		Saturation pressure,	Liquid density	Specific volume of vapour v''	Specific enthalpy			Specific entropy		Temp
t	T	p	ρ_l		h	λ_v	H	s'	s''	t
°C	K	bar	kg/m³	m³/kg	kJ/kg			kJ/(kg·K)		°C
66	339.15	19.328	1019.9	0.00968	297.48	130.48	427.95	1.3138	1.6988	**66**
67	340.15	19.780	1014.2	0.00941	299.17	128.96	428.13	1.3187	1.6982	**67**
68	341.15	20.24	1008.5	0.00914	300.87	127.43	428.31	1.3236	1.6974	**68**
69	342.15	20.70	1003.0	0.00889	302.57	125.91	428.48	1.3285	1.6965	**69**
70	343.15	21.18	997.2	0.00864	304.27	124.39	428.66	1.3335	1.6956	**70**
71	344.15	21.64	990.5	0.00839	305.96	122.87	428.83	1.3384	1.6947	**71**
72	345.15	22.14	984.7	0.00816	307.66	121.35	429.01	1.3433	1.6938	**72**
73	346.15	22.63	977.8	0.00793	309.48	119.49	428.97	1.3482	1.6929	**73**
74	347.15	23.15	972.0	0.00771	311.32	117.68	429.00	1.3531	1.6920	**74**
75	348.15	23.65	965.3	0.00750	313.16	115.86	429.02	1.3580	1.6911	**75**
76	349.15	24.19	957.6	0.00729	315.00	114.02	429.02	1.3629	1.6902	**76**
77	350.15	24.69	950.0	0.00708	316.85	112.17	429.02	1.3682	1.6884	**77**
78	351.15	25.25	942.3	0.00685	318.69	110.28	428.97	1.3736	1.6871	**78**
79	352.15	25.76	934.7	0.00663	320.53	108.39	428.92	1.3789	1.6858	**79**
80	353.15	26.35	927.0	0.00642	322.37	106.55	428.92	1.3842	1.6845	**80**
81	354.15	26.90	919.4	0.00622	324.21	104.48	428.69	1.3896	1.6832	**81**
82	355.15	27.50	911.7	0.00602	326.06	102.41	428.47	1.3949	1.6819	**82**
83	356.15	28.05	904.1	0.00585	328.14	100.10	428.24	1.4002	1.6806	**83**
84	357.15	28.64	896.4	0.00566	330.22	97.80	428.02	1.4055	1.6793	**84**
85	358.15	29.24	888.8	0.00548	332.29	95.50	427.79	1.4109	1.6769	**85**
86	359.15	29.86	881.1	0.00531	334.37	93.20	427.57	1.4162	1.6746	**86**
87	360.15	30.48	868.6	0.00514	336.44	90.60	427.04	1.4215	1.6733	**87**
88	361.15	31.12	857.0	0.00498	338.52	87.96	426.48	1.4269	1.6710	**88**
89	362.15	31.78	848.5	0.00481	340.59	85.32	425.91	1.4331	1.6686	**89**
90	363.15	32.44	837.5	0.00461	342.66	82.69	425.35	1.4395	1.6663	**90**
91	364.15	33.11	826.6	0.00446	345.17	79.61	424.79	1.4459	1.6636	**91**
92	365.15	33.80	815.6	0.00427	347.66	76.56	424.23	1.4522	1.6612	**92**
93	366.15	34.49	801.6	0.00409	350.15	72.93	423.09	1.4586	1.6572	**93**
94	367.15	35.20	788.3	0.00392	352.64	69.08	421.72	1.4649	1.6527	**94**
95	368.15	35.92	775.1	0.00375	355.13	65.22	420.36	1.4713	1.6482	**95**
96	369.15	36.65	761.8	0.00359	357.62	61.37	418.99	1.4777	1.6437	**96**
97	370.15	37.40	736.8	0.00337	360.38	57.24	417.63	1.4879	1.6392	**97**
98	371.15	38.15	713.2	0.00315	365.44	50.82	416.26	1.5012	1.6347	**98**
99	372.15	38.93	689.6	0.00296	369.56	42.12	411.68	1.5059	1.6224	**99**
100	373.15	39.71	645.7	0.00255	374.61	32.31	406.92	1.5192	1.6082	**100**
101	374.15	40.52	580.0	0.00226	381.59	16.92	398.51	1.5410	1.5860	**101**
101.06*	**374.21**	**40.593**	**511.95**	**0.00195**	**389.63**	**0.00**	**389.63**	**1.5620**	**1.5620**	**101.06**

*Critical point

Reference State:

Specific enthalpy of saturated liquid at 0°C (273.15 K) = 200.00 kJ/kg

Specific entropy of saturated liquid at 0°C (273.15 K) = 1.0000 kJ/(kg·K)

TABLE: 6.3 Thermodynamic Properties of 1,1,1,2–Tetrafluoroethane (Refrigerant R–134a) (Superheat Region)

Temperature		Absolute Pressure (Saturation Temperature)									Temp.
T	*t*	0.1 bar (–66.86°C/206.29K)			0.3 bar (–49.69°C/223.46K)			0.6 bar (–36.94°C/236.21K)			*T*
K	°C	*v*	*i*	*s*	*v*	*i*	*s*	*v*	*i*	*s*	K
Satd.	Cond.	1.6667	356.99	1.8177	0.59581	367.85	1.7800	0.31124	375.91	1.7600	
215	–58.15	1.7395	362.91	1.8167	–	–	–	–	–	–	215
220	–53.15	1.7812	366.37	1.8618	–	–	–	–	–	–	220
225	–48.15	1.8227	369.87	1.8775	0.60024	368.95	1.7849	–	–	–	225
230	–43.15	1.8642	373.41	1.8931	0.61455	372.58	1.8008	–	–	–	230
240	–33.15	1.9469	380.63	1.9238	0.64300	379.92	1.8321	0.31686	378.81	1.7721	240
250	–23.15	2.0293	388.03	1.9540	0.67119	387.41	1.8627	0.33151	386.47	1.8033	250
260	–13.15	2.1117	395.61	1.9837	0.69915	395.07	1.8927	0.34597	394.24	1.8338	260
270	–3.15	2.1938	403.37	2.0130	0.72706	402.89	1.9222	0.36028	402.16	1.8637	270
280	6.85	2.2759	411.31	2.0419	0.75483	410.88	1.9512	0.37449	410.23	1.8930	280
290	16.85	2.3579	419.43	2.0704	0.78253	419.04	1.9799	0.38861	418.45	1.9219	290
300	26.85	2.4399	427.72	2.0985	0.81011	427.37	2.0081	0.40264	426.84	1.9503	300
320	46.85	2.6036	444.84	2.1537	0.86520	444.54	2.0635	0.43057	444.10	2.0060	320
340	66.85	2.7672	462.64	2.2077	0.92013	462.39	2.1176	0.45834	462.01	2.0603	340
360	86.85	2.9307	481.11	2.2604	0.97494	480.90	2.1705	0.48598	480.58	2.1134	360
380	106.85	3.0941	500.25	2.3122	1.0296	500.06	2.2223	0.51356	499.78	2.1653	380
400	126.85	3.2573	520.03	2.3629	1.0843	519.87	2.2731	0.54104	519.62	2.2161	400
420	146.85	3.4206	540.45	2.4127	1.1389	540.31	2.3229	0.56847	540.09	2.2661	420
440	166.85	3.5838	561.50	2.4617	1.1934	561.37	2.3719	0.59588	561.18	2.3151	440
460	186.85	3.7470	583.16	2.5098	1.2480	583.05	2.4201	0.62325	582.87	2.3633	460

Temperature		Absolute Pressure (Saturation Temperature)									Temp.
T	*t*	0.8 bar (–31.12°C/242.03K)			1.0 bar (–26.37°C/246.78K)			2.0 bar (–10.08°C/263.07K)			*T*
K	ºC	*v*	*i*	*s*	*v*	*i*	*s*	*v*	*i*	*s*	K
Satd.	Cond.	0.23755	379.61	1.7527	0.19256	382.59	1.7476	0.099880	392.61	1.7334	
250	–23.15	0.24655	385.82	1.7779	0.19553	385.14	1.7577	–	–	–	250
260	–13.15	0.25764	393.68	1.8088	0.20461	393.11	1.7889	–	–	–	260
270	–3.15	0.26857	401.67	1.8389	0.21352	401.16	1.8193	0.10329	398.51	1.7555	270
280	6.85	0.27938	409.79	1.8684	0.22232	409.34	1.8491	0.10807	407.03	1.7865	280
290	16.85	0.29012	418.06	1.8975	0.23101	417.66	1.8783	0.11274	415.60	1.8166	290
300	26.85	0.30077	426.48	1.9260	0.23963	426.12	1.9070	0.11731	424.28	1.8460	300
310	36.85	0.31136	435.06	1.9542	0.24819	434.74	1.9352	0.12182	433.07	1.8748	310
320	46.85	0.32191	443.80	1.9819	0.25671	443.50	1.9630	0.12627	441.98	1.9031	320
330	56.85	0.33241	452.70	2.0093	0.26517	452.43	1.9905	0.13068	451.04	1.9309	330
340	66.85	0.34289	461.76	2.0363	0.27361	461.51	2.0176	0.13505	460.23	1.9584	340
350	76.85	0.35333	470.98	2.0631	0.28202	470.75	2.0444	0.13938	469.57	1.9855	350
360	86.85	0.36374	480.36	2.0895	0.29040	480.14	2.0709	0.14369	479.05	2.0122	360
370	96.85	0.37414	489.90	2.1156				0.14798	488.68	2.0385	370
390	106.85	0.38451	499.59	2.1415	0.30710	499.41	2.1229	0.15225	498.46	2.0646	390
400	116.85	0.39487	509.45	2.1671	0.31542	509.27	2.1486	0.15650	508.39	2.0904	400
410	126.85	0.40522	519.46	2.1924	0.32372	519.29	2.1739	0.16073	518.46	2.1159	410
420	146.85	0.42586	539.94	2.2424	0.34030	539.80	2.2239	0.16916	539.06	2.1662	420
440	166.85	0.44647	561.04	2.2914	0.35684	560.91	2.2730	0.17755	560.26	2.2155	440
460	186.85	0.46705	582.75	2.3397	0.37333	582.63	2.3213	0.18590	582.05	2.2639	460

(Contd.)

TABLE: 6.3 Thermodynamic Properties of 1,1,1,2–Tetrafluoroethane (Refrigerant R–134a) (Superheat Region) [Contd.]

Temperature		Absolute Pressure (Saturation Temperature)									Temp.
T	*t*	3.0 bar (0.67°C/273.82K)			4.0 bar (8.93°C/282.08K)			5.0 bar (15.73°C/288.88K)			*T*
K	ºC	*v*	*i*	*s*	*v*	*i*	*s*	*v*	*i*	*s*	K
Satd.	**Cond.**	0.067705	398.99	1.7267	0.051209	403.71	1.7225	0.041123	407.47	1.7196	
280	6.85	0.069862	404.52	1.7467	–	–	–	–	–	–	**280**
290	16.85	0.073223	413.43	1.7779	0.053379	411.09	1.7483	0.041387	408.55	1.7234	**290**
300	26.85	0.076476	422.35	1.8081	0.055994	420.31	1.7796	0.043653	418.16	1.7560	**300**
310	36.85	0.079650	431.34	1.8376	0.058517	429.53	1.8098	0.045800	427.64	1.7871	**310**
320	46.85	0.082754	440.42	1.8664	0.060964	438.80	1.8392	0.047863	437.12	1.8171	**320**
330	56.85	0.085815	449.61	1.8947	0.063363	448.14	1.8680	0.049868	446.63	1.8464	**330**
340	66.85	0.088842	458.92	1.9225	0.065712	457.58	1.8962	0.051822	456.21	1.8750	**340**
350	76.85	0.091827	468.36	1.9499	0.068027	467.13	1.9239	0.053737	465.88	1.9030	**350**
360	86.85	0.094787	477.94	1.9769	0.070314	476.80	1.9511	0.055620	475.65	1.9306	**360**
370	96.85	0.097714	487.65	2.0035	0.072569	486.60	1.9779	0.057475	485.53	1.9576	**370**
380	106.85	0.10062	497.50	2.0297	0.074806	496.52	2.0044	0.059308	495.53	1.9843	**380**
390	116.85	0.10352	507.49	2.0557	0.077024	506.58	2.0305	0.061121	505.66	2.0106	**390**
400	126.85	0.10640	517.62	2.0813	0.079227	516.77	2.0563	0.062917	515.91	2.0366	**400**
410	136.85	0.10926	527.90	2.1067	0.081413	527.10	2.0819	0.064700	526.30	2.0622	**410**
420	146.85	0.11211	438.32	2.1318	0.083591	537.57	2.1071	0.066467	536.82	2.0876	**420**
430	156.85	0.11495	548.89	2.1567	0.085756	548.18	2.1320	0.068227	547.47	2.1126	**430**
440	166.85	0.11779	559.60	2.1813	0.087904	558.93	2.1568	0.069974	558.26	2.1374	**440**
450	176.85	0.12061	570.45	2.2057	0.090050	569.82	2.1812	0.071715	569.19	2.1620	**450**
460	186.85	0.12343	581.45	2.2299	0.092191	580.86	2.2055	0.073443	580.28	2.1863	**460**

Temperature		Absolute Pressure (Saturation Temperature)									Temp.
T	*t*	6.0 bar (21.57°C/294.72K)			7.0 bar (26.71°C/299.86K)			8.0 bar (31.32°C/304.47K)			*T*
K	ºC	*v*	*i*	*s*	*v*	*i*	*s*	*v*	*i*	*s*	K
Satd.	**Cond.**	0.034301	410.57	1.7174	0.029366	413.19	1.7156	0.025625	415.45	1.7140	
300	26.85	0.035368	415.84	1.7352	0.029391	413.33	1.7161	–	–	–	**300**
310	36.85	0.037284	425.66	1.7673	0.031165	423.55	1.7496	0.026536	421.31	1.7320	**310**
320	46.85	0.039102	435.37	1.7982	0.032819	433.54	1.7813	0.028082	431.63	1.7658	**320**
330	56.85	0.040851	445.07	1.8280	0.034393	443.45	1.8118	0.029533	441.71	1.7970	**330**
340	66.85	0.042546	454.80	1.8571	0.035907	453.35	1.8413	0.030916	451.85	1.8271	**340**
350	76.85	0.044199	464.59	1.8855	0.037375	463.28	1.8701	0.032249	461.93	1.8563	**350**
360	86.85	0.045817	474.47	1.8994	0.038806	473.27	1.8983	0.033541	472.05	1.8848	**360**
370	96.85	0.047407	484.45	1.9406	0.040207	483.35	1.9259	0.034803	482.22	1.9127	**370**
380	106.85	0.048972	494.53	1.9675	0.041582	493.51	1.9530	0.036037	492.48	1.9401	**380**
390	116.85	0.050515	504.73	1.9940	0.042937	503.79	1.9797	0.037249	502.83	1.9670	**390**
400	126.85	0.052043	515.05	2.0201	0.044273	514.17	2.0059	0.038442	513.28	1.9934	**400**
410	136.85	0.053556	525.49	2.0459	0.045593	524.67	2.0319	0.039618	523.84	2.0195	**410**
420	146.85	0.055054	536.06	2.0714	0.046900	535.29	2.0575	0:040781	534.52	2.0452	**420**
430	156.85	0.056542	546.76	2.0966	0.048193	546.04	2.0827	0.041930	545.31	2.0706	**430**
440	166.85	0.058018	557.59	2.1215	0.049478	556.91	2.1077	0.043070	556.23	2.0957	**440**
450	176.85	0.059485	568.56	2.1461	0.050751	567.92	2.1325	0.044201	567.27	2.1205	**450**
460	186.85	0.060946	579.66	2.1705	0.052018	579.05	2.1570	0.045323	578.44	2.1451	**460**

(Contd.)

TABLE: 6.3 Thermodynamic Properties of 1,1,1,2–Tetrafluoroethane (Refrigerant R–134a) (Superheat Region) [Contd.]

Temperature		Absolute Pressure (Saturation Temperature)									Temp.
T	t	10.0 bar (39.38°C/312.53K)			12.0 bar (46.31°C/319.46K)			14.0 bar (52.42°C/325.57K)			T
K	ºC	v	i	s	v	i	s	v	i	s	K
Satd.	Cond.	0.020316	419.16	1.7112	0.016718	422.04	1.7087	0.014110	424.29	1.7062	
320	46.85	0.021374	427.46	1.7375	0.016790	422.68	1.7107	–	–	–	320
330	56.85	0.022677	438.20	1.7705	0.018039	434.26	1.7463	0.014646	429.81	1.7230	330
340	66.85	0.023893	448.71	1.8019	0.019165	445.33	1.7794	0.015740	441.64	1.7583	340
350	76.85	0.025046	459.12	1.8321	0.020212	456.15	1.8107	0.016727	452.97	1.7912	350
360	86.85	0.026153	469.51	1.8614	0.021203	466.85	1.8409	0.017645	464.05	1.8224	360
370	96.85	0.027222	479.91	1.8899	0.022151	477.51	1.8701	0.018513	475.00	1.8524	370
380	106.85	0.028262	490.36	1.9177	0.023066	488.17	1.8985	0.019342	485.91	1.8815	380
390	116.85	0.029277	500.88	1.9450	0.023954	498.87	1.9263	0.020142	496.80	1.9098	390
400	126.85	0.030273	511.47	1.9719	0.024819	509.62	1.9535	0.020916	507.72	1.9374	400
410	136.85	0.031250	522.16	1.9983	0.025665	520.44	1.9803	0.021670	518.68	1.9645	410
420	146.85	0.032212	532.95	2.0242	0.026495	531.35	2.0065	0.022407	529.72	1.9911	420
430	156.85	0.033162	543.84	2.0499	0.027312	542.35	2.0324	0.023130	540.83	2.0172	430
440	166.85	0.034099	554.85	2.0752	0.028116	553.45	2.0579	0.023840	552.03	2.0430	440
450	176.85	0.035027	565.97	2.1002	0.028911	564.66	2.0831	0.024540	563.32	2.0683	450
460	186.85	0.035948	577.22	2.1249	0.029696	575.98	2.1080	0.025230	574.72	2.0934	460
Temperature		**Absolute Pressure (Saturation Temperature)**									**Temp.**
T	**t**	**16.0 bar (57.90°C/331.05K)**			**18.0 bar (62.89°C/336.04K)**			**20.0 bar (67.48°C/340.63K)**			**T**
K	**ºC**	**v**	**i**	**s**	**v**	**i**	**s**	**v**	**i**	**s**	**K**
Satd.	Cond.	0.012127	426.04	1.7035	0.010562	427.35	1.7007	0.009292	428.28	1.6978	
340	66.85	0.013114	437.54	1.7378	0.011003	432.87	1.7170	–	–	–	340
350	76.85	0.014079	449.54	1.7726	0.011980	445.79	1.7545	0.010255	441.61	1.7362	350
360	86.85	0.014953	461.08	1.8051	0.012835	457.91	1.7886	0.011114	454.48	1.7724	360
370	96.85	0.015768	472.38	1.8361	0.013616	469.61	1.8207	0.011878	466.68	1.8059	370
380	106.85	0.016538	483.55	1.8658	0.014345	481.09	1.8513	0.012580	478.52	1.8374	380
390	116.85	0.017274	494.66	1.8947	0.015036	492.45	1.8808	0.013237	490.16	1.8677	390
400	126.85	0.017983	505.76	1.9228	0.015696	503.75	1.9094	0.013860	501.68	1.8969	400
410	136.85	0.018670	516.89	1.9503	0.016332	515.05	1.9373	0.014457	513.16	1.9252	410
420	146.85	0.019339	528.05	1.9772	0.016948	526.36	1.9645	0.015033	524.62	1.9528	420
430	156.85	0.019992	539.28	2.0036	0.017548	537.71	1.9913	0.015591	536.11	1.9798	430
440	166.85	0.020631	550.59	2.0296	0.018134	549.12	2.0175	0.016134	547.64	2.0063	440
450	176.85	0.021260	561.98	2.0552	0.018708	560.61	2.0433	0.016666	559.22	2.0324	450
460	186.85	0.021879	573.46	2.0804	0.019272	572.18	2.0687	0.017186	570.88	2.0580	460
Temperature		**Absolute Pressure (Saturation Temperature)**									**Temp.**
T	**t**	**25.0 bar (77.57°C/350.72K)**			**30.0 bar (86.20°C/359.35K)**			**35.0 bar (93.72°C/366.87K)**			**T**
K	**ºC**	**v**	**i**	**s**	**v**	**i**	**s**	**v**	**i**	**s**	**K**
Satd.	Cond.	0.006941	429.00	1.6880	0.005282	427.33	1.6748	0.003968	422.22	1.6544	
360	86.85	0.007881	444.31	1.7311	0.005372	428.92	1.6792	–	–	–	360
370	96.85	0.008672	458.44	1.7698	0.006392	448.06	1.7317	0.004476	432.44	1.6821	370
380	106.85	0.009353	471.50	1.8047	0.007125	463.30	1.7723	0.005427	453.14	1.7374	380
390	116.85	0.009966	484.01	1.8372	0.007739	477.12	1.8083	0.006093	469.18	1.7791	390
400	126.85	0.010534	496.21	1.8681	0.008286	490.23	1.8414	0.006648	483.59	1.8156	400
410	136.85	0.011067	508.22	1.8977	0.008787	502.91	1.8728	0.007137	497.17	1.8491	410
420	146.85	0.011574	520.12	1.9246	0.009255	515.35	1.9027	0.007585	510.27	1.8807	420
430	156.85	0.012061	531.98	1.9543	0.009697	527.64	1.9317	0.008001	523.08	1.9108	430
440	166.85	0.012530	543.82	1.9815	0.010120	539.84	1.9597	0.008394	535.70	1.9398	440
450	176.85	0.012985	555.68	2.0082	0.010527	552.01	1.9871	0.008769	548.21	1.9679	450
460	186.85	0.013428	567.57	2.0343	0.010921	564.17	2.0138	0.009128	560.67	1.9953	460

(Contd.)

TABLE: 6.3 Thermodynamic Properties of 1,1,1,2–Tetrafluoroethane (Refrigerant R–134a) (Superheat Region) [Contd.]

Temperature		Absolute Pressure (Saturation Temperature)									Temp.
T	t	40.0 bar (100.34°C/373.49K)			50.0 bar			60.0 bar			T
K	°C	v	i	s	v	i	s	v	i	s	K
Satd.	**Cond.**	0.002563	405.37	1.6045	**Non-Existent**			**Non-Existent**			
380	106.85	0.003970	438.79	1.6935	0.001412	376.73	1.5239	0.001246	367.79	1.4969	**380**
390	116.85	0.004791	459.64	1.7477	0.002651	429.39	1.6606	0.001525	394.91	1.5673	**390**
400	126.85	0.005385	476.11	1.7894	0.003508	457.15	1.7310	0.002180	430.63	1.6577	**400**
410	136.85	0.005881	490.91	1.8259	0.004071	476.32	1.7783	0.002825	458.21	1.7259	**410**
420	146.85	0.006321	504.85	1.8595	0.004524	492.74	1.8179	0.003311	478.75	1.7754	**420**
430	156.85	0.006722	518.27	1.8911	0.004917	507.84	1.8535	0.003709	496.26	1.8166	**430**
440	166.85	0.007095	531.38	1.9212	0.005270	522.17	1.8864	0.004054	512.21	1.8533	**440**
450	176.85	0.007448	544.28	1.9502	0.005595	536.02	1.9175	0.004365	527.25	1.8871	**450**
460	186.85	0.007783	557.07	1.9783	0.005900	549.57	1.9473	0.004650	541.71	1.9188	**460**

Adapted from R. Tillner–Roth and H. D. Baehr, *J. Phys. Chem. Ref. Data,* **23** (5), 1994, p. 657 with the Permission of the American Institute of Physics, USA.

TABLE: 7.1 Thermodynamic Properties of Saturated Ammonia (Refrigerant R–717) up to 1 Atmosphere (Temperature Basis)

Temperature		Saturation pressure, p		Liquid density	Specific volume of	Specific enthalpy			Specific entropy		Temp
t	T			ρ_l	vapour v''	h	λ_v	H	s'	s''	t
°C	K	bar	torr	kg/m³	m³/kg	kJ/kg			kJ/(kg·K)		°C
−76	197.15	0.069	51.75	732.00	3.8162	−139.31	1485.27	1345.96	−0.4521	7.0816	**−76**
−75	198.15	0.075	56.25	730.90	2.8397	−134.62	1482.45	1347.83	−0.4284	7.0531	**−75**
−74	199.15	0.081	60.75	729.78	1.9417	−129.92	1479.60	1349.68	−0.4047	7.0249	**−74**
−73	200.15	0.087	65.26	728.68	1.1169	−125.33	1476.86	1351.53	−0.3817	6.9971	**−73**
−72	201.15	0.094	70.51	727.56	0.3569	−120.72	1474.10	1353.38	−0.3587	6.9696	**−72**
−71	202.15	0.102	76.51	726.45	9.6577	−116.20	1471.42	1355.22	−0.3364	6.9425	**−71**
−70	203.15	0.109	81.76	725.32	9.0122	−111.66	1468.71	1357.05	−0.3140	6.9157	**−70**
−69	204.15	0.118	88.51	724.20	8.4172	−107.21	1466.09	1358.88	−0.2921	6.8893	**−69**
−68	205.15	0.126	94.51	723.07	7.8671	−102.72	1463.42	1360.70	−0.2702	6.8632	**−68**
−67	206.15	0.136	102.01	721.94	7.3590	−98.31	1460.82	1362.51	−0.2487	6.8375	**−67**
−66	207.15	0.146	109.51	720.80	6.8885	−93.87	1458.19	1364.32	−0.2272	6.8120	**−66**
−65	208.15	0.156	117.01	719.66	6.4533	−89.49	1455.61	1366.12	−0.2061	6.7869	**−65**
−64	209.15	0.167	125.26	718.51	6.0496	−85.07	1452.98	1367.91	−0.1850	6.7621	**−64**
−63	210.15	0.179	134.26	717.37	5.6755	−80.72	1450.42	1369.70	−0.1642	6.7376	**−63**
−62	211.15	0.192	144.01	716.20	5.3280	−76.32	1447.80	1371.48	−0.1434	6.7134	**−62**
−61	212.15	0.205	153.76	715.05	5.0055	−71.98	1445.23	1373.25	−0.1229	6.6894	**−61**
−60	213.15	0.219	164.26	713.88	4.7055	−67.60	1442.61	1375.01	−0.1023	6.6658	**−60**
−59	214.15	0.234	175.51	712.72	4.4266	−63.27	1440.04	1376.77	−0.0820	6.6424	**−59**
−58	215.15	0.249	186.77	711.54	4.1668	−58.90	1437.41	1378.51	−0.0617	6.6193	**−58**
−57	216.15	0.266	199.52	710.37	3.9249	−54.58	1434.83	1380.25	−0.0416	6.5965	**−57**
−56	217.15	0.283	212.27	709.18	3.6993	−50.21	1432.19	1381.98	−0.0215	6.5739	**−56**
−55	218.15	0.302	226.52	708.00	3.4890	−45.88	1429.59	1383.71	−0.0016	6.5516	**−55**
−54	219.15	0.321	240.77	706.81	3.2925	−41.52	1426.94	1385.42	0.0183	6.5296	**−54**
−53	220.15	0.341	255.77	705.62	3.1091	−37.19	1424.31	1387.12	0.0380	6.5078	**−53**
−52	221.15	0.363	272.27	704.42	2.9375	−32.82	1421.64	1388.82	0.0578	6.4862	**−52**
−51	222.15	0.385	288.77	703.22	2.7772	−28.49	1418.99	1390.50	0.0774	6.4649	**−51**
−50	223.15	0.409	306.78	702.02	2.6270	−24.12	1416.30	1392.18	0.0970	6.4438	**−50**
−49	224.15	0.433	324.78	700.82	2.4864	−19.78	1413.63	1393.85	0.1164	6.4230	**−49**
−48	225.15	0.459	344.28	699.61	2.3546	−15.40	1410.91	1395.51	0.1358	6.4023	**−48**
−47	226.15	0.487	365.28	698.40	2.2311	−11.05	1408.21	1397.16	0.1551	6.3820	**−47**
−46	227.15	0.515	386.28	697.18	2.1152	−6.66	1405.46	1398.80	0.1744	6.3618	**−46**
−45	228.15	0.545	408.78	695.97	2.0064	−2.31	1402.73	1400.42	0.1935	6.3418	**−45**
−44	229.15	0.577	432.79	694.75	1.9041	2.09	1399.95	1402.04	0.2127	6.3221	**−44**
−43	230.15	0.609	456.79	693.53	1.8080	6.46	1397.19	1403.65	0.2317	6.3026	**−43**
−42	231.15	0.644	483.04	692.31	1.7176	10.86	1394.39	1405.25	0.2508	6.2832	**−42**
−41	232.15	0.680	510.04	691.09	1.6326	15.24	1391.60	1406.84	0.2697	6.2641	**−41**
−40	233.15	0.717	537.79	689.86	1.5526	19.64	1388.78	1408.42	0.2886	6.2452	**−40**
−39	234.15	0.757	567.80	688.63	1.4772	24.03	1385.96	1409.99	0.3074	6.2265	**−39**
−38	235.15	0.797	597.80	687.40	1.4061	28.45	1383.09	1411.54	0.3262	6.2079	**−38**
−37	236.15	0.840	630.05	686.17	1.3391	32.85	1380.24	1413.09	0.3448	6.1896	**−37**
−36	237.15	0.885	663.80	684.94	1.2758	37.28	1377.35	1414.63	0.3635	6.1714	**−36**
−35	238.15	0.931	698.31	683.70	1.2161	41.70	1374.46	1416.16	0.3820	6.1534	**−35**
−34	239.15	0.980	735.06	682.46	1.1598	46.13	1371.53	1417.66	0,4006	6.1356	**−34**
−33.33*	**239.82**	**1.01325**	**760.00**	**681.62**	**1.1241**	**49.08**	**1369.59**	**1418.67**	**0.4129**	**6.1240**	**−33.33**

*Normal boiling point

TABLE: 7.2 Thermodynamic Properties of Saturated Ammonia (Refrigerant R–717) above 1 Atmosphere (Temperature Basis)

Temperature		Saturation pressure,	Liquid density	Specific volume of	Specific enthalpy			Specific entropy		Temp
t	T	p	ρ_l	vapour v''	h	λ_v	H	s'	s''	t
°C	K	bar	kg/m³	m³/kg	kJ/kg			kJ/(kg·K)		°C
–33	240.15	1.030	681.22	1.1065	50.56	1368.61	1419.17	0.4190	6.1180	**–33**
–32	241.15	1.083	679.98	1.0561	55.01	1365.65	1420.66	0.4375	6.1006	**–32**
–31	242.15	1.138	678.73	1.0085	59.44	1362.70	1422.14	0.4558	6.0833	**–31**
–30	243.15	1.195	677.48	0.96339	63.90	1359.71	1423.61	0.4742	6.0662	**–30**
–29	244.15	1.254	676.24	0.92072	68.35	1356.72	1425.07	0.4924	6.0493	**–29**
–28	245.15	1.316	674.98	0.88028	72.81	1353.70	1426.51	0.5106	6.0325	**–28**
–27	246.15	1.380	673.73	0.84198	77.27	1350.67	1427.94	0.5287	6.0159	**–27**
–26	247.15	1.446	672.47	0.80565	81.75	1347.62	1429.37	0.5468	5.9994	**–26**
–25	248.15	1.515	671.21	0.77120	86.22	1344.56	1430.78	0.5648	5.9832	**–25**
–24	249.15	1.587	669.95	0.73850	90.71	1341.47	1432.18	0.5828	5.9670	**–24**
–23	250.15	1.662	668.69	0.70746	95.19	1338.37	1433.56	0.6007	5.9510	**–23**
–22	251.15	1.739	667.42	0.67798	99.69	1335.24	1434.93	0.6186	5.9352	**–22**
–21	252.15	1.819	666.15	0.64997	104.18	1332.11	1436.29	0.6365	5.9195	**–21**
–20	253.15	1.902	664.88	0.62334	108.69	1328.95	1437.64	0.6543	5.9039	**–20**
–19	254.15	1.988	663.60	0.59802	113.20	1325.78	1438.98	0.6720	5.8885	**–19**
–18	255.15	2.077	662.32	0.57393	117.72	1322.58	1440.30	0.6897	5.8732	**–18**
–17	256.15	2.169	661.03	0.55101	122.23	1319.38	1441.61	0.7073	5.8581	**–17**
–16	257.15	2.264	659.75	0.52917	126.77	1316.14	1442.91	0.7249	5.8431	**–16**
–15	258.15	2.363	658.46	0.50838	131.29	1312.90	1444.19	0.7424	5.8282	**–15**
–14	259.15	2.465	657.16	0.48857	135.84	1309.62	1445.46	0.7599	5.8134	**–14**
–13	260.15	2.570	655.87	0.46968	140.38	1306.34	1446.72	0.7773	5.7988	**–13**
–12	261.15	2.679	654.57	0.45166	144.93	1303.03	1447.96	0.7947	5.7843	**–12**
–11	262.15	2.792	653.26	0.43448	149.48	1299.72	1449.20	0.8121	5.7700	**–11**
–10	263.15	2.908	651.95	0.41808	154.05	1296.36	1450.41	0.8294	5.7557	**–10**
–9	264.15	3.029	650.64	0.40243	158.61	1293.01	1451.62	0.8466	5.7416	**–9**
–8	265.15	3.153	649.32	0.38747	163.19	1289.61	1452.80	0.8638	5.7276	**–8**
–7	266.15	3.281	648.00	0.37319	167.77	1286.21	1453.98	0.8810	5.7137	**–7**
–6	267.15	3.413	646.68	0.35953	172.36	1282.78	1455.14	0.8981	5.6999	**–6**
–5	268.15	3.549	645.35	0.34648	176.94	1279.35	1456.29	0.9152	5.6862	**–5**
–4	269.15	3.689	644.01	0.33400	181.55	1275.87	1457.42	0.9322	5.6726	**–4**
–3	270.15	3.834	642.67	0.32206	186.15	1272.38	1458.53	0.9492	5.6591	**–3**
–2	271.15	3.983	641.33	0.31062	190.76	1268.88	1459.64	0.9662	5.6458	**–2**
–1	272.15	4.137	639.98	0.29968	195.38	1265.34	1460.72	0.9831	5.6325	**–1**
0	273.15	4.295	638.63	0.28920	200.00	1261.80	1461.80	1.0000	5.6194	**0**
1	274.15	4.458	637.27	0.27916	204.63	1258.22	1462.85	1.0168	5.6063	**1**
2	275.15	4.626	635.91	0.26954	209.27	1254.62	1463.89	1.0336	5.5934	**2**
3	276.15	4.799	634.54	0.26032	213.91	1251.01	1464.92	1.0503	5.5805	**3**
4	277.15	4.976	633.17	0.25147	218.56	1247.37	1465.93	1.0670	5.5677	**4**
5	278.15	5.159	631.79	0.24299	223.22	1243.78	1467.00	1.0837	5.5550	**5**
6	279.15	5.347	630.41	0.23485	227.89	1240.01	1467.90	1.1003	5.5424	**6**
7	280.15	5.540	629.01	0.22703	232.55	1236.31	1468.86	1.1169	5.5299	**7**
8	281.15	5.738	627.62	0.21953	237.23	1232.57	1469.80	1.1335	5.5175	**8**
9	282.15	5.942	626.22	0.21233	241.92	1228.81	1470.73	1.1500	5.5052	**9**
10	283.15	6.152	624.81	0.20541	246.61	1225.03	1471.64	1.1665	5.4929	**10**

(Contd.)

TABLE: 7.2 Thermodynamic Properties of Saturated Ammonia (Refrigerant R–717) above 1 Atmosphere (Temperature Basis) [Contd.]

Temperature t	T	Saturation pressure, p	Liquid density ρ_l	Specific volume of vapour v''	Specific enthalpy h	λ_v	H	Specific entropy s'	s''	Temp t
°C	K	bar	kg/m³	m³/kg	kJ/kg			kJ/(kg·K)		°C
11	284.15	6.367	623.40	0.19876	251.31	1221.23	1472.54	1.1829	5.4807	**11**
12	285.15	6.588	621.98	0.19237	256.02	1217.39	1473.41	1.1993	5.4686	**12**
13	286.15	6.814	620.56	0.18622	260.73	1213.54	1474.27	1.2157	5.4566	**13**
14	287.15	7.047	619.13	0.18031	265.45	1209.66	1475.11	1.2320	5.4447	**14**
15	288.15	7.286	617.70	0.17462	270.18	1205.75	1475.93	1.2483	5.4328	**15**
16	289.15	7.531	616.25	0.16915	274.91	1201.83	1476.74	1.2646	5.4210	**16**
17	290.15	7.782	614.81	0.16388	279.66	1197.86	1477.52	1.2808	5.4093	**17**
18	291.15	8.040	613.35	0.15881	284.41	1193.88	1478.29	1.2971	5.3976	**18**
19	292.15	8.304	611.89	0.15393	289.17	1189.86	1479.03	1.3132	5.3860	**19**
20	293.15	8.575	610.42	0.14922	293.93	1185.83	1479.76	1.3294	5.3745	**20**
21	294.15	8.852	608.95	0.14469	298.71	1181.76	1480.47	1.3455	5.3630	**21**
22	295.15	9.136	607.47	0.14032	303.49	1177.67	1481.16	1.3615	5.3516	**22**
23	296.15	9.427	605.98	0.13611	308.28	1173.55	1481.83	1.3776	5.3403	**23**
24	297.15	9.726	604.49	0.13204	313.08	1169.40	1482.48	1.3936	5.3290	**24**
25	298.15	10.031	602.98	0.12813	317.89	1165.21	1483.10	1.4096	5.3177	**25**
26	299.15	10.344	601.47	0.12435	322.70	1161.01	1483.71	1.4255	5.3065	**26**
27	300.15	10.664	599.96	0.12070	327.53	1156.76	1484.29	1.4414	5.2954	**27**
28	301.15	10.991	598.44	0.11718	332.36	1152.50	1484.86	1.4573	5.2843	**28**
29	302.15	11.327	596.90	0.11378	337.20	1148.20	1485.40	1.4732	5.2733	**29**
30	303.15	11.669	595.37	0.11049	342.05	1143.87	1485.92	1.4890	5.2623	**30**
31	304.15	12.020	593.82	0.10732	346.91	1139.51	1486.42	1.5048	5.2514	**31**
32	305.15	12.379	592.27	0.10426	351.78	1135.11	1486.89	1.5206	5.2405	**32**
33	306.15	12.745	590.71	0.10130	356.65	1130.69	1487.34	1.5364	5.2296	**33**
34	307.15	13.120	589.14	0.09844	361.54	1126.23	1487.77	1.5521	5.2188	**34**
35	308.15	13.503	587.56	0.09567	366.44	1121.73	1488.17	1.5678	5.2080	**35**
36	309.15	13.895	585.98	0.09299	371.34	1117.21	1488.55	1.5835	5.1973	**36**
37	310.15	14.295	584.39	0.09040	376.26	1112.65	1488.91	1.5991	5.1866	**37**
38	311.15	14.704	582.79	0.08790	381.18	1108.06	1489.24	1.6148	5.1759	**38**
39	312.15	15.122	581.18	0.08548	386.12	1103.42	1489.54	1.6304	5.1653	**39**
40	313.15	15.548	579.56	0.08313	391.06	1098.76	1489.82	1.6459	5.1547	**40**
41	314.15	15.984	577.93	0.08086	396.02	1094.05	1490.07	1.6615	5.1441	**41**
42	315.15	16.428	576.30	0.07866	400.98	1089.32	1490.30	1.6770	5.1335	**42**
43	316.15	16.882	574.66	0.07654	405.96	1084.54	1490.50	1.6925	5.1230	**43**
44	317.15	17.346	573.00	0.07447	410.95	1079.73	1490.68	1.7080	5.1125	**44**
45	318.15	17.819	571.34	0.07248	415.94	1074.87	1490.81	1.7235	5.1020	**45**
46	319.15	18.301	569.67	0.07054	420.95	1069.98	1490.93	1.7390	5.0916	**46**
47	320.15	18.793	567.99	0.06866	425.98	1065.04	1491.02	1.7544	5.0811	**47**
48	321.15	19.296	566.30	0.06684	431.01	1060.07	1491.08	1.7698	5.0707	**48**
49	322.15	19.808	564.60	0.06508	436.05	1055.06	1491.11	1.7852	5.0603	**49**
50	323.15	20.330	562.89	0.06337	441.11	1050.00	1491.11	1.8006	5.0499	**50**
51	324.15	20.863	561.17	0.06171	446.18	1044.89	1491.07	1.8160	5.0395	**51**
52	325.15	21.406	559.44	0.06010	451.26	1039.75	1491.01	1.8314	5.0291	**52**
53	326.15	21.960	557.70	0.05854	456.36	1034.56	1490.92	1.8467	5.0187	**53**
54	327.15	22.524	555.94	0.05702	461.47	1029.32	1490.79	1.8620	5.0084	**54**
55	328.15	23.099	554.18	0.05555	466.59	1024.05	1490.64	1.8773	4.9980	**55**

(Contd.)

TABLE: 7.2 Thermodynamic Properties of Saturated Ammonia (Refrigerant R–717) above 1 Atmosphere (Temperature Basis) [Contd.]

Temperature		Saturation pressure,	Liquid density	Specific volume of	Specific enthalpy			Specific entropy		Temp
t	T	p	ρ_l	vapour v''	h	λ_v	H	s'	s''	t
°C	K	bar	kg/m^3	m^3/kg	kJ/kg			kJ/(kg·K)		°C
56	329.15	23.685	552.40	0.05412	471.72	1018.72	1490.44	1.8926	4.9876	**56**
57	330.15	24.282	550.62	0.05273	476.88	1013.34	1490.22	1.9079	4.9773	**57**
58	331.15	24.891	548.82	0.05138	482.04	1007.92	1489.96	1.9232	4.9669	**58**
59	332.15	25.511	547.01	0.05008	487.22	1002.45	1489.67	1.9385	4.9566	**59**
60	333.15	26.142	545.19	0.04880	492.42	996.92	1489.34	1.9538	4.9462	**60**
61	334.15	26.785	543.35	0.04757	497.63	991.34	1488.97	1.9690	4.9358	**61**
62	335.15	27.440	541.51	0.04636	502.86	985.71	1488.57	1.9843	4.9254	**62**
63	336.15	28.106	539.65	0.04520	508.10	980.03	1488.13	1.9996	4.9150	**63**
64	337.15	28.785	537.77	0.04406	513.36	974.29	1487.65	2.0148	4.9046	**64**
65	338.15	29.476	535.89	0.04296	518.64	968.50	1487.14	2.0301	4.8942	**65**
66	339.15	30.179	533.99	0.04188	523.94	962.64	1486.58	2.0453	4.8837	**66**
67	340.15	30.895	532.07	0.04084	529.25	956.74	1485.99	2.0606	4.8733	**67**
68	341.15	31.623	530.14	0.03982	534.59	950.76	1485.35	2.0758	4.8628	**68**
69	342.15	32.364	528.19	0.03883	539.94	944.74	1484.68	2.0911	4.8523	**69**
70	343.15	33.118	526.23	0.03787	545.32	938.64	1483.96	2.1064	4.8417	**70**
71	344.15	33.885	524.25	0.03693	550.71	932.48	1483.19	2.1216	4.8312	**71**
72	345.15	34.666	522.26	0.03601	556.13	926.25	1482.38	2.1369	4.8206	**72**
73	346.15	35.459	520.25	0.03512	561.57	919.96	1481.53	2.1522	4.8099	**73**
74	347.15	36.266	518.22	0.03426	567.03	913.60	1480.63	2.1675	4.7992	**74**
75	348.15	37.087	516.17	0.03341	572.51	907.17	1479.68	2.1828	4.7885	**75**
76	349.15	37.922	514.11	0.03259	578.02	900.67	1478.69	2.1982	4.7778	**76**
77	350.15	38.771	512.02	0.03179	583.56	894.08	1477.64	2.2135	4.7670	**77**
78	351.15	39.634	509.92	0.03101	589.12	887.43	1476.55	2.2289	4.7561	**78**
79	352.15	40.511	507.79	0.03025	594.70	880.70	1475.40	2.2443	4.7452	**79**
80	353.15	41.402	505.65	0.02951	600.32	873.89	1474.21	2.2597	4.7343	**80**
81	354.15	42.309	503.48	0.02878	605.96	866.99	1472.95	2.2752	4.7233	**81**
82	355.15	43.230	501.29	0.02808	611.64	860.00	1471.64	2.2907	4.7122	**82**
83	356.15	44.166	499.08	0.02739	617.34	852.94	1470.28	2.3062	4.7011	**83**
84	357.15	45.117	496.85	0.02671	623.07	845.78	1468.85	2.3217	4.6899	**84**
85	358.15	46.083	494.58	0.02606	628.84	838.53	1467.37	2.3373	4.6786	**85**
86	359.15	47.065	492.30	0.02542	634.64	831.18	1465.82	2.3529	4.6672	**86**
87	360.15	48.062	489.99	0.02479	640.48	823.73	1464.21	2.3686	4.6558	**87**
88	361.15	49.075	487.65	0.02418	646.35	816.19	1462.54	2.3843	4.6443	**88**
89	362.15	50.105	485.28	0.02358	652.26	808.54	1460.80	2.4001	4.6326	**89**
90	363.15	51.150	482.88	0.02300	658.21	800.78	1458.99	2.4159	4.6209	**90**
91	364.15	52.212	480.45	0.02243	664.20	792.90	1457.10	2.4317	4.6091	**91**
92	365.15	53.290	477.99	0.02187	670.24	784.91	1455.15	2.4477	4.5972	**92**
93	366.15	54.385	475.49	0.02133	676.32	776.79	1453.11	2.4637	4.5852	**93**
94	367.15	55.497	472.96	0.02080	682.44	768.56	1451.00	2.4797	4.5730	**94**
95	368.15	56.626	470.40	0.02028	688.61	760.20	1448.81	2.4959	4.5608	**95**
96	369.15	57.772	467.79	0.01977	694.83	751.70	1446.53	2.5121	4.5484	**96**
97	370.15	58.935	465.15	0.01927	701.11	743.05	1444.16	2.5284	4.5358	**97**
98	371.15	60.117	462.46	0.01878	707.44	734.26	1441.70	2.5448	4.5231	**98**
99	372.15	61.316	459.73	0.01831	713.82	725.33	1439.15	2.5612	4.5103	**99**
100	373.15	62.533	456.96	0.01784	720.27	716.23	1436.50	2.5778	4.4972	**100**

(Contd.)

TABLE: 7.2 Thermodynamic Properties of Saturated Ammonia (Refrigerant R–717) above 1 Atmosphere (Temperature Basis) [Contd.]

Temperature t °C	Temperature T K	Saturation pressure, p bar	Liquid density ρ_l kg/m³	Specific volume of vapour v'' m³/kg	Specific enthalpy h kJ/kg	Specific enthalpy λ_v kJ/kg	Specific enthalpy H kJ/kg	Specific entropy s' kJ/(kg·K)	Specific entropy s'' kJ/(kg·K)	Temp t °C
101	374.15	63.769	454.14	0.01738	726.77	706.97	1433.74	2.5945	4.4840	**101**
102	375.15	65.023	451.27	0.01693	733.35	697.52	1430.87	2.6113	4.4706	**102**
103	376.15	66.296	448.34	0.01649	739.99	687.90	1427.89	2.6282	4.4570	**103**
104	377.15	67.588	445.36	0.01606	746.71	678.09	1424.80	2.6453	4.4432	**104**
105	378.15	68.899	442.32	0.01564	753.50	668.07	1421.57	2.6625	4.4292	**105**
106	379.15	70.230	439.22	0.01522	760.37	657.85	1418.22	2.6799	4.4149	**106**
107	380.15	71.580	436.05	0.01481	767.33	647.39	1414.72	2.6974	4.4004	**107**
108	381.15	72.951	432.82	0.01441	774.38	636.70	1411.08	2.7151	4.3855	**108**
109	382.15	74.341	429.50	0.01402	781.52	625.76	1407.28	2.7329	4.3704	**109**
110	383.15	75.753	426.11	0.01363	788.77	614.55	1403.32	2.7510	4.3550	**110**
111	384.15	77.185	422.64	0.01325	796.13	603.05	1399.18	2.7693	4.3391	**111**
112	385.15	78.638	419.07	0.01287	803.61	591.24	1394.85	2.7879	4.3230	**112**
113	386.15	80.112	415.41	0.01250	811.21	579.11	1390.32	2.8067	4.3064	**113**
114	387.15	81.609	411.64	0.01214	818.94	566.62	1385.56	2.8257	4.2893	**114**
115	388.15	83.127	407.75	0.01178	826.83	553.75	1380.58	2.8451	4.2717	**115**
116	389.15	84.668	403.74	0.01142	834.87	540.46	1375.33	2.8648	4.2537	**116**
117	390.15	86.232	399.60	0.01107	843.09	526.72	1369.81	2.8849	4.2350	**117**
118	391.15	87.820	395.31	0.01072	851.50	512.48	1363.98	2.9054	4.2156	**118**
119	392.15	89.431	390.85	0.01038	860.12	497.69	1357.81	2.9264	4.1955	**119**
120	393.15	91.066	386.21	0.01003	868.98	482.29	1351.27	2.9479	4.1746	**120**
121	394.15	92.727	381.36	0.00969	878.09	466.22	1344.31	2.9699	4.1528	**121**
122	395.15	94.412	376.29	0.00935	887.51	449.36	1336.87	2.9927	4.1299	**122**
123	396.15	96.124	370.95	0.00901	897.27	431.63	1328.90	3.0162	4.1057	**123**
124	397.15	97.862	365.30	0.00867	907.42	412.89	1320.31	3.0406	4.0802	**124**
125	398.15	99.628	359.30	0.00833	918.04	392.95	1310.99	3.0660	4.0530	**125**
126	399.15	101.423	352.87	0.00798	929.20	371.62	1300.82	3.0928	4.0238	**126**
128	401.15	105.101	338.33	0.00726	953.75	323.36	1277.11	3.1515	3.9576	**128**
130	403.15	108.908	320.32	0.00650	982.92	263.70	1246.62	3.2211	3.8752	**130**
132	405.15	112.858	294.86	0.00560	1022.26	180.42	1202.68	3.3153	3.7606	**132**

Reference State:
Specific enthalpy of saturated liquid at 0°C (273.15 K) = 200.00 kJ/kg
Specific entropy of saturated liquid at 0°C (273.15 K) = 1.0000 kJ/(kg·K)

TABLE: 7.3 Thermodynamic Properties of Ammonia (Refrigerant R–717) (Superheat Region)

Temperature		Absolute Pressure (Saturation Temperature)									Temp.
T	*t*	0.1 bar (–71.22°C/201.93K)			0.15 bar (–65.59°C/207.56K)			0.20 bar (–61.37°C/211.78K)			*T*
K	°C	*v*	*i*	*s*	*v*	*i*	*s*	*v*	*i*	*s*	K
Satd.	cond.	9.8039	1354.82	6.9486	6.7054	1365.06	6.8017	5.1226	1372.59	6.6983	
210	–63.15	10.2040	1371.03	7.0270	6.7866	1370.00	6.8253	–	–		210
215	–58.15	10.4517	1381.11	7.0745	6.9530	1380.14	6.8731	5.2035	1379.17	6.7291	215
220	–53.15	10.6991	1391.21	7.1209	7.1190	1390.31	6.9198	5.3289	1389.40	6.7762	220
225	–48.15	10.9462	1401.33	7.1664	7.2848	1400.49	6.9656	5.4540	1399.64	6.8222	225
230	–43.15	11.1931	1411.47	7.2110	7.4503	1410.69	7.0104	5.5789	1409.90	6.8673	230
235	–38.15	11.4397	1421.63	7.2547	7.6156	1420.90	7.0543	5.7034	1420.17	6.9115	235
240	–33.15	11.6861	1431.80	7.2975	7.7806	1431.13	7.0974	5.8278	1430.45	6.9547	240
245	–28.15	11.9323	1442.00	7.3396	7.9454	1441.37	7.1396	5.9519	1440.74	6.9972	245
250	–23.15	12.1782	1452.22	7.3808	8.1100	1451.63	7.1811	6.0759	1451.05	7.0388	250
255	–18.15	12.4241	1462.46	7.4214	8.2745	1461.92	7.2218	6.1996	1461.37	7.0797	255
260	–13.15	12.6697	1472.72	7.4613	8.4388	1472.22	7.2618	6.3233	1471.71	7.1199	260
265	–8.15	12.9152	1483.01	7.5004	8.6029	1482.54	7.3011	6.4467	1482.07	7.1593	265
270	–3.15	13.1606	1493.32	7.5390	8.7669	1492.88	7.3398	6.5701	1492.45	7.1981	270
275	1.85	13.4059	1503.66	7.5769	8.9308	1503.26	7.3779	6.6933	1502.85	7.2363	275
280	6.85	13.6510	1514.03	7.6143	9.0946	1513.65	7.4153	6.8164	1513.27	7.2739	280
290	16.85	14.1411	1534.86	7.6874	9.4220	1534.53	7.4886	7.0624	1534.20	7.3473	290
300	26.85	14.6309	1555.83	7.7585	9.7490	1555.54	7.5598	7.3081	1555.24	7.4186	300
310	36.85	15.1204	1576.93	7.8277	10.0758	1576.67	7.6291	7.5535	1576.41	7.4881	310
320	46.85	15.6097	1598.19	7.8952	10.4024	1597.95	7.6967	7.7988	1597.72	7.5557	320
330	56.85	16.0989	1619.60	7.9611	10.7289	1619.39	7.7626	8.0439	1619.18	7.6217	330
340	66.85	16.5880	1641.17	8.0255	11.0552	1640.98	7.8271	8.2889	1640.79	7.6862	340
350	76.85	17.0769	1662.92	8.0885	11.3815	1662.74	7.8902	8.5337	1662.57	7.7494	350
360	86.85	17.5658	1684.83	8.1502	11.7076	1684.67	7.9520	8.7785	1684.51	7.8112	360
370	96.85	18.0546	1706.93	8.2108	12.0337	1706.78	8.0125	9.0232	1706.63	7.8718	370

Temperature		Absolute Pressure (Saturation Temperature)									Temp.
T	*t*	0.25 bar (–57.96°C/215.19K)			0.30 bar (–55.08°C/218.07K)			0.35 bar (–52.58°C/220.57K)			*T*
K	°C	*v*	*i*	*s*	*v*	*i*	*s*	*v*	*i*	*s*	K
Satd.	cond.	4.1576	1378.58	6.6185	3.5060	1383.56	6.5535	3.0355	1387.84	6.4987	
220	–53.15	4.2548	1388.49	6.6640	3.5387	1387.57	6.5718	–	–	–	220
225	–48.15	4.3555	1398.79	6.7103	3.6232	1397.94	6.6184	3.1000	1397.08	6.5401	225
230	–43.15	4.4560	1409.11	6.7557	3.7073	1408.32	6.6640	3.1726	1407.52	6.5860	230
235	–38.15	4.5561	1419.44	6.8001	3.7913	1418.70	6.7086	3.2449	1417.96	6.6309	235
240	–33.15	4.6561	1429.77	6.8436	3.8750	1429.09	6.7524	3.3170	1428.40	6.6749	240
245	–28.15	4.7558	1440.11	6.8863	3.9584	1439.48	6.7952	3.3888	1438.84	6.7180	245
250	–23.15	4.8554	1450.47	6.9281	4.0417	1449.88	6.8373	3.4605	1449.29	6.7602	250
255	–18.15	4.9547	1460.83	6.9691	4.1248	1460.29	6.8785	3.5320	1459.74	6.8016	255
260	–13.15	5.0540	1471.21	7.0094	4.2077	1470.70	6.9189	3.6033	1470.20	6.8422	260
265	–8.15	5.1530	1481.60	7.0490	4.2905	1481.13	6.9587	3.6745	1480.66	6.8820	265
270	–3.15	5.2520	1492.01	7.0880	4.3732	1491.57	6.9977	3.7455	1491.14	6.9212	270
275	1.85	5.3508	1502.44	7.1262	4.4558	1502.03	7.0361	3.8165	1501.63	6.9597	275
280	6.85	5.4495	1512.89	7.1639	4.5382	1512.51	7.0739	3.8873	1512.13	6.9976	280
285	11.85	5.5481	1523.37	7.2010	4.6206	1523.01	7.1110	3.9580	1522.66	7.0348	285
290	16.85	5.6467	1533.87	7.2375	4.7028	1533.53	7.1476	4.0287	1533.20	7.0715	290
300	26.85	5.8435	1554.95	7.3090	4.8672	1554.66	7.2192	4.1697	1554.36	7.1432	300
310	36.85	6.0402	1576.15	7.3785	5.0312	1575.89	7.2889	4.3106	1575.63	7.2130	310
320	46.85	6.2366	1597.49	7.4462	5.1951	1597.25	7.3567	4.4512	1597.02	7.2809	320
330	56.85	6.4329	1618.97	7.5123	5.3589	1618.76	7.4228	4.5917	1618.55	7.3471	330
340	66.85	6.6290	1640.60	7.5769	5.5225	1640.41	7.4875	4.7321	1640.22	7.4118	340
350	76.85	6.8251	1662.39	7.6401	5.6860	1662.22	7.5507	4.8723	1662.04	7.4751	350
360	86.85	7.0210	1684.35	7.7019	5.8494	1684.19	7.6126	5.0125	1684.03	7.5370	360
370	96.85	7.2169	1706.48	7.7626	6.0127	1706.33	7.6733	5.1526	1706.19	7.5977	370
380	106.85	7.4127	1728.79	7.8221	6.1760	1728.66	7.7328	5.2926	1728.52	7.6573	380

(Contd.)

TABLE: 7.3 Thermodynamic Properties of Ammonia (Refrigerant R–717) (Superheat Region) [Contd.]

Temperature		Absolute Pressure (Saturation Temperature)									Temp.
T	*t*	0.40 bar (–50.36°C/222.79K)			0.45 bar (–48.35°C/224.80K)			0.50 bar (–46.52°C/226.63K)			*T*
K	°C	*v*	*i*	*s*	*v*	*i*	*s*	*v*	*i*	*s*	K
Satd.	**Cond.**	2.6794	1391.58	6.4513	2.4001	1394.92	6.4096	2.1751	1397.94	6.3723	
225	–48.15	2.7076	1396.22	6.4720	2.4024	1395.35	6.4115	–	–	–	**225**
230	–43.15	2.7715	1406.72	6.5181	2.4595	1405.91	6.4579	2.2100	1405.11	6.4037	**230**
235	–38.15	2.8351	1417.22	6.5633	2.5164	1416.47	6.5033	2.2614	1415.73	6.4494	**235**
240	–33.15	2.8985	1427.70	6.6075	2.5730	1427.02	6.5478	2.3126	1426.33	6.4941	**240**
245	–28.15	2.9616	1438.21	6.6508	2.6294	1437.57	6.5912	2.3635	1436.93	6.5378	**245**
250	–23.15	3.0246	1448.70	6.6932	2.6855	1448.11	6.6338	2.4143	1447.51	6.5805	**250**
255	–18.15	3.0874	1459.19	6.7347	2.7415	1458.64	6.6756	2.4649	1458.09	6.6224	**255**
260	–13.15	3.1500	1469.69	6.7755	2.7974	1469.18	6.7165	2.5153	1468.67	6.6635	**260**
265	–8.15	3.2124	1480.19	6.8155	2.8530	1479.72	6.7566	2.5655	1479.24	6.7038	**265**
270	–3.15	3.2748	1490.70	6.8548	2.9086	1490.26	6.7960	2.6157	1489.82	6.7433	**270**
275	1.85	3.3370	1501.22	6.8934	2.9640	1500.81	6.8347	2.6657	1500.40	6.7822	**275**
280	6.85	3.3991	1511.75	6.9313	3.0194	1511.37	6.8728	2.7156	1510.98	6.8203	**280**
285	11.85	3.4611	1522.30	6.9687	3.0746	1521.94	6.9102	2.7654	1521.58	6.8578	**285**
290	16.85	3.5230	1532.87	7.0054	3.1298	1532.53	6.9471	2.8151	1532.20	6.8948	**290**
300	26.85	3.6467	1554.07	7.0773	3.2399	1553.77	7.0191	2.9144	1553.48	6.9669	**300**
310	36.85	3.7701	1575.37	7.1472	3.3497	1575.11	7.0890	3.0134	1574.85	7.0370	**310**
320	46.85	3.8933	1596.79	7.2152	3.4594	1596.55	7.1571	3.1122	1596.32	7.1051	**320**
330	56.85	4.0164	1618.34	7.2815	3.5689	1618.12	7.2235	3.2108	1617.91	7.1716	**330**
340	66.85	4.1393	1640.03	7.3462	3.6782	1639.83	7.2883	3.3094	1639.64	7.2364	**340**
350	76.85	4.2621	1661.87	7.4095	3.7875	1661.69	7.3517	3.4078	1661.52	7.2999	**350**
360	86.85	4.3848	1683.87	7.4715	3.8966	1683.71	7.4137	3.5061	1683.55	7.3619	**360**
370	96.85	4.5075	1706.04	7.5322	4.0058	1705.89	7.4744	3.6044	1705.74	7.4227	**370**
380	106.85	4.6301	1728.38	7.5918	4.1148	1728.24	7.5341	3.7025	1728.10	7.4824	**380**
390	116.85	4.7526	1750.90	7.6503	4.2238	1750.77	7.5926	3.8007	1750.64	7.5409	**390**

Temperature		Absolute Pressure (Saturation Temperature)									Temp.
T	*t*	0.60 bar (–43.28°C/229.87K)			0.70 bar (–40.45°C/232.70 K)			0.80 bar (–37.94°C/235.21K)			*T*
K	°C	*v*	*i*	*s*	*v*	*i*	*s*	*v*	*i*	*s*	K
Satd.	**Cond.**	1.8344	1403.20	6.3080	1.5883	1407.71	6.2537	1.4020	1411.64	6.2068	
235	–38.15	1.8788	1414.22	6.3554	1.6056	1412.71	6.2751	–	–	–	**235**
240	–33.15	1.9219	1424.94	6.4005	1.6428	1423.54	6.3207	1.4335	1422.13	6.2510	**240**
245	–28.15	1.9648	1435.64	6.4447	1.6799	1434.34	6.3653	1.4662	1433.04	6.2960	**245**
250	–23.15	2.0074	1446.32	6.4878	1.7167	1445.12	6.4088	1.4987	1443.91	6.3399	**250**
255	–18.15	2.0498	1456.99	6.5301	1.7534	1455.88	6.4514	1.5310	1454.76	6.3829	**255**
260	–13.15	2.0921	1467.64	6.5715	1.7899	1466.61	6.4931	1.5632	1465.58	6.4249	**260**
265	–8.15	2.1343	1478.29	6.6120	1.8262	1477.33	6.5340	1.5951	1476.37	6.4660	**265**
270	–3.15	2.1763	1488.93	6.6518	1.8624	1488.04	6.5740	1.6270	1487.15	6.5063	**270**
275	1.85	2.2182	1499.57	6.6909	1.8985	1498.75	6.6133	1.6587	1497.92	6.5458	**275**
280	6.85	2.2599	1510.22	6.7292	1.9344	1509.45	6.6519	1.6903	1508.67	6.5846	**280**
285	11.85	2.3016	1520.87	6.7669	1.9703	1520.15	6.6897	1.7218	1519.43	6.6226	**285**
290	16.85	2.3432	1531.53	6.8040	2.0061	1530.85	6.7270	1.7533	1530.18	6.6600	**290**
295	21.85	2.3847	1542.20	6.8405	2.0418	1541.57	6.7636	1.7846	1540.93	6.6968	**295**
300	26.85	2.4262	1552.89	6.8764	2.0775	1552.29	6.7997	1.8159	1551.70	6.7330	**300**
310	36.85	2.5089	1574.32	6.9467	2.1486	1573.80	6.8702	1.8783	1573.27	6.8037	**310**
320	46.85	2.5915	1595.85	7.0150	2.2195	1595.38	6.9387	1.9405	1594.91	6.8724	**320**
330	56.85	2.6738	1617.49	7.0816	2.2903	1617.07	7.0054	2.0026	1616.64	6.9393	**330**
340	66.85	2.7561	1639.26	7.1466	2.3609	1638.87	7.0705	2.0645	1638.49	7.0045	**340**
350	76.85	2.8382	1661.17	7.2101	2.4314	1660.81	7.1341	2.1263	1660.46	7.0682	**350**
360	86.85	2.9203	1683.22	7.2723	2.5018	1682.90	7.1963	2.1880	1682.58	7.1305	**360**
370	96.85	3.0023	1705.44	7.3331	2.5722	1705.14	7.2573	2.2496	1704.84	7.1915	**370**
380	106.85	3.0842	1727.82	7.3928	2.6425	1727.55	7.3170	2.3112	1727.27	7.2513	**380**
390	116.85	3.1660	1750.38	7.4514	2.7127	1750.12	7.3757	2.3727	1749.86	7.3100	**390**
400	126.85	3.2479	1773.12	7.5090	2.7829	1772.87	7.4333	2.4342	1772.63	7.3676	**400**

(Contd.)

TABLE: 7.3 Thermodynamic Properties of Ammonia (Refrigerant R–717) (Superheat Region) [Contd.]

Temperature		Absolute Pressure (Saturation Temperature)									Temp.
T	*t*	0.90 bar (–35.67°C/237.48K)			1.00 bar (–33.60°C/239.55K)			1.20 bar (–29.91°C/243.24K)			*T*
K	ºC	*v*	*i*	*s*	*v*	*i*	*s*	*v*	*i*	*s*	K
Satd.	Cond.	1.2558	1415.13	6.1655	1.1380	1418.27	6.1286	0.9596	1423.74	6.0647	
240	–33.15	1.2707	1420.70	6.1588	1.1404	1419.27	6.1327	–	–	–	240
245	–28.15	1.3000	1431.72	6.2343	1.1670	1430.40	6.1786	0.9675	1427.73	6.0811	245
250	–23.15	1.3291	1442.70	6.2786	1.1935	1441.48	6.2234	0.9899	1439.01	6.1267	250
255	–18.15	1.3581	1453.64	6.3219	1.2197	1452.51	6.2671	1.0121	1450.23	6.1711	255
260	–13.15	1.3868	1464.54	6.3643	1.2457	1463.49	6.3097	1.0341	1461.39	6.2144	260
265	–8.15	1.4154	1475.41	6.4057	1.2716	1474.44	6.3514	1.0559	1472.49	6.2567	265
270	–3.15	1.4439	1486.26	6.4462	1.2974	1485.36	6.3922	1.0776	1483.55	6.2981	270
275	1.85	1.4722	1497.08	6.4860	1.3230	1496.25	6.4322	1.0992	1494.56	6.3385	275
280	6.85	1.5005	1507.90	6.5249	1.3485	1507.12	6.4714	1.1207	1505.55	6.3781	280
285	11.85	1.5286	1518.70	6.5632	1.3740	1517.97	6.5098	1.1420	1516.51	6.4169	285
290	16.85	1.5566	1529.50	6.6007	1.3993	1528.82	6.5475	1.1633	1527.46	6.4550	290
295	21.85	1.5846	1540.30	6.6377	1.4246	1539.66	6.5846	1.1845	1538.39	6.4923	295
300	26.85	1.6125	1551.10	6.6740	1.4498	1550.51	6.6211	1.2056	1549.31	6.5290	300
305	31.85	1.6403	1561.92	6.7097	1.4749	1561.36	6.6569	1.2267	1560.23	6.5651	305
310	36.85	1.6681	1572.74	6.7449	1.5000	1572.21	6.6922	1.2477	1571.15	6.6007	310
320	46.85	1.7236	1594.44	6.8138	1.5500	1593.96	6.7613	1.2896	1593.02	6.6701	320
330	56.85	1.7788	1616.22	6.8808	1.5998	1615.79	6.8285	1.3313	1614.94	6.7375	330
340	66.85	1.8339	1638.10	6.9462	1.6495	1637.72	6.8939	1.3729	1636.95	6.8032	340
350	76.85	1.8890	1660.11	7.0100	1.6997	1659.76	6.9578	1.4143	1659.05	6.8673	350
360	86.85	1.9439	1682.25	7.0723	1.7486	1681.93	7.0203	1.4557	1681.28	6.9299	360
370	96.85	1.9988	1704.54	7.1334	1.7981	1704.24	7.0814	1.4970	1703.65	6.9912	370
380	106.85	2.0536	1726.99	7.1933	1.8474	1726.71	7.1413	1.5383	1726.16	7.0512	380
390	116.85	2.1083	1749.60	7.2520	1.8968	1749.34	7.2001	1.5794	1748.82	7.1101	390
400	126.85	2.1630	1772.39	7.3097	1.9460	1772.14	7.2578	1.6206	1771.66	7.1679	400

Temperature		Absolute Pressure (Saturation Temperature)									Temp.
T	*t*	1.40 bar (–26.69°C/246.46K)			1.60 bar (–23.83°C/249.32K)			1.80 bar (–21.23°C/251.92K)			*T*
K	ºC	*v*	*i*	*s*	*v*	*i*	*s*	*v*	*i*	*s*	K
Satd.	Cond.	0.83063	1428.38	6.0108	0.73297	1432.42	5.9642	0.65635	1435.98	5.9231	
250	–23.15	0.84443	1436.52	6.0436	0.73530	1433.99	5.9705	–	–	–	250
255	–18.15	0.86373	1447.93	6.0888	0.75245	1445.60	6.0165	0.66586	1443.24	5.9518	255
260	–13.15	0.88285	1459.26	6.1328	0.76941	1457.11	6.0612	0.68115	1454.94	5.9972	260
265	–8.15	0.90181	1470.52	6.1757	0.78620	1468.53	6.1047	0.69627	1466.53	6.0413	265
270	–3.15	0.92062	1481.72	6.2176	0.80285	1479.88	6.1471	0.71123	1478.02	6.0843	270
275	1.85	0.93930	1492.87	6.2585	0.81937	1491.16	6.1885	0.72607	1489.44	6.1262	275
280	6.85	0.95787	1503.97	6.2985	0.83577	1502.39	6.2290	0.74078	1500.79	6.1671	280
285	11.85	0.97634	1515.04	6.3377	0.85207	1513.56	6.2685	0.75540	1512.07	6.2070	285
290	16.85	0.99472	1526.08	6.3761	0.86827	1524.70	6.3073	0.76991	1523.31	6.2461	290
295	21.85	1.0130	1537.10	6.4138	0.88440	1535.81	6.3453	0.78435	1534.51	6.2844	295
300	26.85	1.0313	1548.10	6.4508	0.90045	1546.89	6.3825	0.79871	1545.67	6.3219	300
305	31.85	1.0494	1559.10	6.4871	0.91644	1557.96	6.4191	0.81301	1556.81	6.3588	305
310	36.85	1.0675	1570.08	6.5228	0.93237	1569.01	6.4550	0.82725	1567.94	6.3949	310
320	46.85	1.1036	1592.07	6.5926	0.96408	1591.11	6.5252	0.85558	1590.15	6.4655	320
330	56.85	1.1395	1614.08	6.6604	0.99563	1613.23	6.5933	0.88374	1612.36	6.5338	330
340	66.85	1.1753	1636.17	6.7263	1.0270	1635.39	6.6594	0.91176	1634.61	6.6002	340
350	76.85	1.2109	1658.35	6.7906	1.0583	1657.64	6.7239	0.93967	1656.92	6.6649	350
360	86.85	1.2465	1680.63	6.8534	1.0896	1679.98	6.7869	0.96750	1679.33	6.7280	360
370	96.85	1.2820	1703.04	6.9148	1.1207	1702.44	6.8484	0.99524	1701.84	6.7897	370
380	106.85	1.3174	1725.60	6.9749	1.1518	1725.04	6.9087	1.0229	1724.48	6.8501	380
390	116.85	1.3528	1748.30	7.0339	1.1828	1747.78	6.9677	1.0506	1747.26	6.9093	390
400	126.85	1.3881	1771.17	7.0918	1.2138	1770.68	7.0257	1.0781	1770.20	6.9673	400
410	136.85	1.4234	1794.21	7.1487	1.2447	1793.75	7.0827	1.1057	1793.29	7.0244	410
420	146.85	1.4587	1817.42	7.2046	1.2756	1816.98	7.1387	1.1332	1816.55	7.0804	420

(Contd.)

TABLE: 7.3 Thermodynamic Properties of Ammonia (Refrigerant R–717) (Superheat Region) [Contd.]

Temperature		Absolute Pressure (Saturation Temperature)									Temp.
T	*t*	2.0 bar (–18.86°C/254.29K)			2.50 bar (–13.66°C/259.49K)			3.0 bar (–9.24°C/263.91K)			*T*
K	ºC	*v*	*i*	*s*	*v*	*i*	*s*	*v*	*i*	*s*	K
Satd.	**Cond.**	0.59456	1439.17	5.8863	0.48211	1445.89	5.8085	0.40605	1451.33	5.7449	
265	–8.15	0.62429	1464.50	5.9839	0.49467	1459.35	5.8598	0.40816	1454.06	5.7552	**265**
270	–3.15	0.63792	1476.15	6.0275	0.50590	1471.39	5.9049	0.41780	1466.53	5.8018	**270**
275	1.85	0.65141	1487.70	6.0699	0.51698	1483.30	5.9486	0.42729	1478.81	5.8469	**275**
280	6.85	0.66478	1499.17	6.1112	0.52794	1495.09	5.9911	0.43665	1490.93	5.8906	**280**
285	11.85	0.67805	1510.57	6.1516	0.53878	1506.78	6.0324	0.44589	1502.92	5.9330	**285**
290	16.85	0.69122	1521.91	6.1910	0.54953	1518.38	6.0728	0.45503	1514.79	5.9743	**290**
295	21.85	0.70430	1533.20	6.2296	0.56019	1529.90	6.1122	0.46408	1526.55	6.0145	**295**
300	26.85	0.71732	1544.45	6.2674	0.57078	1541.36	6.1507	0.47305	1538.23	6.0538	**300**
305	31.85	0.73026	1555.67	6.3045	0.58129	1552.77	6.1884	0.48195	1549.84	6.0921	**305**
310	36.85	0.74315	1566.86	6.3409	0.59175	1564.13	6.2254	0.49079	1561.38	6.1297	**310**
315	41.85	0.75598	1578.03	6.3767	0.60215	1575.46	6.2616	0.49957	1572.87	6.1664	**315**
320	46.85	0.76877	1589.19	6.4118	0.61250	1586.77	6.2972	0.50831	1584.32	6.2025	**320**
325	51.85	0.78152	1600.34	6.4464	0.62281	1598.05	6.3322	0.51700	1595.74	6.2379	**325**
330	56.85	0.79422	1611.50	6.4805	0.63308	1609.33	6.3667	0.52565	1607.14	6.2727	**330**
340	66.85	0.81954	1633.83	6.5471	0.65353	1631.87	6.4339	0.54284	1629.89	6.3406	**340**
350	76.85	0.84474	1656.21	6.6120	0.67385	1654.12	6.4993	0.55992	1652.62	6.4065	**350**
360	86.85	0.86985	1678.67	6.6753	0.69409	1677.03	6.5630	0.57691	1675.38	6.4707	**360**
370	96.85	0.89489	1701.24	6.7371	0.71425	1699.72	6.6252	0.59381	1698.20	6.5332	**370**
380	106.85	0.91986	1723.92	6.7976	0.73434	1722.52	6.6860	0.61066	1721.11	6.5943	**380**
390	116.85	0.94478	1746.74	6.8569	0.75438	1745.43	6.7455	0.62744	1744.12	6.6540	**390**
400	126.85	0.96969	1769.71	6.9150	0.77437	1768.48	6.8038	0.64418	1767.26	6.7126	**400**
410	136.85	0.99449	1792.83	6.9721	0.79433	1791.68	6.8611	0.66088	1790.53	6.7701	**410**
420	146.85	1.0193	1816.12	7.0282	0.81425	1815.04	6.9174	0.67755	1813.96	6.8265	**420**
430	156.85	1.0441	1839.58	7.0834	0.83413	1838.56	6.9728	0.69418	1837.54	6.8820	**430**

Temperature		Absolute Pressure (Saturation Temperature)									Temp.
T	*t*	4.0 bar (–1.89°C/271.26K)			5.0 bar (4.13°C/277.29K)			6.0 bar (9.28°C/282.43K)			*T*
K	ºC	*v*	*i*	*s*	*v*	*i*	*s*	*v*	*i*	*s*	K
Satd.	**Cond.**	0.30940	1459.76	5.6443	0.25033	1466.06	5.5660	0.21037	1470.99	5.5017	
280	6.85	0.32240	1482.36	5.7264	0.25370	1473.43	5.5925	–	–	–	**280**
285	11.85	0.32966	1494.99	5.7710	0.25980	1486.75	5.6396	0.21310	1478.18	5.5271	**285**
290	16.85	0.33681	1507.42	5.8143	0.26577	1499.81	5.6851	0.21831	1491.91	5.5749	**290**
295	21.85	0.34386	1519.70	5.8563	0.27164	1512.63	5.7289	0.22340	1505.34	5.6208	**295**
300	26.85	0.35083	1531.84	5.8971	0.27741	1525.26	5.7714	0.22840	1518.49	5.6650	**300**
305	31.85	0.35772	1543.86	5.9368	0.28311	1537.72	5.8125	0.23331	1531.42	5.7077	**305**
310	36.85	0.36454	1555.77	5.9756	0.28874	1550.03	5.8526	0.23815	1544.15	5.7491	**310**
315	41.85	0.37131	1567.60	6.0134	0.29430	1562.22	5.8916	0.24292	1556.71	5.7893	**315**
320	46.85	0.37802	1579.36	6.0504	0.29981	1574.29	5.9296	0.24763	1569.12	5.8284	**320**
325	51.85	0.38469	1591.05	6.0867	0.30527	1586.28	5.9668	0.25229	1581.41	5.8665	**325**
330	56.85	0.39132	1602.70	6.1223	0.31069	1598.19	6.0031	0.25691	1593.60	5.9037	**330**
335	61.85	0.39791	1614.31	6.1572	0.31607	1610.03	6.0388	0.26148	1605.69	5.9401	**335**
340	66.85	0.40446	1625.89	6.1915	0.32141	1621.83	6.0737	0.26602	1617.71	5.9757	**340**
345	71.85	0.41099	1637.45	6.2253	0.32673	1633.59	6.1080	0.27053	1629.67	6.0106	**345**
350	76.85	0.41749	1648.99	6.2585	0.33201	1645.31	6.1418	0.27501	1641.59	6.0449	**350**
360	86.85	0.43042	1672.06	6.3235	0.34251	1668.70	6.2077	0.28389	1665.30	6.1117	**360**
370	96.85	0.44326	1695.14	6.3867	0.35292	1692.05	6.2717	0.29268	1688.93	6.1765	**370**
380	106.85	0.45604	1718.27	6.4484	0.36326	1715.41	6.3340	0.30140	1712.53	6.2394	**380**
390	116.85	0.46877	1741.48	6.5087	0.37355	1738.82	6.3948	0.31007	1736.15	6.3008	**390**
400	126.85	0.48144	1764.79	6.5677	0.38379	1762.31	6.4542	0.31868	1759.82	6.3607	**400**
410	136.85	0.49408	1788.22	6.6255	0.39399	1785.90	6.5125	0.32726	1782.96	6.4193	**410**
420	146.85	0.50667	1811.78	6.6823	0.40415	1809.60	6.5696	0.33579	1807.41	6.4768	**420**
430	156.85	0.51924	1835.50	6.7381	0.41427	1833.44	6.6257	0.34429	1831.38	6.5332	**430**
440	166.85	0.53178	1859.36	6.7930	424.37000	1857.42	6.6808	0.35277	1855.48	6.5886	**440**

(Contd.)

TABLE: 7.3 Thermodynamic Properties of Ammonia (Refrigerant R–717) (Superheat Region) [Contd.]

Temperature		Absolute Pressure (Saturation Temperature)									Temp.
T	*t*	8.0 bar (17.85°C/291.00K)			10.0 bar (24.90°C/298.05K)			12.0 bar (30.94°C/304.09K)			*T*
K	°C	*v*	*i*	*s*	*v*	*i*	*s*	*v*	*i*	*s*	K
Satd.	**Cond.**	0.15957	1478.17	5.3994	0.12852	1483.04	5.3189	0.10750	1486.39	5.2520	
295	21.85	0.16290	1489.96	5.4396	–	–	–	–	–	–	**295**
300	26.85	0.16696	1504.30	5.4878	0.12988	1489.11	5.3391	–	–	–	**300**
305	31.85	0.17092	1518.26	5.5340	0.13330	1504.29	5.3893	0.10805	1489.35	5.2617	**305**
310	36.85	0.17479	1531.92	5.5784	0.13662	1518.99	5.4372	0.11104	1505.28	5.3135	**310**
315	41.85	0.17858	1545.29	5.6212	0.13986	1533.30	5.4829	0.11392	1520.65	5.3627	**315**
320	46.85	0.18232	1558.44	5.6626	0.14302	1547.27	5.5269	0.11672	1535.54	5.4096	**320**
325	51.85	0.18599	1571.38	5.7027	0.14612	1560.94	5.5693	0.11945	1550.02	5.4545	**325**
330	56.85	0.18961	1584.16	5.7418	0.14916	1574.36	5.6103	0.12212	1564.16	5.4977	**330**
335	61.85	0.19319	1596.78	5.7797	0.15215	1587.57	5.6500	0.12473	1578.01	5.5393	**335**
340	66.85	0.19673	1609.29	5.8168	0.15510	1600.59	5.6886	0.12730	1591.60	5.5796	**340**
345	71.85	0.20024	1621.68	5.8530	0.15802	1613.46	5.7262	0.12982	1604.98	5.6187	**345**
350	76.85	0.20371	1633.99	5.8884	0.16090	1626.19	5.7628	0.13231	1618.17	5.6566	**350**
355	81.85	0.20716	1646.22	5.9231	0.16374	1638.81	5.7986	0.13477	1631.21	5.6936	**355**
360	86.85	0.21058	1658.40	5.9571	0.16657	1651.34	5.8337	0.13719	1644.11	5.7297	**360**
370	96.85	0.21736	1682.61	6.0235	0.17214	1676.17	5.9017	0.14197	1669.59	5.7995	**370**
380	106.85	0.22406	1706.71	6.0877	0.17763	1700.78	5.9674	0.14667	1694.76	5.8667	**380**
390	116.85	0.23070	1730.75	6.1502	0.18306	1725.27	6.0310	0.15129	1719.72	5.9315	**390**
400	126.85	0.23729	1754.79	6.2111	0.18844	1749.70	6.0928	0.15586	1744.54	5.9943	**400**
410	136.85	0.24383	1778.86	6.2705	0.19377	1774.11	6.1531	0.16038	1769.31	6.0555	**410**
420	146.85	0.25034	1803.00	6.3287	0.19906	1798.55	6.2120	0.16486	1794.06	6.1151	**420**
430	156.85	0.25681	1827.23	6.3857	0.20431	1823.05	6.2696	0.16931	1818.84	6.1734	**430**
440	166.85	0.26325	1851.57	6.4416	0.20954	1847.63	6.3261	0.17372	1843.67	6.2305	**440**
450	176.85	0.26967	1876.04	6.4966	0.21474	1872.33	6.3816	0.17811	1868.59	6.2865	**450**
460	186.85	0.27606	1900.64	6.5507	0.21991	1897.14	6.4362	0.18247	1893.62	6.3415	**460**

Temperature		Absolute Pressure (Saturation Temperature)									Temp.
T	*t*	14.0 bar (36.26°C/309.41K)			16.0 bar (41.04°C/314.19K)			18.0 bar (45.38°C/318.53K)			*T*
K	°C	*v*	*i*	*s*	*v*	*i*	*s*	*v*	*i*	*s*	K
Satd.	**Cond.**	0.092301	1488.65	5.1943	0.080780	1490.08	5.1437	0.071736	1490.86	5.0981	
310	36.85	0.092619	1490.64	5.2009	–	–	–	–	–	–	**310**
315	41.85	0.095281	1507.24	5.2540	0.081182	1492.95	5.1528	–	–	–	**315**
320	46.85	0.097844	1523.19	5.3042	0.083590	1510.12	5.2069	0.072404	1496.23	5.1149	**320**
325	51.85	0.100326	1538.59	5.3520	0.085902	1526.57	5.2579	0.074605	1513.88	5.1696	**325**
330	56.85	0.102737	1553.53	5.3976	0.088135	1542.41	5.3063	0.076714	1530.75	5.2211	**330**
335	61.85	0.105088	1568.08	5.4414	0.090301	1557.75	5.3524	0.078746	1546.96	5.2699	**335**
340	66.85	0.107388	1582.30	5.4835	0.092409	1572.65	5.3966	0.080714	1562.63	5.3163	**340**
345	71.85	0.109641	1596.23	5.5242	0.094467	1587.19	5.4390	0.082627	1577.83	5.3607	**345**
350	76.85	0.111856	1609.92	5.5636	0.096482	1601.42	5.4800	0.084492	1592.65	5.4033	**350**
355	81.85	0.114035	1623.40	5.6018	0.098459	1615.38	5.5196	0.086316	1607.13	5.4444	**355**
360	86.85	0.116183	1636.71	5.6391	0.100402	1629.12	5.5580	0.088104	1621.33	5.4842	**360**
365	91.85	0.118304	1649.86	5.6754	0.102316	1642.66	5.5954	0.089861	1635.29	5.5227	**365**
370	96.85	0.120400	1662.89	5.7108	0.104204	1656.04	5.6318	0.091589	1649.05	5.5601	**370**
375	101.85	0.122474	1675.81	5.7455	0.106068	1669.28	5.6673	0.093293	1662.63	5.5965	**375**
380	106.85	0.124528	1688.64	5.7795	0.107912	1682.40	5.7021	0.094974	1676.06	5.6321	**380**
390	116.85	0.128585	1714.08	5.8456	0.111543	1708.36	5.7695	0.098279	1702.56	5.7010	**390**
400	126.85	0.132582	1739.33	5.9095	0.115113	1734.05	5.8345	0.101518	1728.70	5.7671	**400**
410	136.85	0.136529	1764.46	5.9715	0.118631	1759.55	5.8975	0.104704	1754.59	5.8311	**410**
420	146.85	0.140433	1789.53	6.0319	0.122105	1784.95	5.9587	0.107844	1780.34	5.8931	**420**
430	156.85	0.144302	1814.59	6.0909	0.125541	1810.31	6.0184	0.110946	1805.99	5.9535	**430**
440	166.85	0.148138	1839.68	6.1486	0.128945	1835.66	6.0767	0.114014	1831.62	6.0124	**440**
450	176.85	0.151946	1864.84	6.2051	0.132320	1861.06	6.1338	0.117052	1857.25	6.0700	**450**
460	186.85	0.155729	1890.08	6.2606	0.135669	1886.51	6.1897	0.120065	1882.93	6.1265	**460**
470	196.85	0.159490	1915.42	6.3151	0.138996	1912.06	6.2446	0.123054	1908.68	6.1818	**470**

(Contd.)

TABLE: 7.3 Thermodynamic Properties of Ammonia (Refrigerant R–717) (Superheat Region) [Contd.]

Temperature		Absolute Pressure (Saturation Temperature)									Temp.
T	*t*	20.0 bar (49.37°C/322.52K)			22.0 bar (53.07°C/326.22K)			24.0 bar (56.53°C/329.68K)			*T*
K	°C	*v*	*i*	*s*	*v*	*i*	*s*	*v*	*i*	*s*	K
Satd.	Cond.	0.064441	1491.11	5.0564	0.058427	1490.91	5.0180	0.053381	1490.33	4.9822	
325	51.85	0.065484	1500.42	5.0852	–	–	–	–	–	–	
330	56.85	0.067510	1518.46	5.1403	0.059910	1505.46	5.0623	0.053503	1491.63	4.9861	330
335	61.85	0.069448	1535.67	5.1920	0.061785	1523.80	5.1175	0.055341	1511.29	5.0452	335
340	66.85	0.071313	1552.18	5.2410	0.063577	1541.28	5.1693	0.057083	1529.85	5.1002	340
345	71.85	0.073117	1568.13	5.2875	0.065300	1558.04	5.2182	0.058747	1547.52	5.1518	345
350	76.85	0.074869	1583.58	5.3320	0.066964	1574.20	5.2647	0.060346	1564.47	5.2006	350
355	81.85	0.076576	1598.63	5.3747	0.068580	1589.87	5.3092	0.061891	1580.81	5.2470	355
360	86.85	0.078243	1613.34	5.4158	0.070153	1605.11	5.3518	0.063389	1596.64	5.2913	360
365	91.85	0.079877	1627.74	5.4556	0.071689	1620.00	5.3929	0.064848	1612.04	5.3337	365
370	96.85	0.081480	1641.90	5.4941	0.073193	1634.58	5.4326	0.066272	1627.08	5.3747	370
375	101.85	0.083057	1655.84	5.5315	0.074669	1648.90	5.4710	0.067666	1641.82	5.4142	375
380	106.85	0.084611	1669.59	5.5679	0.076120	1663.00	5.5084	0.069033	1656.28	5.4525	380
385	111.85	0.086143	1683.19	5.6035	0.077549	1676.91	5.5447	0.070376	1670.52	5.4898	385
390	116.85	0.087657	1696.66	5.6383	0.078957	1690.67	5.5802	0.071699	1684.58	5.5260	390
400	126.85	0.090635	1723.28	5.7056	0.081723	1717.78	5.6489	0.074290	1712.21	5.5960	400
420	146.85	0.096431	1775.68	5.8335	0.087088	1770.97	5.7787	0.079299	1766.22	5.7278	420
440	166.85	0.102066	1827.54	5.9541	0.092287	1823.44	5.9007	0.084136	1819.30	5.8513	440
460	186.85	0.107579	1879.33	6.0692	0.097362	1875.71	6.0169	0.088846	1872.07	5.9685	460
480	206.85	0.112998	1931.32	6.1799	0.102340	1928.10	6.1284	0.093457	1924.87	6.0809	480
500	226.85	0.118337	1983.69	6.2867	0.107238	1980.82	6.2360	0.097989	1977.95	6.1892	500
520	246.85	0.123611	2036.53	6.3904	0.112070	2033.98	6.3402	0.102452	2031.43	6.2941	520
540	266.85	0.128826	2089.92	6.4911	0.116844	2087.66	6.4415	0.106858	2085.39	6.3959	540
560	286.85	0.133991	2143.88	6.5892	0.121366	2141.89	6.5401	0.111212	2139.89	6.4950	560
580	306.85	0.139109	2198.45	6.6850	0.126242	2196.70	6.6363	0.115520	2194.95	6.5916	580

Temperature		Absolute Pressure (Saturation Temperature)									Temp.
T	*t*	26.0 bar (59.78°C/332.93K)			28.0 bar (62.84°C/335.99K)			30.0 bar (65.75°C/338.90K)			*T*
K	°C	*v*	*i*	*s*	*v*	*i*	*s*	*v*	*i*	*s*	K
Satd.	Cond.	0.049084	1489.41	4.9485	0.045379	1488.20	4.9167	0.042149	1486.73	4.8864	
335	61.85	0.049826	1498.02	4.9743	–	–	–	–	–	–	335
340	66.85	0.051540	1517.83	5.0330	0.046736	1505.13	4.9667	0.042517	1491.63	4.9008	340
345	71.85	0.053163	1536.53	5.0876	0.048337	1525.01	5.0248	0.044111	1512.89	4.9629	345
350	76.85	0.054715	1554.35	5.1389	0.049855	1543.81	5.0789	0.045609	1532.78	5.0201	350
360	86.85	0.057645	1587.91	5.2334	0.052699	1578.89	5.1777	0.048390	1569.55	5.1238	360
380	106.85	0.063024	1649.42	5.3998	0.057863	1642.42	5.3496	0.053379	1635.26	5.3015	380
400	126.85	0.067994	1706.57	5.5464	0.062591	1700.83	5.4994	0.057902	1695.02	5.4548	400
420	146.85	0.072703	1761.42	5.6802	0.067047	1756.57	5.6354	0.062140	1751.66	5.5930	420
440	166.85	0.077236	1815.13	5.8051	0.071319	1810.94	5.7619	0.066189	1806.71	5.7210	440
460	186.85	0.081638	1868.40	5.9235	0.075459	1864.72	5.8814	0.070102	1861.01	5.8417	460
480	206.85	0.085940	1921.63	6.0368	0.079496	1918.36	5.9956	0.073910	1915.08	5.9568	480
500	226.85	0.090161	1975.06	6.1459	0.083451	1972.16	6.1054	0.077636	1969.25	6.0674	500
520	246.85	0.094314	2028.86	6.2514	0.087338	2026.28	6.2115	0.081292	2023.69	6.1741	520
540	266.85	0.098409	2083.11	6.3537	0.091166	2080.82	6.3144	0.084889	2078.52	6.2776	540
560	286.85	0.102451	2137.88	6.4533	0.094943	2135.86	6.4145	0.088435	2133.83	6.3781	560
580	306.85	0.106448	2193.19	6.5504	0.098673	2191.42	6.5120	0.091934	2189.64	6.4761	580
600	326.85	0.110403	2249.06	6.6451	0.102361	2247.53	6.6071	0.095392	2245.98	6.5716	600
620	346.85	0.114321	2305.51	6.7376	0.106012	2304.19	6.7000	0.098812	2302.86	6.6648	620
640	366.85	0.118203	2362.53	6.8281	0.109628	2361.42	6.7908	0.102198	2360.30	6.7560	640
660	386.85	0.112055	2420.12	6.9167	0.113213	2419.21	6.8797	0.105552	2418.29	6.8452	660
680	406.85	0.125877	2478.29	7.0036	0.116769	2477.56	6.9668	0.108877	2476.83	6.9326	680
700	426.85	0.129672	2537.03	7.0887	0.120298	2536.47	7.0522	0.112176	2535.91	7.0182	700
720	446.85	0.133443	2596.33	7.1722	0.123803	2595.94	7.1360	0.115450	2595.55	7.1022	720
740	466.85	0.137192	2656.19	7.2542	0.127285	2655.96	7.2182	0.118702	2655.72	7.1847	740

(Contd.)

TABLE: 7.3 Thermodynamic Properties of Ammonia (Refrigerant R–717) (Superheat Region) [Contd.]

Temperature		Absolute Pressure (Saturation Temperature)									Temp.
T	*t*	35.0 bar (72.42°C/345.57K)			40.0 bar (78.42°C/351.57K)			45.0 bar (83.88°C/357.03K)			*T*
K	°C	*v*	*i*	*s*	*v*	*i*	*s*	*v*	*i*	*s*	°C
Satd.	Cond.	0.035635	1482.03	4.8161	0.030688	1476.08	4.7516	0.026795	1469.03	4.6912	
350	76.85	0.036963	1502.69	4.8755	–	–	–	–	–	–	350
355	81.85	0.038362	1524.36	4.9370	0.031666	1493.47	4.8008	–	–	–	355
360	86.85	0.039676	1544.65	4.9937	0.032992	1516.93	4.8664	0.027611	1485.35	4.7368	360
370	96.85	0.042114	1582.12	5.0964	0.035386	1559.04	4.9819	0.030047	1533.73	4.8693	370
380	106.85	0.044366	1616.63	5.1884	0.037543	1596.80	5.0826	0.032168	1575.54	4.9809	380
390	116.85	0.046484	1649.09	5.2728	0.039538	1631.67	5.1731	0.034090	1613.27	5.0789	390
400	126.85	0.048500	1680.08	5.3512	0.041416	1664.53	5.2563	0.035874	1648.28	5.1675	400
420	146.85	0.052314	1739.18	5.4954	0.044925	1726.34	5.4072	0.039161	1713.11	5.3258	420
440	166.85	0.055921	1795.98	5.6276	0.048208	1785.04	5.5437	0.042198	1773.87	5.4671	440
460	186.85	0.059382	1851.64	5.7513	0.051335	1842.13	5.6706	0.045070	1832.47	5.5974	460
480	206.85	0.062735	1906.82	5.8687	0.054349	1898.45	5.7905	0.047823	1889.98	5.7198	480
500	226.85	0.066002	1961.91	5.9812	0.057274	1954.50	5.9049	0.050484	1947.01	5.8362	500
520	246.85	0.069199	2017.17	6.0895	0.060128	2010.59	6.0149	0.053071	2003.94	5.9478	520
540	266.85	0.072336	2072.74	6.1944	0.062920	2066.90	6.1211	0.055597	2061.00	6.0555	540
560	286.85	0.075420	2128.71	6.2962	0.065660	2123.55	6.2241	0.058070	2118.33	6.1597	560
580	306.85	0.078459	2185.15	6.3952	0.068354	2180.60	6.3243	0.060496	2176.01	6.2609	580
600	326.85	0.081456	2242.07	6.4917	0.071006	2238.11	6.4217	0.062881	2234.09	6.3594	600
620	346.85	0.084415	2299.51	6.5858	0.073621	2296.09	6.5168	0.065228	2292.62	6.4553	620
640	366.85	0.087340	2357.46	6.6778	0.076201	2354.56	6.6096	0.067541	2351.60	6.5490	640
660	386.85	0.090234	2415.93	6.7678	0.078750	2413.52	6.7003	0.069823	2411.05	6.6404	660
680	406.85	0.093098	2474.94	6.8559	0.081271	2472.98	6.7891	0.072077	2470.97	6.7299	680
700	426.85	0.095937	2534.46	6.9421	0.083765	2532.95	6.8760	0.074304	2531.37	6.8174	700
720	446.85	0.098751	2594.51	7.0267	0.086235	2593.41	6.9611	0.076507	2592.25	6.9031	720
740	466.85	0.101542	2655.08	7.1097	0.088682	2654.37	7.0447	0.078688	2653.60	6.9872	740

Temperature		Absolute Pressure (Saturation Temperature)									Temp.
T	*t*	50.0 bar (88.90°C/362.05K)			55.0 bar (93.56°C/366.71K)			60.0 bar (97.90°C/371.05K)			*T*
K	°C	*v*	*i*	*s*	*v*	*i*	*s*	*v*	*i*	*s*	K
Satd.	Cond.	0.023644	1460.98	4.6338	0.021034	1451.95	4.5785	0.018831	1441.95	4.5244	
365	91.85	0.024431	1478.49	4.6820	–	–	–	–	–	–	365
370	96.85	0.025650	1505.46	4.7554	0.021894	1473.01	4.6356	–	–	–	370
375	101.85	0.026762	1529.91	4.8211	0.023069	1501.59	4.7124	0.019844	1469.06	4.5970	375
380	106.85	0.027794	1552.51	4.8809	0.024130	1527.27	4.7804	0.020975	1499.08	4.6766	380
385	111.85	0.028764	1573.69	4.9363	0.025110	1550.84	4.8420	0.021989	1525.83	4.7465	385
390	116.85	0.029683	1593.73	4.9880	0.026026	1572.83	4.8988	0.022921	1550.26	4.8096	390
400	126.85	0.031407	1631.24	5.0830	0.027718	1613.30	5.0013	0.024607	1594.31	4.9211	400
420	146.85	0.034532	1699.46	5.2495	0.030727	1685.36	5.1772	0.027539	1670.76	5.1077	420
440	166.85	0.037381	1762.45	5.3960	0.033430	1750.77	5.3293	0.030128	1738.81	5.2661	440
460	186.85	0.040052	1822.65	5.5299	0.035941	1812.67	5.4669	0.032509	1802.52	5.4077	460
480	206.85	0.042598	1881.40	5.6549	0.038320	1872.71	5.5947	0.034751	1863.91	5.5384	480
500	226.85	0.045049	1939.44	5.7734	0.040600	1931.79	5.7153	0.036891	1924.06	5.6611	500
520	246.85	0.047424	1997.23	5.8867	0.042804	1990.46	5.8303	0.038952	1983.62	5.7779	520
540	266.85	0.049738	2055.05	5.9958	0.044943	2049.04	5.9409	0.040948	2042.98	5.8899	540
560	286.85	0.051997	2113.05	6.1013	0.047030	2107.73	6.0476	0.042890	2102.36	5.9979	560
580	306.85	0.054210	2171.36	6.2036	0.049069	2166.66	6.1510	0.044785	2161.92	6.1024	580
600	326.85	0.056382	2230.03	6.3030	0.051066	2225.91	6.2514	0.046638	2221.74	6.2038	600
620	346.85	0.058516	2289.09	6.3998	0.053027	2285.52	6.3492	0.048454	2281.89	6.3025	620
640	366.85	0.060616	2348.59	6.4943	0.054953	2345.52	6.4444	0.050236	2342.40	6.3985	640
660	386.85	0.062686	2408.52	6.5865	0.056849	2405.93	6.5374	0.051987	2403.30	6.4922	660
680	406.85	0.064726	2468.90	6.6766	0.058716	2466.77	6.6282	0.053711	2464.59	6.5837	680
700	426.85	0.066741	2529.73	6.7648	0.060557	2528.03	6.7170	0.055408	2526.28	6.6731	700
720	446.85	0.068731	2591.02	6.8511	0.062374	2589.73	6.8039	0.057081	2588.38	6.7606	720
740	466.85	0.070700	2652.76	6.9357	0.064169	2651.86	6.8890	0.058733	2650.90	6.8462	740

TABLE: 8.1 Thermodynamic Properties of Saturated Nitrogen (Refrigerant R–728) up to 1 Atmosphere (Temperature Basis)

Temperature		Saturation pressure, p		Liquid density	Specific volume of vapour v''	Specific enthalpy			Specific entropy		Temp
t	T			ρ_l		h	λ_v	H	s'	s''	t
°C	K	bar	torr	kg/m³	m³/kg	kJ/kg			kJ/(kg·K)		°C
–209.999[a]	**63.151**	**0.1252**	**93.93**	**867.21**	**1.4831**	**–150.88**	**215.66**	**64.78**	**2.4257**	**5.8383**	**–209.999**
–209	64.15	0.1498	112.34	863.12	1.2574	–148.73	214.33	65.59	2.4531	5.8188	**–209**
–208	65.15	0.1786	133.95	858.95	1.0714	–146.73	213.24	66.51	2.4830	5.7856	**–208**
–207	66.15	0.2113	158.52	854.80	0.91575	–144.72	212.15	67.42	2.5129	5.7524	**–207**
–206	67.15	0.2462	184.66	850.66	0.78797	–142.71	211.05	68.34	2.5427	5.7192	**–206**
–205	68.15	0.2917	218.78	846.41	0.68117	–140.70	209.96	69.25	2.5726	5.6860	**–205**
–204	69.15	0.3400	255.02	842.16	0.59168	–138.69	208.86	70.17	2.6025	5.6528	**–204**
–203	70.15	0.3944	295.85	837.90	0.51615	–136.67	207.75	71.09	2.6324	5.6196	**–203**
–202	71.15	0.4544	340.84	833.54	0.45367	–134.64	206.64	72.00	2.6656	5.5760	**–202**
–201	72.15	0.5224	391.87	829.23	0.39885	–132.62	205.44	72.82	2.6929	5.5480	**–201**
–200	73.15	0.5983	448.76	824.85	0.35217	–130.59	204.23	73.64	2.7203	5.5207	**–200**
–199	74.15	0.6823	511.73	820.44	0.31217	–128.56	203.07	74.51	2.7476	5.4941	**–199**
–198	75.15	0.7751	581.41	815.98	0.27752	–126.53	201.85	75.32	2.7750	5.4682	**–198**
–197	76.15	0.8775	658.16	811.54	0.24760	–124.49	200.63	76.14	2.8023	5.4431	**–197**
–196	77.15	0.9877	740.86	807.03	0.22152	–122.44	199.35	76.91	2.8297	5.4185	**–196**
–195.795[b]	**77.355**	**1.01325**	**760.00**	**806.09**	**0.21682**	**–121.53**	**198.69**	**77.16**	**2.8342**	**5.4092**	**–195.795**

[a]Triple point
[b]Normal boiling point

TABLE: 8.2 Thermodynamic Properties of Saturated Nitrogen (Refrigerant R–728) above 1 Atmosphere (Temperature Basis)

Temperature		Saturation pressure,	Liquid density	Specific volume of vapour v''	Specific enthalpy			Specific entropy		Temp
t	T	p	ρ_l		h	λ_v	H	s'	s''	t
°C	K	bar	kg/m³	m³/kg	kJ/kg			kJ/(kg·K)		°C
–195	78.15	1.118	802.45	0.19914	–120.39	198.16	77.77	2.8552	5.3956	**–195**
–194	79.15	1.245	797.85	0.17938	–118.33	196.79	78.46	2.8808	5.3737	**–194**
–193	80.15	1.391	793.22	0.16182	–116.26	195.42	79.16	2.9064	5.3518	**–193**
–192	81.15	1.550	788.57	0.14638	–114.20	194.05	79.85	2.9320	5.3299	**–192**
–191	82.15	1.722	783.85	0.13277	–112.12	192.67	80.55	2.9575	5.3080	**–191**
–190	83.15	1.907	779.11	0.12072	–110.04	191.28	81.24	2.9823	5.2878	**–190**
–189	84.15	2.108	774.28	0.10995	–107.98	189.84	81.85	3.0067	5.2684	**–189**
–188	85.15	2.324	769.44	0.10036	–105.85	188.28	82.43	3.0311	5.2489	**–188**
–187	86.15	2.552	764.46	0.09191	–103.74	186.74	83.00	3.0555	5.2295	**–187**
–186	87.15	2.800	759.49	0.08424	–101.61	185.17	83.56	3.0800	5.2100	**–186**
–185	88.15	3.066	754.48	0.07734	–99.48	183.58	84.10	3.1037	5.1923	**–185**
–184	89.15	3.349	749.37	0.07113	–97.34	181.93	84.59	3.1272	5.1746	**–184**
–183	90.15	3.652	744.26	0.06551	–95.21	180.23	85.02	3.1507	5.1569	**–183**
–182	91.15	3.974	739.04	0.06044	–93.02	178.46	85.45	3.1742	5.1392	**–182**
–181	92.15	4.316	733.78	0.05583	–90.78	176.64	85.86	3.1977	5.1215	**–181**
–180	93.15	4.675	728.42	0.05166	–88.54	174.81	86.27	3.2207	5.1048	**–180**
–179	94.15	5.060	722.89	0.04787	–86.30	173.17	86.87	3.2435	5.0883	**–179**
–178	95.15	5.466	717.52	0.04441	–84.06	171.25	87.18	3.2663	5.0717	**–178**
–177	96.15	5.897	711.81	0.04123	–81.82	169.28	87.45	3.2892	5.0552	**–177**
–176	97.15	6.352	706.15	0.03833	–79.58	167.27	87.69	3.3120	5.0387	**–176**
–175	98.15	6.829	700.36	0.03566	–77.35	165.23	87.88	3.3345	5.0226	**–175**
–174	99.15	7.336	694.48	0.03321	–75.11	163.15	88.04	3.3570	5.0066	**–174**
–173	100.15	7.867	2242.93	0.03096	–72.87	160.95	88.08	3.3794	4.9906	**–173**
–172	101.15	8.419	682.25	0.02888	–70.54	158.70	88.16	3.4019	4.9746	**–172**
–171	102.15	9.004	676.00	0.02697	–68.14	156.34	88.19	3.4243	4.9586	**–171**
–170	103.15	9.619	669.69	0.02519	–65.74	153.93	88.19	3.4466	4.9426	**–170**
–169	104.15	10.263	663.17	0.02355	–63.34	151.44	88.10	3.4691	4.9266	**–169**
–168	105.15	10.937	656.52	0.02202	–60.94	148.90	87.96	3.4916	4.9106	**–168**
–167	106.15	11.642	649.71	0.02060	–58.45	146.21	87.76	3.5140	4.8946	**–167**
–166	107.15	12.379	642.69	0.01928	–55.85	143.34	87.49	3.5365	4.8786	**–166**
–165	108.15	13.143	635.26	0.01805	–53.33	140.48	87.15	3.5591	4.8598	**–165**
–164	109.15	13.946	627.84	0.01691	–50.66	137.32	86.66	3.5821	4.8427	**–164**
–163	110.15	14.785	620.21	0.01582	–47.30	133.47	86.17	3.6052	4.8255	**–163**
–162	111.15	15.659	612.29	0.01481	–45.38	130.97	85.59	3.6282	4.8084	**–162**
–161	112.15	16.569	2852.85	0.01386	–42.58	127.50	84.91	3.6512	4.7912	**–161**
–160	113.15	17.517	595.61	0.01297	–39.78	123.92	84.14	3.6742	4.7791	**–160**
–159	114.15	18.504	586.73	0.01214	–36.89	120.28	83.39	3.6986	4.7594	**–159**
–158	115.15	19.530	576.99	0.01134	–33.97	116.35	82.38	3.7237	4.7397	**–158**
–157	116.15	20.593	567.33	0.01059	–30.87	112.07	81.20	3.7489	4.7200	**–157**
–156	117.15	21.702	556.94	0.00988	–27.70	107.57	79.86	3.7741	4.7003	**–156**
–155	118.15	22.855	546.07	0.00918	–24.39	102.75	78.36	3.7992	4.6688	**–155**

(Contd.)

TABLE: 8.2 Thermodynamic Properties of Saturated Nitrogen (Refrigerant R–728) above 1 Atmosphere (Temperature Basis) [Contd.]

Temperature		Saturation pressure,	Liquid density	Specific volume of	Specific enthalpy			Specific entropy		Temp
t	T	p	ρ_l	vapour v''	h	λ_v	H	s'	s''	t
°C	K	bar	kg/m³	m³/kg	kJ/kg			kJ/(kg·K)		°C
–154	119.15	24.053	534.33	0.00853	–20.93	97.45	76.52	3.8268	4.6454	**–154**
–153	120.15	25.298	521.69	0.00791	–17.32	91.50	74.18	3.8568	4.6199	**–153**
–152	121.15	26.592	506.97	0.00730	–13.30	85.10	71.80	3.8867	4.5944	**–152**
–151	122.15	27.935	491.47	0.00670	–9.17	78.56	69.39	3.9167	4.5625	**–151**
–150	123.15	29.334	473.68	0.00610	–4.57	70.51	65.95	3.9529	4.5270	**–150**
–149	124.15	30.788	452.14	0.00568	0.66	60.94	61.60	3.9977	4.4828	**–149**
–148	125.15	32.907	451.56	0.00482	7.19	48.00	55.19	4.0425	4.4386	**–148**
–147	126.15	33.890	369.42	0.00334	20.07	22.90	42.97	4.1323	4.3404	**–147**
–146.958*	**126.192**	**33.958**	**313.30**	**0.00319**	**29.23**	**0.00**	**29.23**	**4.2148**	**4.2148**	**–146.958**

*Critical point

Reference State:

Specific enthalpy of gaseous nitrogen at 25°C (298.15 K) = 8670 kJ/kmol = 309.494 kJ/kg

Specific entropy of gaseous nitrogen at 25°C (298.15 K) and 1.01325 bar a = 191.5 kJ/(kmol×K) = 6.8360 kJ/(kg×K)

TABLE: 8.3 Thermodynamic Properties of Nitrogen (Refrigerant R–728) (Superheat Region)

Temperature		Absolute Pressure (Saturation Temperature)									Temp.
T	t	1.0 bar (–195.906°C/77.244K)			2.0 bar (–189.524°C/83.626K)			5.0 bar (–179.155°C/93.995K)			T
K	°C	v	i	s	v	i	s	v	i	s	K
Satd. Cond.		0.21947	77.07	5.4117	0.114545	81.52	5.2725	0.04435	86.52	5.0840	
80	–193.15	0.22835	80.15	5.4510	–	–	–	–	–	–	**80**
85	–188.15	0.24425	85.67	5.5177	0.11780	83.13	5.2918	–	–	–	**85**
90	–183.15	0.25996	91.13	5.5802	0.12618	88.91	5.3578	–	–	–	**90**
95	–178.15	0.27553	96.54	5.6387	0.13439	94.57	5.4192	0.049225	87.86	5.0983	**95**
100	–173.15	0.29098	101.90	5.6937	0.14247	100.14	5.4763	0.053027	94.30	5.1643	**100**
105	–168.15	0.30636	107.25	5.7458	0.15045	105.66	5.5299	0.056668	100.47	5.2243	**105**
110	–163.15	0.32165	112.57	5.7954	0.15835	111.12	5.5809	0.060194	106.47	5.2803	**110**
115	–158.15	0.33689	117.88	5.8426	0.16619	116.54	5.6291	0.063634	112.32	5.3324	**115**
120	–153.15	0.35211	123.17	5.8875	0.17398	121.94	5.6751	0.067009	118.08	5.3814	**120**
125	–148.15	0.35726	128.45	5.9307	0.18172	127.31	5.7190	0.070332	123.76	5.4278	**125**
130	–143.15	0.38238	133.72	5.9721	0.18944	132.66	5.7608	0.073615	129.38	5.4717	**130**
135	–138.15	0.39748	138.98	6.0118	0.19711	137.99	5.8012	0.076861	134.95	5.5138	**135**
140	–133.15	0.41255	144.24	6.0500	0.20478	143.31	5.8397	0.080078	140.47	5.5541	**140**
145	–128.15	0.42760	149.49	6.0867	0.21241	148.62	5.8772	0.083270	145.96	5.5923	**145**
150	–123.15	0.44264	154.73	6.1224	0.22003	153.91	5.9129	0.086442	151.42	5.6294	**150**
160	–113.15	0.47266	165.21	6.1899	0.23521	164.48	5.9811	0.092734	162.27	5.6994	**160**
170	–103.15	0.50263	175.67	6.2524	0.25035	175.02	6.0450	0.098972	173.04	5.7647	**170**
180	–93.15	0.53256	186.12	6.3131	0.26545	185.53	6.1049	0.105165	183.75	5.8261	**180**
200	–73.15	0.59234	206.97	6.4230	0.29553	206.52	6.2156	0.117452	205.04	5.9382	**200**
220	–53.15	0.65202	227.86	6.5226	0.32553	227.45	6.3155	0.129648	226.22	6.0393	**220**
240	–33.15	0.71164	248.71	6.6133	0.35548	248.36	6.4062	0.141779	247.32	6.1310	**240**
260	–13.15	0.77121	269.54	6.6968	0.38537	269.25	6.4901	0.153861	268.35	6.2152	**260**
280	6.85	0.83075	290.37	6.7739	0.41522	290.11	6.5672	0.165910	289.35	6.2931	**280**
300	26.85	0.89025	311.20	6.8457	0.44505	310.98	6.6393	0.177935	310.32	6.3652	**300**
350	76.85	1.03895	363.29	7.0063	0.51953	363.11	6.8000	0.207892	362.68	6.5265	**350**
400	126.85	1.18753	415.44	7.1455	0.59393	415.34	6.9395	0.237775	415.05	6.6665	**400**
450	176.85	1.33607	467.81	7.2690	0.66829	467.74	7.0630	0.267595	467.53	6.7903	**450**
500	226.85	1.48460	520.47	7.3797	0.74261	520.43	7.1741	0.297402	520.29	6.9013	**500**
600	326.85	1.78156	626.99	7.5739	0.89116	626.99	7.3683	0.356936	626.99	7.0959	**600**
700	426.85	2.07844	735.61	7.7413	1.03967	735.65	7.5357	0.416386	735.72	7.2633	**700**
800	526.85	2.37538	846.63	7.8898	1.18816	846.67	7.6838	0.475823	846.81	7.4118	**800**
900	626.85	2.67215	960.04	8.0233	1.33662	960.11	7.8173	0.535239	960.29	7.5453	**900**
1000	726.85	2.96908	1075.74	8.1450	1.48503	1075.81	7.9394	0.594636	1076.02	7.6670	**1000**

(Contd.)

TABLE: 8.3 Thermodynamic Properties of Nitrogen (Refrigerant R–728) (Superheat Region) [Contd.]

Temperature		Absolute Pressure (Saturation Temperature)									Temp.
T	*t*	10.0 bar(–169.403°C/103.747K)			15.0 bar(–162.751°C/110.399K)			20.0 bar(–157.551°C/115.599K)			*T*
K	°C	*v*	*i*	*s*	*v*	*i*	*s*	*v*	*i*	*s*	K
Satd.	**Cond.**	0.024195	87.73	4.9259	0.015565	85.62	4.8156	0.010997	81.24	4.7192	
105	–168.15	0.024805	89.77	4.9451	–	–	–	–	–	–	**105**
110	–163.15	0.027076	97.33	5.0158	–	–	–	–	–	–	**110**
115	–158.15	0.029174	104.32	5.0779	0.017320	94.34	4.8930	–	–	–	**115**
120	–153.15	0.031158	110.94	5.1343	0.018972	102.52	4.9630	0.012575	91.81	4.8088	**120**
125	–148.15	0.033059	117.32	5.1861	0.020477	109.98	5.0237	0.014012	101.29	4.8866	**125**
130	–143.15	0.034901	123.51	5.2346	0.021887	116.98	5.0786	0.015271	109.58	4.9516	**130**
135	–138.15	0.036696	129.55	5.2803	0.023230	123.68	5.1293	0.016425	117.19	5.0090	**135**
140	–133.15	0.038454	135.49	5.3235	0.024524	130.14	5.1761	0.017510	124.35	5.0611	**140**
145	–128.15	0.040181	141.33	5.3646	0.024778	136.42	5.2204	0.018544	131.19	5.1090	**145**
150	–123.15	0.041883	147.10	5.4038	0.027002	142.57	5.2621	0.019539	137.79	5.1536	**150**
160	–113.15	0.045226	158.47	5.4770	0.029378	154.55	5.3392	0.021444	150.48	5.2357	**160**
170	–103.15	0.048507	169.67	5.5448	0.031680	166.22	5.4099	0.023266	162.68	5.3096	**170**
180	–93.15	0.051740	180.73	5.6084	0.033933	177.66	5.4756	0.025031	174.54	5.3774	**180**
190	–83.15	0.054935	191.69	5.6676	0.036143	188.94	5.5366	0.026752	186.16	5.4403	**190**
200	–73.15	0.058100	202.58	5.7233	0.038322	200.09	5.5938	0.028439	197.59	5.4988	**200**
220	–53.15	0.064363	224.16	5.8261	0.042609	222.11	5.6987	0.031739	220.05	5.6059	**220**
240	–33.15	0.070559	245.58	5.9193	0.046826	243.85	5.7933	0.034966	242.13	5.7019	**240**
260	–13.15	0.076705	266.88	6.0046	0.050994	265.41	5.8797	0.038144	263.95	5.7894	**260**
280	6.85	0.082816	288.08	6.0832	0.055125	286.83	5.9589	0.041285	285.58	5.8697	**280**
300	26.85	0.088899	309.23	6.1560	0.059229	308.15	6.0325	0.044398	307.08	5.9436	**300**
350	76.85	0.104019	361.93	6.3188	0.069401	361.18	6.1960	0.052095	360.43	6.1081	**350**
400	126.85	0.119058	414.52	6.4591	0.079491	414.02	6.3370	0.059711	413.52	6.2499	**400**
450	176.85	0.134049	467.20	6.5833	0.089532	466.88	6.4616	0.067276	466.53	6.3748	**450**
500	226.85	0.148999	520.07	6.6947	0.099538	519.89	6.5733	0.074807	519.72	6.4869	**500**
600	326.85	0.178844	626.99	6.8896	0.119485	626.99	6.7685	0.089804	626.99	6.6825	**600**
700	426.85	0.208621	735.86	7.0573	0.139371	735.97	6.9363	0.104748	736.11	6.8507	**700**
800	526.85	0.238379	847.02	7.2058	0.159235	847.27	7.0852	0.119661	847.49	6.9992	**800**
900	626.85	0.268118	960.58	7.3393	0.179068	960.86	7.2187	0.134554	961.18	7.1334	**900**
1000	726.85	0.297824	1076.34	7.4614	0.198892	1076.70	7.3408	0.149429	1077.06	7.2551	**1000**

(Contd.)

TABLE: 8.3 Thermodynamic Properties of Nitrogen (Refrigerant R–728) (Superheat Region) [Contd.]

Temperature		Absolute Pressure (Saturation Temperature)									Temp.
T	t	25.0 bar(–153.234°C/119.916K)			30.0 bar(–149.534°C/123.616K)			35.0 bar			T
K	°C	v	i	s	v	i	s	v	i	s	K
Satd. Cond.		0.008047	74.35	4.6206	0.005793	63.04	4.4996	Non-Existent			
120	–153.15	0.008089	74.74	4.6239	–	–	–	–	–	–	120
125	–148.15	0.009891	90.21	4.7506	0.006638	72.72	4.5775	–	–	–	125
130	–143.15	0.011175	100.88	4.8341	0.008275	89.98	4.7131	0.005911	74.29	4.57	130
135	–138.15	0.012268	109.89	4.9023	0.009414	101.46	4.7998	0.007274	91.28	4.69	135
140	–133.15	0.013256	118.02	4.9616	0.010373	110.99	4.8695	0.008265	103.07	4.78	140
150	–123.15	0.015042	132.73	5.0629	0.018028	127.36	4.9823	0.009861	121.64	4.91	150
160	–113.15	0.016676	146.25	5.1504	0.013494	141.86	5.0761	0.011216	137.30	5.01	160
170	–103.15	0.018217	159.05	5.2279	0.014850	155.33	5.1575	0.012446	151.52	5.10	170
180	–93.15	0.019692	171.36	5.2982	0.016128	168.14	5.2307	0.013598	164.87	5.17	180
190	–83.15	0.021120	183.35	5.3631	0.017370	180.51	5.2978	0.014695	177.64	5.24	190
200	–73.15	0.022513	195.07	5.4231	0.018567	192.54	5.3596	0.015753	190.01	5.30	200
220	–53.15	0.025221	217.99	5.5324	0.020882	215.94	5.4710	0.017785	213.89	5.42	220
240	–33.15	0.027856	240.41	5.6302	0.023120	238.71	5.5702	0.019742	237.02	5.52	240
260	–13.15	0.030440	262.50	5.7183	0.025306	261.07	5.6594	0.021645	259.64	5.61	260
280	6.85	0.032986	284.35	5.7994	0.027457	283.13	5.7412	0.023511	281.93	5.69	280
300	26.85	0.035505	306.03	5.8743	0.029578	304.99	5.8169	0.025350	303.96	5.77	300
350	76.85	0.041715	359.72	6.0396	0.034496	359.01	5.9832	0.029860	358.33	5.94	350
400	126.85	0.047845	413.02	6.1820	0.039936	412.55	6.1264	0.034288	412.09	6.08	400
500	226.85	0.059971	519.54	6.4198	0.050082	519.36	6.3645	0.043019	519.18	6.32	500
600	326.85	0.071999	626.99	6.6154	0.060130	626.99	6.5608	0.051651	627.02	6.51	600
700	426.85	0.083974	736.25	6.7839	0.070125	736.40	6.7293	0.060233	736.54	6.68	700
800	526.85	0.095916	847.74	6.9328	0.080089	847.95	6.8785	0.068783	848.20	6.83	800
900	626.85	0.107840	961.47	7.0666	0.090033	961.75	7.0124	0.077314	962.08	6.97	900
1000	726.85	0.119749	1077.41	7.1887	0.099964	1077.77	7.1344	0.085831	1078.13	7.09	1000

Temperature		Absolute Pressure									Temp.
T	t	40.0 bar			50.0 bar			60.0 bar			T
K	°C	v	i	s	v	i	s	v	i	s	K
125	–148.15	0.002002	–5.42	3.9292	0.001880	–9.95	3.8778	0.001808	–12.46	3.8428	125
130	–143.15	0.003334	39.59	4.2805	0.002172	9.30	4.0284	0.001998	3.19	3.9656	130
135	–138.15	0.005538	78.21	4.5732	0.002987	40.47	4.2633	0.002325	22.66	4.1123	135
140	–133.15	0.006636	93.94	4.6878	0.004256	70.86	4.4846	0.002935	47.15	4.2904	140
145	–128.15	0.007489	105.63	4.7699	0.005185	88.84	4.6110	0.003717	70.03	4.4511	145
150	–123.15	0.008225	116.23	4.8370	0.005921	102.02	4.7006	0.004418	87.21	4.5678	150
160	–113.15	0.009508	94.54	4.9473	0.007120	122.60	4.8334	0.005553	112.12	4.7288	160
170	–103.15	0.010646	147.63	5.0387	0.008136	139.61	4.9366	0.006484	131.39	4.8455	170
180	–93.15	0.011699	161.56	5.1183	0.009052	154.83	5.0237	0.007307	148.04	4.9408	180
190	–83.15	0.012693	174.76	5.1897	0.009904	168.97	5.1001	0.008061	163.16	5.0226	190
200	–73.15	0.013647	187.46	5.2546	0.010711	182.38	5.1690	0.008769	177.33	5.0954	200
220	–53.15	0.015468	211.85	5.3710	0.012235	207.80	5.2903	0.010093	203.81	5.2214	220
240	–33.15	0.017211	235.33	5.4731	0.013679	232.01	5.3956	0.011335	228.76	5.3303	240
260	–13.15	0.018901	258.24	5.5648	0.015069	255.46	5.4895	0.012524	252.76	5.4263	260
280	6.85	0.020555	280.73	5.6484	0.016422	278.40	5.5745	0.013676	276.12	5.5127	280
300	26.85	0.022179	302.94	5.7248	0.017748	300.95	5.6523	0.014810	299.02	5.5920	300
350	76.85	0.026158	357.61	5.8936	0.020979	356.28	5.8229	0.017531	354.99	5.7644	350
400	126.85	0.030056	411.62	6.0378	0.024131	410.73	5.9682	0.020186	409.88	5.9111	400
450	176.85	0.033904	465.31	6.1642	0.027237	464.78	6.0957	0.022794	464.24	6.0393	450
500	226.85	0.037722	519.00	6.2774	0.030308	518.72	6.2092	0.025369	518.43	6.1531	500
600	326.85	0.045294	627.02	6.4744	0.036395	627.09	6.4069	0.030464	627.16	6.3516	600
700	426.85	0.052814	736.68	6.6432	0.052429	736.97	6.5761	0.035505	737.29	6.5212	700
800	526.85	0.060304	848.42	6.7925	0.048434	848.92	6.7257	0.040521	849.38	6.6707	800
900	626.85	0.067774	962.36	6.9267	0.054420	963.00	6.8599	0.045517	963.61	6.8053	900
1000	726.85	0.075231	1078.48	7.0488	0.060392	1079.20	6.9824	0.050500	1079.91	6.9278	1000

(Contd.)

TABLE: 8.3 Thermodynamic Properties of Nitrogen (Refrigerant R–728) (Superheat Region) [Contd.]

Temperature		Absolute Pressure									Temp.
T	*t*	80.0 bar			100.0 bar			150.0 bar			*T*
K	°C	*v*	*i*	*s*	*v*	*i*	*s*	*v*	*i*	*s*	K
125	−148.15	0.001717	−15.21	3.7928	0.001657	−16.52	3.7553	0.001560	−17.12	3.6861	**125**
130	−143.15	0.001838	−2.22	3.8946	0.001749	−4.74	3.8474	0.001622	−6.73	3.7678	**130**
135	−138.15	0.001999	11.94	4.0013	0.001862	7.62	3.9410	0.001691	3.85	3.8474	**135**
140	−133.15	0.002221	27.58	4.1152	0.002001	20.66	4.0356	0.001768	14.61	3.9260	**140**
145	−128.15	0.002527	44.58	4.2344	0.002174	34.36	4.1320	0.001855	25.55	4.0027	**145**
150	−123.15	0.002911	61.78	4.3511	0.002386	48.52	4.2280	0.001953	36.64	4.0777	**150**
160	−113.15	0.003745	91.75	4.5446	0.002906	76.44	4.4082	0.002182	59.03	4.2223	**160**
170	−103.15	0.004510	115.21	4.6870	0.003472	101.27	4.5589	0.002451	81.14	4.3565	**170**
180	−93.15	0.005191	134.69	4.7984	0.004020	122.63	4.6810	0.002748	102.28	4.4771	**180**
190	−83.15	0.005811	151.83	4.8912	0.004533	141.40	4.7824	0.003056	122.11	4.5846	**190**
200	−73.15	0.006386	167.49	4.9715	0.005014	158.37	4.8695	0.003365	140.57	4.6792	**200**
220	−53.15	0.007448	196.10	5.1079	0.005902	188.92	5.0151	0.003965	174.11	4.8391	**220**
240	−33.15	0.008432	222.50	5.2229	0.006721	216.67	5.1361	0.004533	204.38	4.9708	**240**
260	−13.15	0.009365	247.58	5.3232	0.007495	242.75	5.2403	0.005073	232.47	5.0833	**260**
280	6.85	0.010263	271.77	5.4131	0.008235	267.73	5.3332	0.005590	259.08	5.1822	**280**
300	26.85	0.011134	295.34	5.4942	0.008950	291.94	5.4167	0.006089	284.63	5.2703	**300**
350	76.85	0.013234	352.55	5.6709	0.010668	350.29	5.5966	0.007279	345.52	5.4581	**350**
400	126.85	0.015262	408.27	5.8197	0.012317	406.84	5.7476	0.008415	403.84	5.6137	**400**
450	176.85	0.017247	463.24	5.9493	0.013925	462.39	5.8786	0.009514	460.74	5.7480	**450**
500	226.85	0.019201	517.93	6.0642	0.015504	572.51	6.0996	0.010590	516.90	5.8661	**500**
600	326.85	0.023053	627.34	6.2638	0.018609	627.59	6.4094	0.012693	628.48	6.0696	**600**
700	426.85	0.026856	737.97	6.4344	0.021667	738.65	6.3666	0.014756	740.61	6.2427	**700**
800	526.85	0.030631	850.38	6.5843	0.024701	851.41	6.5172	0.016794	854.16	6.3941	**800**
900	626.85	0.034390	964.86	6.7193	0.027713	966.14	6.6522	0.018816	969.46	6.5301	**900**
1000	726.85	0.038134	1081.38	6.8421	0.030715	1082.84	6.7750	0.020827	1086.59	6.6532	**1000**

Adapted from R. Span, E. W. Lemmon, R. T. Jackobsen, W. Wagner and A. Yokozeki, *J. Phys. Chem. Ref. Data*, **29** (6), 2000, p. 1361 with the Permission of the American Institute of Physics, USA.

TABLE: 9.1 Thermodynamic Properties of Carbon Dioxide (Refrigerant R–744) on Sublimation Curve (Temperature Basis)

Temperature		Saturation pressure,	Solid density	Specific volume of	Specific enthalpy			Specific entropy		Temp
T	t	p	ρ_s	vapour v''	h_s	λ_s	H	s_s	s''	t
K	°C	bar	kg/m^3	m^3/kg	kJ/kg			kJ/(kg·K)		°C
–102	171.15	0.1136	1597	2.837	–176.92	586.60	409.68	–0.6962	2.7298	**–102**
–100	173.15	0.1397	1595	2.327	–174.75	585.50	410.75	–0.6832	2.6978	**–100**
–98	175.15	0.1715	1593	1.916	–172.48	584.20	411.72	–0.6692	2.6658	**–98**
–96	177.15	0.2095	1591	1.583	–170.18	583.10	412.92	–0.6562	2.6348	**–96**
–94	179.15	0.2549	1588	1.314	–167.84	581.90	414.06	–0.6422	2.6048	**–94**
–92	181.15	0.3089	1585	1.095	–165.46	580.60	415.14	–0.6292	2.5758	**–92**
–90	183.15	0.3727	1582	0.9173	–163.04	579.40	416.36	–0.6162	2.5468	**–90**
–88	185.15	0.4476	1579	0.7719	–160.59	578.00	417.41	–0.6032	2.5188	**–88**
–86	187.15	0.5353	1576	0.6513	–158.10	576.70	418.60	–0.5902	2.4918	**–86**
–84	189.15	0.6377	1573	0.5507	–155.57	575.30	419.73	–0.5772	2.4648	**–84**
–82	191.15	0.7572	1569	0.4671	–153.01	573.80	420.79	–0.5632	2.4388	**–82**
–80	193.15	0.8962	1565	0.3977	–150.41	572.30	421.89	–0.5502	2.4118	**–80**
–78.46[a]	**194.69**	**1.01325**	**1562**	**0.3553**	**–148.39**	**571.00**	**422.61**	**–0.5400**	**2.3950**	**–78.46**
–78	195.15	1.057	1561	0.3398	–147.77	570.70	422.93	–0.5372	2.3868	**–78**
–76	197.15	1.242	1558	0.2911	–145.10	569.10	424.00	–0.5252	2.3618	**–76**
–74	199.15	1.456	1554	0.2499	–142.38	567.50	425.12	–0.5122	2.3378	**–74**
–72	201.15	1.700	1549	0.2151	–139.63	565.80	426.17	–0.4992	2.3138	**–72**
–70	203.15	1.981	1545	0.1857	–136.85	564.00	427.15	–0.4862	2.2908	**–70**
–68	205.15	2.302	1541	0.1608	–134.02	561.40	427.38	–0.4732	2.2668	**–68**
–66	207.15	2.670	1536	0.1395	–131.16	559.70	428.54	–0.4602	2.2418	**–66**
–64	209.15	3.089	1532	0.1211	–128.27	557.30	429.03	–0.4472	2.2168	**–64**
–62	211.15	3.567	1527	0.1051	–125.33	554.70	429.37	–0.4342	2.1928	**–62**
–60	213.15	4.098	1522	0.09123	–122.36	552.00	429.64	–0.4212	2.1698	**–60**
–58	215.15	4.671	1517	0.08100	–119.35	549.30	429.95	–0.4082	2.1508	**–58**
–56.56[b]	**216.59**	**5.1795**	**1513**	**0.07267**	**–117.17**	**547.60**	**430.43**	**–0.3892**	**2.1388**	**–56.56**

[a]Sublimation temperature at 1 standard atmosphere
[b]Triple point

TABLE: 9.2 Thermodynamic Properties of Carbon Dioxide (Refrigerant R–744) Saturation Region (Temperature Basis)

Temperature		Saturation pressure,	Liquid density	Specific volume of	Specific enthalpy			Specific entropy		Temp
t	T	p	ρ_l	vapour v''	h	λ_v	H	s'	s''	t
°C	K	bar	kg/m^3	m^3/kg	kJ/kg			kJ/(kg·K)		°C
–56.56*	**216.59**	**5.1795**	**1178.5**	**0.07267**	**80.05**	**350.38**	**430.43**	**0.5211**	**2.1388**	**–56.56**
–56	217.15	5.3151	1176.5	0.07093	81.14	349.51	430.65	0.5262	2.1354	**–56**
–55	218.15	5.5364	1172.8	0.06810	83.11	347.88	430.99	0.5350	2.1297	**–55**
–54	219.15	5.7902	1168.9	0.06540	85.07	346.25	431.33	0.5439	2.1241	**–54**
–53	220.15	6.0441	1165.3	0.06282	87.04	344.63	431.67	0.5527	2.1184	**–53**
–52	221.15	6.2979	1161.5	0.06035	89.01	343.00	432.01	0.5615	2.1128	**–52**
–51	222.15	6.5518	1157.9	0.05799	90.98	341.37	432.35	0.5704	2.1071	**–51**
–50	223.15	6.8212	1154.0	0.05573	92.95	339.74	432.69	0.5792	2.1015	**–50**
–49	224.15	7.0811	1150.3	0.05357	94.91	338.12	433.03	0.5881	2.0958	**–49**
–48	225.15	7.3922	1146.5	0.05156	96.88	336.50	433.38	0.5968	2.0904	**–48**
–47	226.15	7.7033	1142.6	0.04958	98.89	334.74	433.63	0.6054	2.0852	**–47**
–46	227.15	8.0144	1138.8	0.04778	100.89	332.98	433.87	0.6140	2.0799	**–46**
–45	228.15	8.3255	1134.9	0.04596	102.90	331.22	434.12	0.6225	2.0746	**–45**
–44	229.15	8.6480	1131.3	0.04434	104.90	329.46	434.36	0.6311	2.0694	**–44**
–43	230.15	8.9592	1127.4	0.04266	106.90	327.71	434.61	0.6397	2.0641	**–43**
–42	231.15	9.3167	1123.6	0.04117	108.91	325.95	434.86	0.6483	2.0588	**–42**
–41	232.15	9.6781	1119.7	0.03963	110.91	324.19	435.10	0.6569	2.0535	**–41**
–40	233.15	10.067	1115.9	0.03821	112.92	322.43	435.35	0.6655	2.0483	**–40**
–39	234.15	10.426	1112.0	0.03679	114.92	320.67	435.59	0.6741	2.0430	**–39**
–38	235.15	10.815	1107.8	0.03553	116.91	318.93	435.84	0.6826	2.0378	**–38**
–37	236.15	11.204	1103.7	0.03430	118.97	317.00	435.97	0.6909	2.0328	**–37**
–36	237.15	11.605	1099.8	0.03311	121.02	315.07	436.09	0.7014	2.0278	**–36**
–35	238.15	11.994	1095.7	0.03197	123.07	313.15	436.22	0.7097	2.0228	**–35**
–34	239.15	12.458	1091.7	0.03088	125.13	311.22	436.35	0.7163	2.0177	**–34**
–33	240.15	12.871	1087.5	0.02982	127.18	309.29	436.47	0.7246	2.0127	**–33**
–32	241.15	13.349	1083.4	0.02881	129.24	307.36	436.60	0.7333	2.0077	**–32**
–31	242.15	13.807	1079.3	0.02784	131.29	305.43	436.72	0.7416	2.0027	**–31**
–30	243.15	14.285	1075.2	0.02690	133.34	303.51	436.85	0.7497	1.9977	**–30**
–29	244.15	14.763	1071.0	0.02599	135.40	301.58	436.98	0.7580	1.9926	**–29**
–28	245.15	15.265	1066.9	0.02519	137.44	299.60	437.05	0.7665	1.9878	**–28**
–27	246.15	15.773	1062.4	0.02436	139.57	297.46	437.02	0.7748	1.9828	**–27**
–26	247.15	16.301	1058.2	0.02356	141.69	295.31	437.00	0.7829	1.9779	**–26**
–25	248.15	16.836	1054.1	0.02279	143.82	293.25	437.07	0.7912	1.9730	**–25**
–24	249.15	17.388	1049.6	0.02205	145.95	291.10	437.05	0.7995	1.9680	**–24**
–23	250.15	17.940	1045.1	0.02134	148.07	288.98	437.05	0.8078	1.9631	**–23**
–22	251.15	18.510	1040.7	0.02065	150.20	286.83	437.03	0.8161	1.9582	**–22**
–21	252.15	19.095	1036.2	0.01999	152.32	284.66	436.98	0.8243	1.9532	**–21**
–20	253.15	19.699	1031.7	0.01935	154.45	282.51	436.96	0.8326	1.9483	**–20**
–19	254.15	20.332	1027.2	0.01873	156.58	280.26	436.83	0.8409	1.9433	**–19**
–18	255.15	20.965	1022.3	0.01806	158.70	278.11	436.81	0.8491	1.9385	**–18**
–17	256.15	21.599	1017.8	0.01748	160.93	275.66	436.59	0.8574	1.9335	**–17**
–16	257.15	22.232	1013.1	0.01693	163.15	273.22	436.38	0.8657	1.9284	**–16**
–15	258.15	22.884	1008.1	0.01639	165.38	270.89	436.27	0.8740	1.9234	**–15**
–14	259.15	23.605	1003.4	0.01588	167.61	268.45	436.06	0.8823	1.9184	**–14**
–13	260.15	24.300	998.42	0.01538	169.84	266.06	435.90	0.8906	1.9133	**–13**
–12	261.15	25.021	993.45	0.01490	172.06	263.62	435.69	0.8989	1.9083	**–12**
–11	262.15	25.743	988.48	0.01444	174.29	261.18	435.47	0.9072	1.9032	**–11**
–10	263.15	26.490	983.51	0.01399	176.52	258.74	435.26	0.9155	1.8982	**–10**

*Triple point

(Contd.)

TABLE: 9.2 Thermodynamic Properties of Carbon Dioxide (Refrigerant R–744) in Saturation Region (Temperature Basis) [Contd.]

Temperature		Saturation pressure,	Liquid density	Specific volume of	Specific enthalpy			Specific entropy		Temp
t	T	p	ρ_l	vapour v''	h	λ_v	H	s'	s''	t
°C	K	bar	kg/m³	m³/kg	kJ/kg			kJ/(kg·K)		°C
−9	264.15	27.245	978.27	0.01356	178.75	256.11	434.86	0.9238	1.8932	**−9**
−8	265.15	28.035	973.12	0.01320	180.96	253.69	434.65	0.9322	1.8882	**−8**
−7	266.15	28.825	967.69	0.01278	183.34	250.84	434.18	0.9407	1.8827	**−7**
−6	267.15	29.641	961.99	0.01238	185.72	247.98	433.70	0.9491	1.8773	**−6**
−5	268.15	30.458	956.29	0.01199	188.10	245.30	433.40	0.9576	1.8719	**−5**
−4	269.15	31.301	950.59	0.01162	190.48	242.44	432.92	0.9661	1.8665	**−4**
−3	270.15	32.166	944.89	0.01125	192.86	239.65	432.51	0.9746	1.8611	**−3**
−2	271.15	33.038	939.19	0.01091	195.24	236.79	432.03	0.9830	1.8557	**−2**
−1	272.15	33.939	933.49	0.01057	197.62	233.86	431.48	0.9915	1.8503	**−1**
0	273.15	34.859	927.79	0.01024	200.00	231.00	431.00	1.0000	1.8449	**0**
1	274.15	35.779	922.09	0.00993	202.38	227.94	430.32	1.0085	1.8395	**1**
2	275.15	36.733	915.96	0.00961	204.81	225.05	429.86	1.0167	1.8341	**2**
3	276.15	37.685	909.48	0.00929	207.44	221.53	428.97	1.0257	1.8278	**3**
4	277.15	38.679	903.14	0.00899	209.93	218.44	428.37	1.0346	1.8224	**4**
5	278.15	39.673	896.53	0.00870	212.55	214.93	427.48	1.0436	1.8161	**5**
6	279.15	40.716	889.92	0.00841	215.18	211.53	426.71	1.0526	1.8099	**6**
7	280.15	41.763	883.04	0.00814	217.81	208.01	425.82	1.0615	1.8036	**7**
8	281.15	42.821	876.07	0.00788	220.43	204.50	424.93	1.0705	1.7974	**8**
9	282.15	43.920	868.98	0.00763	223.06	200.98	424.04	1.0794	1.7911	**9**
10	283.15	45.033	861.81	0.00738	225.83	196.90	422.73	1.0884	1.7840	**10**
11	284.15	46.147	853.78	0.00715	228.46	193.38	421.84	1.0973	1.7778	**11**
12	285.15	47.289	846.12	0.00690	231.02	190.00	421.02	1.1063	1.7715	**12**
13	286.15	48.457	838.19	0.00667	234.16	185.17	419.33	1.1164	1.7633	**13**
14	287.15	49.659	829.83	0.00642	237.30	181.01	418.30	1.1267	1.7550	**14**
15	288.15	50.861	821.18	0.00618	240.03	176.58	416.61	1.1358	1.7486	**15**
16	289.15	52.106	812.11	0.00595	243.16	171.99	415.15	1.1460	1.7403	**16**
17	290.15	53.351	803.34	0.00573	246.30	167.17	413.47	1.1557	1.7328	**17**
18	291.15	54.640	793.84	0.00552	249.43	162.20	411.63	1.1659	1.7244	**18**
19	292.15	55.944	783.89	0.00532	252.57	157.38	409.94	1.1765	1.7155	**19**
20	293.15	57.287	773.13	0.00513	256.11	152.14	408.26	1.1883	1.7071	**20**
21	294.15	58.735	763.77	0.00494	259.25	147.32	406.57	1.1986	1.6964	**21**
22	295.15	60.018	750.67	0.00475	261.87	142.71	404.58	1.2077	1.6899	**22**
23	296.15	61.466	733.15	0.00453	266.64	133.80	400.44	1.2229	1.6741	**23**
24	297.15	62.854	725.14	0.00431	271.41	127.08	398.49	1.2337	1.6650	**24**
25	298.15	64.302	707.65	0.00409	274.98	119.37	394.36	1.2489	1.6492	**25**
26	299.15	65.809	694.35	0.00389	279.74	110.86	390.61	1.2626	1.6347	**26**
27	300.15	67.257	676.88	0.00367	284.51	101.97	386.47	1.2778	1.6159	**27**
28	301.15	68.890	656.09	0.00347	289.27	92.16	381.43	1.2951	1.6001	**28**
29	302.15	70.504	638.61	0.00325	295.85	81.45	377.30	1.3102	1.5788	**29**
30	303.15	72.117	593.93	0.00286	305.32	58.72	364.04	1.3457	1.5491	**30**
30.98*	**304.13**	**73.773**	**467.60**	**0.00214**	**332.26**	**0.00**	**332.26**	**1.4334**	**1.4334**	**30.98**

*Critical point

Reference State:

Specific enthalpy of saturated liquid at 0°C (273.15 K) = 200.00 kJ/kg

Specific entropy of saturated liquid at 0°C (273.15 K) = 1.0000 kJ/(kg·K)

TABLE: 9.3 Thermodynamic Properties of Carbon Dioxide (Refrigerant R–744) (Superheat Region)

Temperature		Absolute Pressure (Saturation Temperature)									Temp.
T	*t*	0.5 bar (–86.71°C/186.44K)			1.0 bar (–76.62°C/194.53K)			1.01325 bar(–78.46°C/194.69K)			*T*
K	°C	*v*	*i*	*s*	*v*	*i*	*s*	*v*	*i*	*s*	K
Subl.	**Cond.**	0.69589	417.85	2.5012	0.35976	422.53	2.3969	0.35528	422.61	2.3950	
190	–83.15	0.70977	420.51	2.5153	–	–	–	–	–	–	**190**
200	–73.15	0.74856	428.00	2.5538	0.37064	426.74	2.4183	0.36570	426.71	2.4157	**200**
210	–63.15	0.78715	435.57	2.5907	0.39037	434.47	2.4560	0.38516	434.44	2.4534	**210**
220	–53.15	0.82563	443.21	2.6262	0.40994	442.25	2.4922	0.40450	442.22	2.4896	**220**
230	–43.15	0.86393	450.95	2.6606	0.42941	450.09	2.5271	0.42371	450.07	2.5245	**230**
240	–33.15	0.90220	458.78	2.6940	0.44879	458.01	2.5608	0.44285	457.99	2.5582	**240**
250	–23.15	0.94038	466.71	2.7263	0.46810	466.01	2.5934	0.46192	466.00	2.5909	**250**
260	–13.15	0.97857	474.74	2.7578	0.48735	474.11	2.6252	0.48093	474.09	2.6227	**260**
270	–3.15	1.0167	482.88	2.7885	0.50656	482.30	2.6561	0.49990	482.28	2.6536	**270**
280	6.85	1.0547	491.12	2.8185	0.52573	490.58	2.6862	0.51883	490.57	2.6837	**280**
290	16.85	1.0928	499.46	2.8478	0.54490	498.97	2.7156	0.53772	498.95	2.7131	**290**
300	26.85	1.1308	507.90	2.8764	0.56402	507.44	2.7444	0.55661	507.43	2.7419	**300**
325	51.85	1.2257	529.44	2.9453	0.61170	529.05	2.8136	0.60368	529.04	2.8110	**325**
350	76.85	1.3206	551.57	3.0109	0.65933	551.25	2.8793	0.65066	551.24	2.8768	**350**
375	101.85	1.4153	574.28	3.0736	0.70686	574.00	2.9421	0.69759	573.99	2.9396	**375**
400	126.85	1.5100	597.53	3.1336	0.75432	597.28	3.0022	0.74444	597.28	2.9997	**400**
425	151.85	1.6047	621.29	3.1912	0.80180	621.07	3.0599	0.79126	621.07	3.0574	**425**
450	176.85	1.6993	645.55	3.2467	0.84918	645.35	3.1154	0.83808	645.34	3.1129	**450**
475	201.85	1.7939	670.26	3.3001	0.89654	670.08	3.1689	0.88480	670.08	3.1664	**475**
500	226.85	1.8885	695.42	3.3517	0.94393	695.26	3.2205	0.93153	695.25	3.2180	**500**
525	251.85	1.9831	720.99	3.4016	0.99128	720.85	3.2705	0.97828	720.85	3.2680	**525**
550	276.85	2.0777	746.97	3.4500	1.0386	746.84	3.3188	1.0250	746.83	3.3163	**550**
575	301.85	2.1723	773.33	3.4968	1.0859	773.21	3.3657	1.0717	773.20	3.3632	**575**
600	326.85	2.2668	800.05	3.5423	1.1332	799.94	3.4112	1.1184	799.93	3.4087	**600**

Temperature		Absolute Pressure (Saturation Temperature)									Temp.
T	*t*	2.0 bar (–69.84°C/203.31K)			3.0 bar (–64.35°C/208.8K)			5.0 bar (–57.07°C/216.08K)			*T*
K	°C	*v*	*i*	*s*	*v*	*i*	*s*	*v*	*i*	*s*	K
Subl.	**Cond.**	0.18500	426.82	2.2913	0.12478	428.86	2.2279	0.075290	430.39	2.1447	
210	–63.15	0.19188	432.22	2.3174	0.12564	429.87	2.2327	–	–	–	**210**
220	–53.15	0.20204	440.27	2.3549	0.13268	438.23	2.2716	0.077077	433.92	2.1609	**220**
230	–43.15	0.21208	448.34	2.3907	0.13960	446.54	2.3086	0.081533	442.78	2.2004	**230**
240	–33.15	0.22203	456.44	2.4252	0.14642	454.84	2.3439	0.085866	451.52	2.2375	**240**
250	–23.15	0.23191	464.60	2.4585	0.15316	463.16	2.3779	0.090114	460.20	2.2730	**250**
260	–13.15	0.24173	472.83	2.4908	0.15984	471.52	2.4107	0.094286	468.85	2.3069	**260**
270	–3.15	0.25150	481.13	2.5221	0.16647	479.94	2.4424	0.098416	477.51	2.3396	**270**
280	6.85	0.26123	489.51	2.5526	0.17306	488.42	2.4733	0.10249	486.20	2.3712	**280**
290	16.85	0.27094	497.97	2.5823	0.17961	496.97	2.5033	0.10654	494.93	2.4018	**290**
300	26.85	0.28062	506.52	2.6113	0.18614	505.59	2.5325	0.11055	503.71	2.4316	**300**
325	51.85	0.30470	528.28	2.6809	0.20236	527.50	2.6026	0.12049	525.93	2.5027	**325**
350	76.85	0.32870	550.59	2.7470	0.21848	549.92	2.6691	0.13032	548.59	2.5699	**350**
375	101.85	0.35261	573.42	2.8101	0.23453	572.85	2.7324	0.14007	571.70	2.6336	**375**
400	126.85	0.37648	596.78	2.8703	0.25053	596.28	2.7928	0.14977	595.27	2.6945	**400**
425	151.85	0.40030	620.63	2.9282	0.26648	620.19	2.8508	0.15943	619.30	2.7528	**425**
450	176.85	0.42409	644.95	2.9838	0.28241	644.56	2.9065	0.16905	643.77	2.8087	**450**
475	201.85	0.44787	669.73	3.0374	0.29830	669.38	2.9602	0.17865	668.67	2.8625	**475**
500	226.85	0.47161	694.94	3.0891	0.31418	694.62	3.0120	0.18823	693.98	2.9145	**500**
525	251.85	0.49534	720.56	3.1391	0.33004	720.27	3.0620	0.19779	719.70	2.9647	**525**
550	276.85	0.51908	746.58	3.1875	0.34589	746.32	3.1105	0.20734	745.79	3.0132	**550**
575	301.85	0.54277	772.97	3.2344	0.36172	772.73	3.1575	0.21688	772.25	3.0603	**575**
600	326.85	0.56648	799.72	3.2800	0.37754	799.50	3.2031	0.22641	799.07	3.1059	**600**
700	426.85	0.66120	910.01	3.4499	0.44078	909.86	3.3731	0.26445	909.05	3.2761	**700**
800	526.85	0.75586	1024.92	3.6033	0.50393	1024.81	3.5265	0.30242	1024.60	3.4297	**800**

(Contd.)

TABLE: 9.3 Thermodynamic Properties of Carbon Dioxide (Refrigerant R–744) (Superheat Region) [Contd.]

Temperature		Absolute Pressure (Saturation Temperature)									Temp.
T	t	07.5 bar (–47.64°C/225.51 K)			10.0 bar (–40.12°C/233.03 K)			20.0 bar (–19.50°C/253.65 K)			T
K	°C	v	i	s	v	i	s	v	i	s	K
Satd. Cond.		0.050916	433.40	2.0887	0.038453	435.31	2.0489	0.019033	436.86	1.9459	
230	–43.15	0.052370	437.73	2.1077	–	–	–	–	–	–	230
240	–33.15	0.055500	447.13	2.1477	0.040230	442.39	–	–	–	–	240
250	–23.15	0.058524	456.32	2.1852	0.042671	452.20	–	–	–	–	250
260	–13.15	0.061467	465.38	2.2208	0.045015	461.74	0.020034	0.020034	444.82	1.9769	260
270	–3.15	0.064350	474.38	2.2547	0.047288	471.12	0.021490	0.021490	456.49	2.0209	270
280	6.85	0.067186	483.35	2.2874	0.049507	480.41	0.022846	0.022846	467.49	2.0609	280
290	16.85	0.069979	492.32	2.3188	0.051682	489.64	0.024132	0.024132	478.07	2.0980	290
300	26.85	0.072743	501.31	2.3493	0.053824	498.85	0.025368	0.025368	488.37	2.1330	300
325	51.85	0.079542	523.94	2.4218	0.059060	521.92	0.028302	0.028302	513.48	2.2134	320
350	76.85	0.086229	546.90	2.4898	0.065015	545.20	0.031090	0.031090	538.18	2.2866	340
400	126.85	0.099384	594.01	2.6156	0.074200	592.73	0.036411	0.036411	587.57	2.4185	400
420	146.85	0.10460	613.31	2.6627	0.078149	612.16	0.038478	0.038478	607.53	2.4672	420
440	166.85	0.10979	632.89	2.7082	0.082082	631.85	0.040522	0.040522	627.66	2.5140	440
460	186.85	0.11496	652.73	2.7523	0.085992	651.78	0.042546	0.042546	647.97	2.5591	460
480	206.85	0.12011	672.83	2.7951	0.089888	671.97	0.044555	0.044555	668.49	2.6028	480
500	226.85	0.12526	693.19	2.8366	0.093773	692.39	0.046551	0.046551	689.20	2.6451	500
525	251.85	0.13167	718.98	2.8870	0.098610	718.26	0.049029	0.049029	715.38	2.6962	525
550	276.85	0.13807	745.14	2.9356	0.103439	744.49	0.051496	0.051496	741.89	2.7455	550
575	301.85	0.14446	771.66	2.98279	0.108254	771.07	0.053949	0.053949	768.70	2.7932	575
600	326.85	0.15084	798.53	3.02852	0.113060	797.98	0.056392	0.056392	795.83	2.8393	600
650	376.85	0.16358	853.24	3.1161	0.122647	852.78	0.061256	0.061256	850.99	2.9276	650
700	426.85	0.17629	909.17	3.19899	0.132209	908.79	0.066094	0.066094	907.28	3.0111	700
800	526.85	0.20166	1024.32	3.35269	0.151281	1024.05	0.075717	0.075717	1022.983	3.1655	800
900	626.85	0.22698	1143.28	3.49276	0.170306	1143.08	0.085295	0.085295	1142.333	3.3060	900

Temperature		Absolute Pressure (Saturation Temperature)									Temp.
T	t	30.0 bar (–5.55°C/267.6K)			40.0 bar (5.30°C/278.45K)			50.0 bar (14.28°C/287.43K)			T
K	°C	v	i	s	v	i	s	v	i	s	K
Satd. Cond.		0.012207	433.62	1.8752	0.008640	427.26	1.8143	0.006383	417.67	1.7541	
270	–3.15	0.012530	437.47	1.8895	–	–	–	–	–	–	270
280	6.85	0.013735	451.83	1.9418	0.008843	430.51	1.8259	–	–	–	280
290	16.85	0.014800	464.64	1.9867	0.009953	448.15	1.8879	0.006738	424.90	1.7792	290
300	26.85	0.015779	476.58	2.0272	0.010874	462.86	1.9378	0.007788	445.96	1.8507	300
320	46.85	0.017578	498.99	2.0995	0.012452	488.54	2.0207	0.009327	476.88	1.9506	320
340	66.85	0.019242	520.34	2.1643	0.013844	511.85	2.0914	0.010584	502.75	2.0291	340
360	86.85	0.020820	541.18	2.2238	0.015130	534.03	2.1548	0.011707	526.54	2.0971	360
380	106.85	0.022339	561.78	2.2795	0.016347	555.62	2.2131	0.012749	549.26	2.1585	380
400	126.85	0.023814	582.30	2.3321	0.017515	576.90	2.2677	0.013736	571.39	2.2152	400
420	146.85	0.025256	602.82	2.3822	0.018647	598.04	2.3193	0.014683	593.19	2.2684	420
440	166.85	0.026673	623.42	2.4301	0.019752	619.14	2.3684	0.015602	614.82	2.3188	440
460	186.85	0.028070	644.14	2.4762	0.020836	640.28	2.4154	0.016498	636.40	2.3667	460
480	206.85	0.029450	665.00	2.5206	0.021901	661.50	2.4605	0.017376	657.99	2.4127	480
500	226.85	0.030816	686.01	2.5634	0.022954	682.82	2.5040	0.018240	679.63	3.0207	500
525	251.85	0.032509	712.52	2.6152	0.024252	709.66	2.5564	0.019302	706.81	2.5099	525
550	276.85	0.034186	739.30	2.6650	0.025536	736.72	2.6068	0.020350	734.16	2.5608	550
575	301.85	0.035851	766.36	2.7131	0.026808	764.03	2.6553	0.021385	761.71	2.6097	575
600	326.85	0.037508	793.70	2.7597	0.028070	791.58	2.7022	0.022411	789.48	2.6570	600
625	351.85	0.039154	821.32	2.8048	0.029323	819.39	2.7476	0.023427	817.48	2.7027	625
650	376.85	0.040795	849.21	2.8485	0.030570	847.45	2.7917	0.024437	845.71	2.7470	650
700	426.85	0.044059	905.79	2.9324	0.033045	904.31	2.8759	0.026439	902.86	2.8317	700
800	526.85	0.050533	1021.92	3.0874	0.037943	1020.88	3.0315	0.030391	1019.86	2.9879	800
900	626.85	0.056961	1141.58	3.2283	0.042795	1140.86	3.1728	0.034298	1140.14	3.1296	900
1000	726.85	0.063359	1264.27	3.3575	0.047617	1263.78	3.3023	0.038174	1263.30	3.2593	1000

(Contd.)

TABLE: 9.3 Thermodynamic Properties of Carbon Dioxide (Refrigerant R–744) (Superheat Region) [Contd.]

Temperature		Absolute Pressure (Saturation Temperature)									Temp.
T	t	60 bar (–21.98°C/295.13 K)			70 bar (–28.68°C/301.83 K)			75 bar			T
K	°C	v	i	s	v	i	s	v	i	s	K
Satd.	Cond.	0.004742	403.33	1.6860	0.003289	376.92	1.5842	Non-Existent			
310	36.85	0.006452	445.76	1.8268	0.004748	423.32	1.7364	0.003947	407.28	1.6777	310
320	46.85	0.007189	463.55	1.8833	0.005595	447.76	1.8141	0.004925	438.51	1.7770	320
330	56.85	0.007820	478.92	1.9306	0.006246	466.31	1.8712	0.005602	459.35	1.8411	330
340	66.85	0.008391	492.95	1.9725	0.006805	482.29	1.9189	0.006164	476.59	1.8926	340
360	86.85	0.009417	518.69	2.0461	0.007775	510.44	1.9994	0.007116	506.15	1.9772	360
380	106.85	0.010347	542.70	2.1110	0.008631	535.92	2.0683	0.007944	532.45	2.0483	380
400	126.85	0.011216	565.75	2.1702	0.009418	560.00	2.1301	0.008699	557.09	2.1115	400
410	136.85	0.011634	577.06	2.1981	0.009793	571.72	2.1590	0.009058	569.02	2.1410	410
420	146.85	0.012043	588.27	2.2251	0.010159	583.28	2.1869	0.009406	580.76	2.1693	420
430	156.85	0.012444	599.40	2.2513	0.010516	594.73	2.2138	0.009747	592.37	2.1966	430
440	166.85	0.012838	610.47	2.2767	0.010867	606.08	2.2399	0.010079	603.87	2.2230	440
460	186.85	0.013610	632.51	2.3257	0.011549	628.60	2.2900	0.010727	626.64	2.2736	460
480	206.85	0.014363	654.48	2.3725	0.012213	650.97	2.3376	0.011043	649.21	2.3217	480
500	226.85	0.015100	676.45	2.4173	0.012861	673.27	2.3831	0.011967	671.68	2.3675	500
525	251.85	0.016006	703.97	2.4710	0.013654	701.14	2.4375	0.012714	699.73	2.4223	525
550	276.85	0.016896	731.61	2.5225	0.014431	729.08	2.4895	0.013446	727.82	2.4745	550
575	301.85	0.017773	759.41	2.5719	0.015196	757.13	2.5394	0.014166	756.00	2.5246	575
600	326.85	0.018641	787.40	2.6195	0.015950	785.34	2.5874	0.014875	784.32	2.5728	600
625	351.85	0.019500	815.59	2.6656	0.016697	813.72	2.6337	0.015576	812.79	2.6192	625
650	376.85	0.020352	843.99	2.7101	0.017436	842.29	2.6785	0.016270	841.44	2.6643	650
675	401.85	0.021197	872.60	2.7533	0.018168	871.05	2.7220	0.016957	870.28	2.7078	675
700	426.85	0.022038	901.43	2.7952	0.018896	900.01	2.7641	0.017639	899.31	2.7500	700
800	526.85	0.025358	1018.85	2.9520	0.021765	1017.86	2.9214	0.020328	1017.37	2.9077	800
900	626.85	0.028634	1139.44	3.0740	0.024589	1138.75	3.0638	0.022972	1138.42	3.0502	900
1000	726.85	0.031880	1262.83	3.2240	0.027385	1262.37	3.1940	0.025587	1262.15	3.1805	1000

Temperature		Absolute Pressure									Temp.
T	t	100.0 bar			150.0 bar			200.0 bar			T
K	°C	v	i	s	v	i	s	v	i	s	K
310	36.85	0.001458	297.31	1.3923	0.001246	277.21	1.2193	0.001168	269.99	1.1766	310
320	46.85	0.002231	362.91	1.5135	0.001376	304.61	1.3063	0.001246	292.74	1.2488	320
330	56.85	0.003223	414.74	1.6733	0.001574	335.88	1.4025	0.001345	316.69	1.3225	330
340	66.85	0.003867	443.40	1.7589	0.001867	370.38	1.5055	0.001471	341.92	1.3978	340
350	76.85	0.004371	464.87	1.8212	0.002226	403.40	1.6012	0.001628	368.00	1.4734	350
360	86.85	0.004802	483.04	1.8723	0.002583	431.50	1.6804	0.001813	393.95	1.5465	360
380	106.85	0.005544	514.35	1.9570	0.003210	475.79	1.8003	0.002224	441.80	1.6760	380
400	126.85	0.006191	542.12	2.0283	0.003739	511.23	1.8912	0.002628	482.63	1.7807	400
420	146.85	0.006781	568.00	2.0915	0.004206	542.20	1.9668	0.003001	518.00	1.8671	420
440	166.85	0.007332	592.77	2.1491	0.004630	570.62	2.0329	0.003345	549.80	1.9411	440
460	186.85	0.007855	616.83	2.2026	0.005003	597.48	2.0926	0.003664	579.29	2.0066	460
480	206.85	0.008358	640.45	2.2528	0.005398	623.31	2.1476	0.003965	607.23	2.0661	480
500	226.85	0.008844	663.80	2.3005	0.005755	648.46	2.1989	0.004251	634.10	2.1210	500
525	251.85	0.009433	692.76	2.3570	0.006183	679.27	2.2591	0.004592	666.67	2.1845	525
550	276.85	0.010007	721.60	2.4107	0.006595	709.62	2.3156	0.004919	698.47	2.2437	550
575	301.85	0.010568	750.42	2.4619	0.006994	739.72	2.3691	0.005234	729.77	2.2994	575
600	326.85	0.011118	779.28	2.5111	0.007384	769.67	2.4201	0.005540	760.74	2.3521	600
625	351.85	0.011660	808.24	2.5583	0.007765	799.56	2.4689	0.005838	791.52	2.4023	625
650	376.85	0.012195	837.31	2.6039	0.008139	829.45	2.5158	0.006130	822.19	2.4505	650
675	401.85	0.012724	866.52	2.6480	0.008508	859.38	2.5609	0.006417	852.81	2.4967	675
700	426.85	0.013247	895.88	2.6907	0.008872	889.39	2.6046	0.006699	883.43	2.5412	700
800	526.85	0.015302	1014.99	2.8498	0.010288	1010.52	2.7663	0.007792	1006.46	2.7055	800
900	626.85	0.017313	1136.78	2.9932	0.011663	1133.74	2.9114	0.008846	1131.02	2.8522	900
1000	726.85	0.019296	1261.07	3.1241	0.013011	1259.10	3.0435	0.009875	1257.39	2.9853	1000
1100	826.85	0.021259	1387.57	3.2447	0.014341	1386.45	3.1648	0.010886	1385.53	3.1074	1100

(Contd.)

TABLE: 9.3 Thermodynamic Properties of Carbon Dioxide (Refrigerant R–744) (Superheat Region) [Contd.]

Temperature		Absolute Pressure									Temp.
T	*t*	250.0 bar			300.0 bar			400.0 bar			*T*
K	°C	*v*	*i*	*s*	*v*	*i*	*s*	*v*	*i*	*s*	K
310	36.85	0.001120	266.18	1.1459	0.001085	264.03	1.1212	0.001035	262.43	1.0819	**310**
320	46.85	0.001178	286.94	1.2118	0.001133	283.64	1.1835	0.001071	280.72	1.1400	**320**
330	56.85	0.001247	308.25	1.2774	0.001187	303.54	1.2447	0.001110	299.09	1.1965	**330**
340	66.85	0.001328	330.12	1.3427	0.001248	323.72	1.3049	0.001153	317.54	1.2516	**340**
350	76.85	0.001424	352.48	1.4075	0.001318	344.15	1.3642	0.001199	336.03	1.3052	**350**
360	86.85	0.001534	375.05	1.4711	0.001396	364.74	1.4222	0.001250	354.56	1.3574	**360**
380	106.85	0.001792	419.32	1.5908	0.001576	405.70	1.5329	0.001362	391.52	1.4573	**380**
400	126.85	0.002077	460.49	1.6964	0.001781	445.23	1.6343	0.001488	427.99	1.5508	**400**
420	146.85	0.002363	497.80	1.7874	0.001998	482.48	1.7252	0.001625	463.48	1.6374	**420**
440	166.85	0.002638	531.77	1.8665	0.002217	517.25	1.8061	0.001769	497.72	1.7171	**440**
460	186.85	0.002899	563.23	1.9364	0.002430	549.80	1.8784	0.001915	530.64	1.7902	**460**
480	206.85	0.003147	592.86	1.9995	0.002637	580.58	1.9439	0.002062	562.27	1.8576	**480**
500	226.85	0.003385	621.18	2.0573	0.002837	609.97	2.0039	0.002208	592.75	1.9198	**500**
525	251.85	0.003668	655.27	2.1238	0.003077	645.26	2.0728	0.002386	629.49	1.9915	**525**
550	276.85	0.003939	688.34	2.1854	0.003307	679.38	2.1363	0.002560	665.03	2.0576	**550**
575	301.85	0.004200	720.72	2.2430	0.003530	712.68	2.1955	0.002729	699.64	2.1192	**575**
600	326.85	0.004454	752.63	2.2973	0.003745	745.38	2.2512	0.002894	733.55	2.1769	**600**
625	351.85	0.004700	784.21	2.3488	0.003955	777.67	2.3040	0.003054	766.95	2.2315	**625**
650	376.85	0.004941	815.59	2.3981	0.004160	809.68	2.3542	0.003212	799.95	2.2832	**650**
675	401.85	0.005177	846.84	2.4452	0.004361	841.49	2.4022	0.003366	832.67	2.3326	**675**
700	426.85	0.005408	878.02	2.4906	0.004558	873.17	2.4483	0.003518	865.19	2.3799	**700**
800	526.85	0.006303	1002.81	2.6572	0.005318	999.56	2.6170	0.004101	994.28	2.5523	**800**
900	626.85	0.007162	1128.62	2.8054	0.006044	1126.52	2.7666	0.004659	1123.24	2.7042	**900**
1000	726.85	0.007997	1255.93	2.9395	0.006750	1254.70	2.9016	0.005198	1252.97	2.8409	**1000**
1100	826.85	0.008817	1384.80	3.0623	0.007440	1384.27	3.0251	0.005725	1383.77	2.9655	**1100**

(Contd.)

TABLE: 9.3 Thermodynamic Properties of Carbon Dioxide (Refrigerant R–744) (Superheat Region) [Contd.]

Temperature		Absolute Pressure									Temp.
T	*t*	500.0 bar			750.0 bar			1000.0 bar			*T*
K	°C	*v*	*i*	*s*	*v*	*i*	*s*	*v*	*i*	*s*	K
310	36.85	0.001000	262.85	1.0505	0.000941	268.26	0.9899	0.000902	276.82	0.9433	**310**
320	46.85	0.001029	280.38	1.1062	0.000962	728.79	1.0424	0.000919	292.85	0.9942	**320**
330	56.85	0.001061	297.92	1.1601	0.000984	301.27	1.0931	0.000937	308.80	1.0433	**330**
340	66.85	0.001094	315.46	1.2125	0.001007	317.68	1.1421	0.000955	324.68	1.0907	**340**
350	76.85	0.001130	332.98	1.2633	0.001031	334.03	1.1895	0.000973	340.48	1.1365	**350**
360	86.85	0.001168	350.47	1.3125	0.001056	350.31	1.2353	0.000992	356.21	1.1808	**360**
380	106.85	0.001251	385.29	1.4067	0.001108	382.65	1.3228	0.001031	387.43	1.2652	**380**
400	126.85	0.001341	419.77	1.4951	0.001163	414.67	1.4049	0.001072	418.36	1.3445	**400**
420	146.85	0.001440	453.70	1.5779	0.001221	446.34	1.4821	0.001114	448.98	1.4192	**420**
440	166.85	0.001544	486.90	1.6551	0.001282	477.64	1.5550	0.001158	479.31	1.4898	**440**
460	186.85	0.001651	519.26	1.7270	0.001345	508.56	1.6237	0.001203	509.35	1.5565	**460**
480	206.85	0.001761	550.76	1.7941	0.001410	539.08	1.6886	0.001249	539.12	1.6199	**480**
500	226.85	0.001871	581.43	1.8567	0.001476	569.18	1.7501	0.001296	568.62	1.6801	**500**
525	251.85	0.002009	618.69	1.9294	0.001560	606.24	1.8224	0.001356	605.15	1.7514	**525**
550	276.85	0.002145	654.92	1.9968	0.001645	642.69	1.8902	0.001417	641.29	1.8186	**550**
575	301.85	0.002279	690.29	2.0597	0.001730	678.59	1.9541	0.001478	677.09	1.8823	**575**
600	326.85	0.002411	724.97	2.1188	0.001814	714.01	2.0144	0.001539	712.56	1.9427	**600**
625	351.85	0.002540	759.10	2.1745	0.001898	749.00	2.0715	0.001600	747.74	2.0001	**625**
650	376.85	0.002666	792.81	2.2274	0.001982	783.63	2.1258	0.001661	782.65	2.0549	**650**
675	401.85	0.002791	826.19	2.2778	0.002064	817.95	2.1776	0.001722	817.34	2.1073	**675**
700	426.85	0.002913	859.33	2.3260	0.002146	852.02	2.2272	0.001783	851.83	2.1574	**700**
800	526.85	0.003386	990.53	2.5012	0.002463	986.66	2.4070	0.002020	988.37	2.3398	**800**
900	626.85	0.003838	1121.12	2.6550	0.002769	1120.12	2.5642	0.002251	1123.68	2.4992	**900**
1000	726.85	0.004276	1252.13	2.7930	0.003065	1253.51	2.7048	0.002474	1258.70	2.6414	**1000**
1100	826.85	0.004702	1383.98	2.9187	0.003354	1387.34	2.8323	0.002693	1393.93	2.7703	**1100**

Adapted from R. Span and W. Wagner, *J. Phys. Chem. Ref. Data,* **25**(6), 1996, p. 1509.

TABLE: 10.1 Thermodynamic Properties of Saturated n-Hexane up to 1 Atmosphere (Temperature Basis)

Temperature		Saturation pressure, p		Liquid density	Specific volume of vapour v''	Specific enthalpy			Specific entropy		Temp
t	T			ρ_l		h	λ_v	H	s'	s''	t
°C	K	bar	torr	kg/m^3	m^3/kg	kJ/kg			kJ/(kg·K)		°C
−23	250.15	0.0150	11.24	697.26	15.88	141.17	396.48	537.66	0.7683	2.4020	−23
−22	251.15	0.0160	11.97	696.43	14.92	143.69	395.88	539.58	0.7916	2.4015	−22
−21	252.15	0.0170	12.74	695.61	14.01	146.21	395.28	541.50	0.8010	2.4010	−21
−20	253.15	0.0181	13.56	694.68	13.16	148.74	394.68	543.42	0.8104	2.4006	−20
−19	254.15	0.0192	14.43	693.76	12.37	151.26	394.08	545.34	0.8198	2.4002	−19
−18	255.15	0.0205	15.35	692.84	11.63	153.78	393.48	547.26	0.8292	2.3999	−18
−17	256.15	0.0218	16.33	691.92	10.80	156.71	392.88	549.59	0.8386	2.3996	−17
−16	257.15	0.0231	17.36	691.01	10.23	159.23	392.28	551.51	0.8480	2.3994	−16
−15	258.15	0.0246	18.45	690.09	9.690	161.75	391.67	553.43	0.8573	2.3992	−15
−14	259.15	0.0262	19.61	689.18	9.183	164.28	391.07	555.35	0.8667	2.3991	−14
−13	260.15	0.0278	20.84	688.28	8.704	166.80	390.47	557.27	0.8761	2.3990	−13
−12	261.15	0.0295	22.14	687.37	8.251	169.32	389.87	559.19	0.8855	2.3989	−12
−11	262.15	0.0314	23.52	686.47	7.824	171.84	389.27	561.11	0.8949	2.3989	−11
−10	263.15	0.0333	24.97	685.56	7.338	174.37	388.67	563.04	0.9043	2.3989	−10
−9	264.15	0.0353	26.51	684.67	6.961	176.89	388.07	564.96	0.9155	2.3990	−9
−8	265.15	0.0375	28.14	683.77	6.604	179.41	387.47	566.88	0.9249	2.3991	−8
−7	266.15	0.0398	29.86	682.88	6.267	181.93	386.87	568.80	0.9343	2.3993	−7
−6	267.15	0.0422	31.67	681.98	5.949	184.46	386.26	570.72	0.9437	2.3995	−6
−5	268.15	0.0448	33.60	681.09	5.647	186.98	385.66	572.64	0.9531	2.3997	−5
−4	269.15	0.0475	35.63	680.21	5.362	189.50	385.06	574.56	0.9625	2.4000	−4
−3	270.15	0.0504	37.77	679.32	5.093	192.43	384.46	576.89	0.9718	2.4000	−3
−2	271.15	0.0534	40.04	678.44	4.837	194.95	383.86	578.81	0.9812	2.4004	−2
−1	272.15	0.0566	42.43	677.56	4.596	197.48	383.26	580.74	0.9906	2.4008	−1
0	273.15	0.0610	45.76	676.68	4.367	200.00	382.66	582.66	1.0000	2.4013	0
1	274.15	0.0641	48.09	675.81	4.151	202.52	382.03	584.56	1.0094	2.4018	1
2	275.15	0.0674	50.53	674.93	3.946	205.05	381.38	586.43	1.0150	2.4016	2
3	276.15	0.0708	53.08	674.06	3.751	207.57	380.72	588.30	1.0243	2.4022	3
4	277.15	0.0743	55.75	673.19	3.567	210.10	380.07	590.17	1.0337	2.4028	4
5	278.15	0.0781	58.55	671.60	3.393	212.62	379.42	592.04	1.0431	2.4035	5
6	279.15	0.0820	61.47	670.73	3.228	215.14	378.76	593.91	1.0525	2.4042	6
7	280.15	0.0860	64.53	669.87	3.071	217.67	378.11	595.78	1.0618	2.4043	7
8	281.15	0.0903	67.73	669.01	2.923	220.19	377.45	597.65	1.0712	2.4054	8
9	282.15	0.0948	71.08	668.15	2.755	222.72	376.80	599.52	1.0806	2.4064	9
10	283.15	0.101	75.64	667.29	2.639	225.24	376.15	601.39	1.0900	2.4075	10
11	284.15	0.106	79.35	666.44	2.527	227.76	375.49	603.26	1.0994	2.4086	11
12	285.15	0.111	83.23	665.59	2.421	229.43	374.84	604.27	1.0979	2.4098	12
13	286.15	0.116	87.28	664.74	2.319	231.95	374.19	606.14	1.1072	2.4109	13
14	287.15	0.122	91.52	663.89	2.222	234.48	373.53	608.01	1.1166	2.4121	14
15	288.15	0.128	95.94	663.05	2.130	237.00	372.88	609.87	1.1259	2.4133	15
16	289.15	0.134	100.57	662.33	2.041	239.52	372.39	611.91	1.1353	2.4145	16
17	290.15	0.141	105.40	661.37	1.957	242.04	371.74	613.78	1.1446	2.4157	17
18	291.15	0.147	110.44	660.41	1.876	244.56	371.08	615.64	1.1539	2.4169	18
19	292.15	0.154	115.71	659.46	1.799	247.08	370.43	617.51	1.1633	2.4182	19
20	293.15	0.162	121.22	658.51	1.726	248.01	369.78	617.79	1.1726	2.4195	20
21	294.15	0.169	126.96	657.56	1.655	250.38	369.12	619.51	1.1820	2.4208	21
22	295.15	0.177	132.95	656.62	1.588	252.75	368.47	621.22	1.1913	2.4221	22
23	296.15	0.186	139.21	655.68	1.523	255.12	367.82	622.94	1.1957	2.4234	23
24	297.15	0.194	145.73	654.74	1.462	257.49	367.16	624.66	1.2031	2.4248	24
25	298.15	0.203	152.54	653.80	1.403	259.86	366.51	626.38	1.2105	2.4262	25

(Contd.)

TABLE: 10.1 Thermodynamic Properties of Saturated *n*-Hexane up to 1 Atmosphere (Temperature Basis) [Contd.]

Temperature		Saturation pressure, p		Liquid density	Specific volume of vapour v''	Specific enthalpy			Specific entropy		Temp
t	T			ρ_l		h	λ_v	H	s'	s''	t
°C	K	bar	torr	kg/m³	m³/kg	kJ/kg			kJ/(kg·K)		°C
26	299.15	0.213	159.64	652.87	1.347	262.23	365.86	628.09	1.2179	2.4276	**26**
27	300.15	0.223	167.04	651.94	1.293	264.61	365.20	629.81	1.2253	2.4290	**27**
28	301.15	0.233	174.76	651.01	1.241	266.98	364.55	631.53	1.2327	2.4304	**28**
29	302.15	0.244	182.82	650.08	1.192	269.35	363.90	633.24	1.2401	2.4319	**29**
30	303.15	0.249	187.01	649.16	1.145	271.72	363.24	634.96	1.2475	2.4333	**30**
31	304.15	0.257	192.44	648.24	1.099	274.09	362.79	636.88	1.2549	2.4348	**31**
32	305.15	0.267	200.07	647.32	1.056	276.46	362.04	638.50	1.2654	2.4363	**32**
33	306.15	0.277	207.97	646.41	1.015	278.83	361.29	640.12	1.2728	2.4379	**33**
34	307.15	0.288	216.16	645.49	0.9749	281.20	360.55	641.74	1.2802	2.4394	**34**
35	308.15	0.307	230.12	644.58	0.9369	283.57	359.80	643.37	1.2876	2.4410	**35**
36	309.15	0.319	239.12	643.68	0.9005	285.94	359.05	644.99	1.2950	2.4426	**36**
37	310.15	0.331	248.45	642.77	0.8656	288.31	358.30	646.61	1.3024	2.4442	**37**
38	311.15	0.344	258.11	641.87	0.8322	290.68	357.55	648.23	1.3098	2.4458	**38**
39	312.15	0.357	268.11	640.97	0.8136	293.05	356.80	649.86	1.3172	2.4474	**39**
40	313.15	0.371	278.47	640.07	0.7861	295.42	356.05	651.48	1.3246	2.4491	**40**
41	314.15	0.386	289.19	639.18	0.7597	297.79	355.31	653.10	1.3321	2.4508	**41**
42	315.15	0.400	300.28	638.28	0.7342	300.16	354.56	654.72	1.3401	2.4524	**42**
43	316.15	0.416	311.77	637.39	0.7097	302.54	353.81	656.34	1.3475	2.4542	**43**
44	317.15	0.432	323.65	636.51	0.6860	304.91	353.06	657.97	1.3549	2.4559	**44**
45	318.15	0.448	335.95	635.62	0.6632	307.28	352.31	659.59	1.3623	2.4577	**45**
46	319.15	0.465	348.68	635.41	0.6413	309.65	351.56	661.21	1.3697	2.4594	**46**
47	320.15	0.484	362.89	634.42	0.6201	312.02	350.82	662.83	1.3772	2.4612	**47**
48	321.15	0.502	376.55	633.43	0.5997	314.39	350.07	664.45	1.3846	2.4630	**48**
49	322.15	0.521	390.68	632.45	0.5800	316.76	349.32	666.08	1.3920	2.4649	**49**
50	323.15	0.540	405.30	631.47	0.5610	319.13	348.57	667.70	1.3994	2.4667	**50**
51	324.15	0.560	420.41	630.49	0.5427	321.50	347.82	669.32	1.4068	2.4686	**51**
52	325.15	0.581	436.04	629.51	0.5251	323.87	347.07	670.94	1.4139	2.4705	**52**
53	326.15	0.603	452.19	628.54	0.5081	326.24	346.32	672.57	1.4213	2.4724	**53**
54	327.15	0.625	468.90	627.57	0.4860	328.61	345.58	674.19	1.4287	2.4743	**54**
55	328.15	0.648	486.16	626.61	0.4703	330.98	344.83	675.81	1.4361	2.4762	**55**
56	329.15	0.670	502.55	625.64	0.4552	333.35	344.08	677.43	1.4435	2.4782	**56**
57	330.15	0.695	520.94	624.68	0.4406	335.72	343.33	679.05	1.4509	2.4789	**57**
58	331.15	0.720	539.95	623.73	0.4266	338.09	342.58	680.68	1.4584	2.4809	**58**
59	332.15	0.746	559.58	622.77	0.4130	340.46	341.83	682.30	1.4658	2.4829	**59**
60	333.15	0.763	572.26	621.82	0.3999	342.84	341.08	683.92	1.4732	2.4849	**60**
61	334.15	0.787	590.20	620.87	0.3850	345.21	340.63	685.84	1.4806	2.4870	**61**
62	335.15	0.811	608.64	619.92	0.3728	347.58	339.77	687.35	1.4843	2.4878	**62**
63	336.15	0.837	627.61	618.98	0.3611	350.75	338.91	689.66	1.4917	2.4898	**63**
64	337.15	0.863	647.10	618.04	0.3497	353.13	338.05	691.18	1.4991	2.4919	**64**
65	338.15	0.894	670.65	617.10	0.3388	355.50	337.19	692.69	1.5065	2.4941	**65**
66	339.15	0.922	691.36	616.17	0.3282	357.87	336.33	694.20	1.5139	2.4962	**66**
67	340.15	0.950	712.65	615.24	0.3180	360.25	335.47	695.72	1.5213	2.4987	**67**
68	341.15	0.979	734.52	614.31	0.3082	362.62	334.61	697.23	1.5338	2.5010	**68**
68.736*	**341.886**	**1.01325**	**760.00**	**613.63**	**0.3011**	**364.37**	**333.98**	**698.34**	**1.5388**	**2.5027**	**68.736**

*Normal boiling point

TABLE: 10.2 Thermodynamic Properties of Saturated *n*–Hexane above 1 Atmosphere (Temperature Basis)

Temperature *t* °C	Temperature *T* K	Saturation pressure, *p* bar	Liquid density ρ_l kg/m³	Specific volume of vapour *v″* m³/kg	Specific enthalpy *h* kJ/kg	Specific enthalpy λ_v kJ/kg	Specific enthalpy *H* kJ/kg	Specific entropy *s′* kJ/(kg·K)	Specific entropy *s″* kJ/(kg·K)	Temp *t* °C
69	342.15	1.019	613.38	0.3063	364.99	333.75	698.74	1.5406	2.5033	**69**
70	343.15	1.050	612.46	0.2977	367.36	332.89	700.26	1.5474	2.5057	**70**
71	344.15	1.082	611.54	0.2894	369.74	332.03	701.77	1.5542	2.5080	**71**
72	345.15	1.115	610.62	0.2814	372.11	331.17	703.28	1.5610	2.5103	**72**
73	346.15	1.148	609.19	0.2736	373.62	330.31	703.93	1.5643	2.5126	**73**
74	347.15	1.183	608.27	0.2660	376.09	329.45	705.54	1.5711	2.5149	**74**
75	348.15	1.219	607.37	0.2587	378.56	328.59	707.15	1.5779	2.5173	**75**
76	349.15	1.266	606.46	0.2516	381.03	327.73	708.76	1.5846	2.5196	**76**
77	350.15	1.304	605.47	0.2418	383.50	326.87	710.37	1.5914	2.5219	**77**
78	351.15	1.343	604.38	0.2352	385.97	326.01	711.98	1.5982	2.5242	**78**
79	352.15	1.383	603.29	0.2288	388.44	325.15	713.59	1.6050	2.5265	**79**
80	353.15	1.424	602.21	0.2226	390.91	324.29	715.20	1.6118	2.5289	**80**
81	354.15	1.466	601.13	0.2165	393.38	323.43	716.81	1.6185	2.5312	**81**
82	355.15	1.509	600.05	0.2107	395.85	322.57	718.42	1.6253	2.5335	**82**
83	356.15	1.553	598.98	0.2050	398.32	321.71	720.03	1.6321	2.5358	**83**
84	357.15	1.599	597.91	0.1989	400.79	320.85	721.64	1.6389	2.5381	**84**
85	358.15	1.646	596.84	0.1936	403.26	319.99	723.25	1.6457	2.5405	**85**
86	359.15	1.684	595.78	0.1884	405.73	319.13	724.86	1.6524	2.5428	**86**
87	360.15	1.733	594.72	0.1834	408.20	318.27	726.47	1.6592	2.5451	**87**
88	361.15	1.783	593.67	0.1785	410.67	317.41	728.08	1.6660	2.5474	**88**
89	362.15	1.834	592.62	0.1738	413.14	316.55	729.69	1.6728	2.5498	**89**
90	363.15	1.887	591.57	0.1692	415.61	315.69	731.30	1.6796	2.5521	**90**
91	364.15	1.941	590.53	0.1647	418.08	314.83	732.91	1.6863	2.5544	**91**
92	365.15	1.997	589.49	0.1604	420.55	313.97	734.52	1.6931	2.5567	**92**
93	366.15	2.054	588.46	0.1562	423.02	313.11	736.13	1.6999	2.5590	**93**
94	367.15	2.113	587.43	0.1521	425.49	312.25	737.74	1.7067	2.5614	**94**
95	368.15	2.173	586.40	0.1489	427.96	311.39	739.35	1.7135	2.5637	**95**
96	369.15	2.199	585.37	0.1453	430.43	310.21	740.64	1.7202	2.5660	**96**
97	370.15	2.285	584.35	0.1418	432.90	309.35	742.25	1.7270	2.5683	**97**
98	371.15	2.342	583.34	0.1385	435.37	308.49	743.86	1.7338	2.5706	**98**
99	372.15	2.400	582.32	0.1352	437.84	307.66	745.51	1.7406	2.5730	**99**
100	373.15	2.459	581.31	0.1320	440.31	306.70	747.01	1.7474	2.5753	**100**
101	374.15	2.520	580.30	0.1289	442.78	305.74	748.52	1.7541	2.5776	**101**
102	375.15	2.582	579.30	0.1259	445.25	304.77	750.02	1.7598	2.5799	**102**
103	376.15	2.646	578.30	0.1229	447.72	303.81	751.53	1.7666	2.5822	**103**
104	377.15	2.710	577.30	0.1199	450.19	302.84	753.03	1.7733	2.5846	**104**
105	378.15	2.776	576.31	0.1171	452.66	301.88	754.54	1.7801	2.5869	**105**
106	379.15	2.857	575.32	0.1144	455.13	300.91	756.05	1.7869	2.5892	**106**
107	380.15	2.926	574.34	0.1118	457.60	299.95	757.55	1.7937	2.5915	**107**
108	381.15	2.997	573.11	0.1092	460.07	298.99	759.06	1.8007	2.5938	**108**
109	382.15	3.069	571.89	0.1067	462.54	298.02	760.56	1.8073	2.5962	**109**
110	383.15	3.143	570.67	0.1042	464.17	297.06	761.22	1.8139	2.5985	**110**
111	384.15	3.219	569.46	0.1016	466.90	296.09	762.99	1.8205	2.6008	**111**
112	385.15	3.296	568.26	0.09931	469.62	295.13	764.75	1.8271	2.6031	**112**
113	386.15	3.375	567.06	0.09704	472.35	294.16	766.52	1.8337	2.6055	**113**
114	387.15	3.455	565.86	0.09484	475.08	293.20	768.28	1.8403	2.6078	**114**
115	388.15	3.544	564.67	0.09268	477.81	292.24	770.05	1.8469	2.6101	**115**

(Contd.)

TABLE: 10.2 Thermodynamic Properties of Saturated *n*–Hexane above 1 Atmosphere (Temperature Basis) [Contd.]

Temperature		Saturation pressure,	Liquid density	Specific volume of	Specific enthalpy			Specific entropy		Temp
t	T	p	ρ_l	vapour v''	h	λ_v	H	s'	s''	t
°C	K	bar	kg/m^3	m^3/kg	kJ/kg			kJ/(kg·K)		°C
116	389.15	3.628	563.48	0.09059	480.54	291.27	771.81	1.8535	2.6124	**116**
117	390.15	3.714	562.30	0.08854	483.27	290.31	773.58	1.8601	2.6147	**117**
118	391.15	3.802	561.13	0.08655	486.00	289.34	775.34	1.8667	2.6171	**118**
119	392.15	3.891	559.95	0.08460	488.73	288.38	777.11	1.8733	2.6194	**119**
120	393.15	3.982	558.79	0.08270	491.46	287.41	778.87	1.8799	2.6217	**120**
121	394.15	4.076	557.63	0.08086	494.19	286.45	780.64	1.8865	2.6240	**121**
122	395.15	4.171	556.47	0.07905	496.52	285.49	782.00	1.8931	2.6263	**122**
123	396.15	4.268	555.32	0.07729	499.25	284.52	783.77	1.8997	2.6287	**123**
124	397.15	4.367	554.17	0.07558	501.98	283.56	785.53	1.9063	2.6310	**124**
125	398.15	4.480	553.03	0.07391	504.70	282.59	787.30	1.9129	2.6333	**125**
126	399.15	4.574	551.89	0.07200	507.43	281.63	789.06	1.9195	2.6356	**126**
127	400.15	4.671	550.76	0.07047	510.16	280.66	790.82	1.9261	2.6395	**127**
128	401.15	4.769	549.63	0.06897	512.89	279.70	792.59	1.9327	2.6419	**128**
129	402.15	4.869	548.50	0.06750	515.62	278.74	794.36	1.9393	2.6442	**129**
130	403.15	4.971	547.39	0.06607	518.34	277.77	796.11	1.9459	2.6465	**130**
131	404.15	5.075	546.27	0.06468	521.07	276.81	797.88	1.9525	2.6488	**131**
132	405.15	5.181	545.16	0.06332	523.80	275.84	799.64	1.9591	2.6528	**132**
133	406.15	5.288	544.06	0.06199	526.53	274.41	800.93	1.9657	2.6551	**133**
134	407.15	5.398	542.95	0.06069	529.26	273.44	802.70	1.9723	2.6575	**134**
135	408.15	5.509	541.86	0.05942	531.98	272.48	804.46	1.9789	2.6598	**135**
136	409.15	5.647	540.77	0.05818	534.71	271.51	806.22	1.9855	2.6621	**136**
137	410.15	5.763	539.39	0.05697	537.44	270.54	807.98	1.9921	2.6644	**137**
138	411.15	5.881	538.17	0.05578	540.17	269.58	809.75	1.9987	2.6668	**138**
139	412.15	6.002	536.95	0.05463	542.90	268.61	811.51	2.0053	2.6691	**139**
140	413.15	6.124	535.71	0.05350	545.62	267.65	813.27	2.0119	2.6714	**140**
141	414.15	6.249	534.46	0.05240	548.35	266.68	815.03	2.0185	2.6738	**141**
142	415.15	6.376	533.19	0.05132	551.08	265.72	816.80	2.0251	2.6786	**142**
143	416.15	6.505	531.92	0.05027	553.81	264.75	818.56	2.0317	2.6809	**143**
144	417.15	6.637	530.63	0.04924	556.54	263.06	819.60	2.0383	2.6832	**144**
145	418.15	6.771	529.33	0.04823	559.26	262.82	822.08	2.0449	2.6856	**145**
146	419.15	6.895	528.03	0.04725	561.99	261.86	823.85	2.0515	2.6879	**146**
147	420.15	7.033	526.71	0.04629	564.72	260.89	825.61	2.0581	2.6903	**147**
148	421.15	7.174	525.38	0.04535	567.45	258.12	825.56	2.0647	2.6926	**148**
149	422.15	7.317	524.04	0.04443	570.18	257.15	827.32	2.0713	2.6949	**149**
150	423.15	7.463	522.69	0.04354	572.90	256.18	829.08	2.0779	2.6973	**150**
151	424.15	7.612	521.32	0.04266	575.63	255.21	830.84	2.0845	2.6996	**151**
152	425.15	7.763	519.95	0.04180	578.78	254.24	833.02	2.0911	2.7032	**152**
153	426.15	7.916	518.57	0.04097	581.51	253.27	834.78	2.0977	2.7055	**153**
154	427.15	8.073	517.18	0.04015	584.24	252.30	836.54	2.1043	2.7079	**154**
155	428.15	8.232	515.79	0.03935	586.97	251.33	838.30	2.1109	2.7102	**155**
156	429.15	8.403	514.38	0.03856	589.70	249.90	839.60	2.1175	2.7126	**156**
157	430.15	8.557	512.96	0.03780	592.43	248.65	841.08	2.1241	2.7149	**157**
158	431.15	8.714	511.54	0.03705	595.16	247.40	842.55	2.1014	2.7172	**158**
159	432.15	8.873	510.11	0.03628	597.89	246.14	844.03	2.1090	2.7196	**159**
160	433.15	9.035	508.66	0.03555	600.62	244.89	845.51	2.1167	2.7219	**160**
161	434.15	9.199	507.22	0.03485	604.98	243.64	848.63	2.1243	2.7242	**161**
162	435.15	9.366	505.76	0.03416	607.68	242.39	850.07	2.1320	2.7283	**162**
163	436.15	9.536	504.29	0.03349	610.40	241.14	851.54	2.1396	2.7307	**163**
164	437.15	9.708	502.82	0.03283	613.14	239.89	853.03	2.1473	2.7330	**164**
165	438.15	9.883	501.34	0.03218	615.91	238.64	854.55	2.1549	2.7353	**165**

(Contd.)

TABLE: 10.2 Thermodynamic Properties of Saturated n–Hexane above 1 Atmosphere (Temperature Basis) [Contd.]

Temperature		Saturation pressure,	Liquid density	Specific volume of	Specific enthalpy			Specific entropy		Temp
t	T	p	ρ_l	vapour v''	h	λ_v	H	s'	s''	t
°C	K	bar	kg/m³	m³/kg	kJ/kg			kJ/(kg·K)		°C
166	439.15	10.10	499.86	0.03155	618.69	237.39	856.08	2.1626	2.7377	**166**
167	440.15	10.28	498.29	0.03093	621.50	236.14	857.64	2.1702	2.7400	**167**
168	441.15	10.46	496.84	0.03033	624.33	234.89	859.21	2.1779	2.7424	**168**
169	442.15	10.65	495.36	0.02974	627.18	233.64	860.81	2.1811	2.7447	**169**
170	443.15	10.84	493.85	0.02916	630.05	234.70	864.75	2.1888	2.7471	**170**
171	444.15	11.03	492.32	0.02860	632.95	233.45	866.40	2.1964	2.7494	**171**
172	445.15	11.23	490.75	0.02804	636.26	232.21	868.47	2.2040	2.7517	**172**
173	446.15	11.43	489.16	0.02750	639.20	230.96	870.16	2.2117	2.7541	**173**
174	447.15	11.63	487.55	0.02697	642.16	229.72	871.88	2.2193	2.7564	**174**
175	448.15	11.83	485.90	0.02646	645.14	228.47	873.61	2.2270	2.7588	**175**
176	449.15	12.07	484.23	0.02595	648.15	227.22	875.37	2.2346	2.7611	**176**
177	450.15	12.28	482.54	0.02545	651.18	225.98	877.16	2.2422	2.7635	**177**
178	451.15	12.50	480.82	0.02497	654.23	224.21	878.44	2.2499	2.7658	**178**
179	452.15	12.72	479.08	0.02449	657.30	222.96	880.26	2.2575	2.7681	**179**
180	453.15	12.94	477.32	0.02403	660.39	221.72	882.11	2.2652	2.7705	**180**
181	454.15	13.16	475.53	0.02357	663.51	220.47	883.98	2.2728	2.7728	**181**
182	455.15	13.39	473.72	0.02312	666.64	219.22	885.87	2.2804	2.7752	**182**
183	456.15	13.62	471.89	0.02269	669.80	217.97	887.78	2.2881	2.7775	**183**
184	457.15	13.86	470.03	0.02226	672.98	216.73	889.71	2.2957	2.7798	**184**
185	458.15	14.11	468.16	0.02184	676.19	214.28	890.47	2.3034	2.7822	**185**
186	459.15	14.35	466.27	0.02123	679.41	211.84	891.25	2.3110	2.7845	**186**
187	460.15	14.58	464.36	0.02076	682.66	210.58	893.24	2.3186	2.7869	**187**
188	461.15	14.83	462.43	0.02029	685.93	209.33	895.26	2.3263	2.7892	**188**
189	462.15	15.07	460.48	0.01984	689.22	207.35	896.56	2.3339	2.7916	**189**
190	463.15	15.32	458.51	0.01940	692.53	205.83	898.36	2.3416	2.7939	**190**
191	464.15	15.57	456.53	0.01897	695.86	204.26	900.12	2.3492	2.7962	**191**
192	465.15	15.83	454.53	0.01855	699.22	202.64	901.86	2.3568	2.7986	**192**
193	466.15	16.09	452.52	0.01814	702.60	200.97	903.56	2.3645	2.8009	**193**
194	467.15	16.35	450.49	0.01774	705.99	199.25	905.24	2.3721	2.8033	**194**
195	468.15	16.63	448.44	0.01735	709.42	197.47	906.89	2.3798	2.8056	**195**
196	469.15	16.90	446.38	0.01696	712.86	195.65	908.51	2.3874	2.8079	**196**
197	470.15	17.17	444.31	0.01659	716.32	193.77	910.09	2.3950	2.8103	**197**
198	471.15	17.45	442.23	0.01623	719.81	191.84	911.65	2.4027	2.8126	**198**
199	472.15	17.73	440.13	0.01587	723.32	189.86	913.18	2.4103	2.8150	**199**
200	473.15	18.02	438.02	0.01553	726.85	187.83	914.68	2.4180	2.8173	**200**
201	474.15	18.31	434.73	0.01519	730.40	185.74	916.15	2.4256	2.8197	**201**
202	475.15	18.61	432.60	0.01486	733.98	183.61	917.59	2.4332	2.8220	**202**
203	476.15	18.90	430.47	0.01454	737.57	181.42	918.99	2.4409	2.8243	**203**
204	477.15	19.21	428.32	0.01422	741.19	179.18	920.37	2.4485	2.8267	**204**
205	478.15	19.51	426.17	0.01392	744.83	176.89	921.72	2.4562	2.8290	**205**
206	479.15	19.79	424.00	0.01362	748.49	174.54	923.03	2.4638	2.8314	**206**
207	480.15	20.11	421.83	0.01333	752.18	172.14	924.32	2.4714	2.8331	**207**
208	481.15	20.43	415.82	0.01304	755.88	169.69	925.58	2.4791	2.8360	**208**
209	482.15	20.75	413.65	0.01276	759.61	167.19	926.80	2.4867	2.8388	**209**
210	483.15	21.08	411.48	0.01249	763.36	164.63	927.99	2.4944	2.8415	**210**
211	484.15	21.41	409.30	0.01222	770.62	162.02	932.64	2.5020	2.8441	**211**
212	485.15	21.75	407.11	0.01198	774.43	159.36	933.79	2.5096	2.8408	**212**
213	486.15	22.09	403.18	0.01168	778.26	156.64	934.90	2.5221	2.8406	**213**
214	487.15	22.44	402.17	0.01138	782.11	153.87	935.99	2.5298	2.8405	**214**
215	488.15	22.71	399.96	0.01108	785.99	151.05	937.04	2.5374	2.8404	**215**

(Contd.)

TABLE: 10.2 Thermodynamic Properties of Saturated n–Hexane above 1 Atmosphere (Temperature Basis) [Contd.]

Temperature		Saturation pressure,	Liquid density	Specific volume of	Specific enthalpy			Specific entropy		Temp
t	T	p	ρ_l	vapour v''	h	λ_v	H	s'	s''	t
°C	K	bar	kg/m³	m³/kg	kJ/kg			kJ/(kg·K)		°C
216	489.15	23.07	396.60	0.01077	789.89	148.17	938.06	2.5451	2.8402	**216**
217	490.15	23.42	392.14	0.01047	793.80	145.24	939.05	2.5527	2.8401	**217**
218	491.15	23.78	386.65	0.01017	797.75	142.26	940.01	2.5604	2.8399	**218**
219	492.15	24.15	384.83	0.01027	801.71	139.22	940.93	2.5680	2.8398	**219**
220	493.15	24.52	376.87	0.009968	805.69	136.13	941.82	2.5757	2.8397	**220**
221	494.15	24.89	368.72	0.009666	809.70	124.01	933.71	2.5833	2.8395	**221**
222	495.15	25.27	359.93	0.009364	827.02	120.76	947.78	2.6207	2.8394	**222**
223	496.15	25.66	350.61	0.009062	831.11	117.46	948.57	2.6284	2.8368	**223**
224	497.15	26.05	340.88	0.008760	835.23	114.10	949.33	2.6361	2.8366	**224**
225	498.15	26.32	330.84	0.008458	839.37	110.69	950.05	2.6438	2.8365	**225**
226	499.15	26.72	320.59	0.008156	843.53	107.22	950.75	2.6515	2.8363	**226**
227	500.15	27.12	310.22	0.007854	847.71	89.18	936.89	2.6593	2.8362	**227**
228	501.15	27.53	299.80	0.007552	851.91	85.51	937.43	2.6670	2.8360	**228**
229	502.15	27.95	289.42	0.007250	858.66	78.86	937.52	2.6747	2.8359	**229**
230	503.15	28.37	279.13	0.006533	875.58	75.06	950.64	2.6824	2.8357	**230**
231	504.15	28.79	268.99	0.006230	879.91	71.21	951.12	2.6901	2.8356	**231**
232	505.15	29.22	259.04	0.005927	889.36	64.32	953.68	2.6978	2.8355	**232**
233	506.15	29.66	249.32	0.005624	893.76	60.34	954.10	2.7055	2.8353	**233**
234	507.15	30.10	239.86	0.005322	898.18	56.30	954.48	2.7132	2.8352	**234**
234.53*	**507.68**	**30.40**	**235.11**	**0.004253**	**911.55**	**0.00**	**911.55**	**2.7843**	**2.7843**	**234.53**

*Critical point

Reference State:

Specific enthalpy of saturated liquid at 0°C (273.15 K) = 200.00 kJ/kg

Specific entropy of saturated liquid at 0°C (273.15 K) = 1.0000 kJ/(kg·K)

TABLE: 10.2 Thermodynamic Properties of Saturated n-Hexane (Superheat Region)

Temperature		Absolute Pressure (Saturation Temperature)									Temp.
T	t	1.01325 bar (68.74°C/341.89 K)			2.0 bar (92.05°C/365.2 K)			4.0 bar (120.19°C/393.34 K)			T
K	°C	v	i	s	v	i	s	v	i	s	K
Satd.	**cond.**	0.3011	698.34	2.5027	0.1602	734.60	2.5568	0.08235	779.21	2.6221	
400	126.85	0.3701	810.67	2.8133	0.1818	806.52	2.7400	0.08468	797.32	2.6559	**400**
440	166.85	0.4107	898.88	3.0234	0.2038	895.71	2.9525	0.09733	888.94	2.8741	**440**
480	206.85	0.4505	993.41	3.2289	0.2249	990.92	3.1595	0.1090	985.69	3.0845	**480**
520	246.85	0.4900	1094.14	3.4304	0.2456	1092.13	3.3619	0.1201	1087.96	3.289	**520**
560	286.85	0.5291	1200.86	3.6280	0.2659	1199.21	3.5602	0.1308	1195.80	3.4888	**560**
600	326.85	0.5680	1313.33	3.8220	0.2860	1311.95	3.7546	0.1412	1309.11	3.6842	**600**
640	366.85	0.6067	1431.25	4.0122	0.3059	1430.08	3.9452	0.1515	1427.69	3.8755	**640**
680	406.85	0.6453	1554.33	4.1987	0.3258	1553.33	4.1319	0.1617	1551.28	4.0627	**680**

Temperature		Absolute Pressure (Saturation Temperature)									Temp.
T	t	7.0 bar (146.77°C/419.92 K)			10.0 bar (165.46°C/438.61 K)			20.0 bar (206.66°C/479.81 K)			T
K	°C	v	i	s	v	i	s	v	i	s	K
Satd.	**cond.**	0.04651	825.55	2.6897	0.03189	855.25	2.7364	0.01342	923.89	2.8329	
440	166.85	0.05128	877.57	2.8005	0.03230	863.94	2.7417	–	–	–	**440**
480	206.85	0.05910	977.24	3.0173	0.03894	967.99	2.9680	–	–	–	**480**
520	246.85	0.06617	1081.42	3.2256	0.04452	1074.43	3.1810	0.01880	1046.75	3.0723	**520**
560	286.85	0.07281	1190.52	3.4277	0.04958	1184.99	3.3858	0.02230	1164.72	3.2909	**560**
600	326.85	0.07917	1304.76	3.6247	0.05433	1300.29	3.5846	0.02528	1284.37	3.4972	**600**
640	366.85	0.08535	1424.04	3.8171	0.05887	1420.32	3.7782	0.02796	1407.37	3.6956	**640**
680	406.85	0.09139	1548.18	4.0052	0.06327	1545.04	3.9672	0.03047	1534.25	3.8879	**680**

Temperature		Absolute Pressure									Temp.
T	t	40.0 bar			70.0 bar			100.0 bar			T
K	°C	v	i	s	v	i	s	v	i	s	K
520	246.85	0.00347	1047.84	2.7694	0.00249	868.89	2.6776	0.00229	860.08	2.6470	**520**
560	286.85	0.00807	1247.46	3.1406	0.00336	1022.94	2.9628	0.00271	998.48	2.9033	**560**
600	326.85	0.01061	1386.16	3.3800	0.00476	1182.68	3.2384	0.00336	1146.81	3.1590	**600**
640	366.85	0.01251	1518.31	3.5932	0.00615	1332.49	3.4802	0.00416	1297.09	3.4015	**640**
680	406.85	0.01413	1650.89	3.7941	0.00735	1475.81	3.6975	0.00498	1445.5	3.6265	**680**

Adapted from:

1. B. D. Smith and R. Srivastava, *Thermodynamic Data for Pure Compounds, Part–A, Hydrocarbons and Ketones*, Elsevier Science Publishers B.V., The Netherlands, 1986.
2. N. B. Vargaftik, *Tables on the Thermophysical Properties of Liquids and Gases,* 2nd Edition, Hemisphere Publishing Corporation, USA, 1975 and
3. W. C. Reynolds, *Thermodynamic Properties in SI,* Department of Mechanical Engineering, Stanford University, California, USA, 1979. Reproduced with the Permission of the Department of Mechanical Engineering, Stanford University, California, USA.

Properties of Brines

Table: 11 General Properties of Recommended Brines

Average operating temperature of brine, °C	Recommended brine	Recommended concentration, % by mass	Density, kg/L	Specific heat, kJ/(kg·K)	Thermal coductivity, W/(m·K)	Viscosity, mPa·s or cP	Recommended minimum velocity, m/s (refer Note 1)	Calculated film coefficient, W/(m²·K) (refer Note 2)
0	Sodium Chloride	12	1.092	3.62	0.488	2.3	0.5	5300
	Calcium Chloride	12	1.109	3.465	0.553	2.4	0.5	5500
	Methanol	15	0.981	4.18	0.490	2.8	0.8	4400
	Ethanol	20	0.975	4.35	0.440	5.4	1.4	3500
	Ethylene Glycol	25	1.038	3.82	0.500	3.7	0.9	4400
	Propylene Glycol	30	1.030	3.82	0.462	7.5	1.9	3000
–10	Sodium Chloride	21	1.168	3.35	0.428	4.2	0.9	4000
	Calcium Chloride	20	1.196	2.98	0.527	4.6	1.0	4200
	Methanol	22	0.976	4.06	0.450	5.7	1.4	3400
	Ethanol	25	0.977	4.25	0.430	8.2	2.1	2900
	Ethylene Glycol	35	1.055	3.50	0.460	6.8	1.6	3300
	Propylene Glycol	40	1.046	3.73	0.465	25.0	>3	590
–20	Calcium Chloride	25	1.254	2.81	0.507	9.7	2.1	2900
	Methanol	35	0.958	3.77	0.405	9.9	2.6	560
	Ethanol	36	0.971	3.95	0.381	13.5	>3	550
	Ethylene Glycol	45	1.076	3.18	0.420	17.2	>3	590
	Propylene Glycol	50	1.051	3.54	0.388	85.0	>3	560
–35	Calcium Chloride	30	1.318	2.65	0.482	27.0	>3	630
	Methanol	45	0.955	3.40	0.380	21.0	>3	520
	Ethanol	52	0.955	3.40	0.329	20.2	>3	470
	Ethylene Glycol	55	1.096	2.78	0.415	78.0	>3	530
	Propylene Glycol	60	1.073	3.16	0.360	860	>3	530

Note 1 : Minimum brine velocity, recommended at Re = 3500 for 14 mm ID tubing

Note 2 : Film coefficient between brine and surface at 2 m/s velocity for 14 mm ID tubing

(Data, presented in Table 11 and Figs. 12.1 to 12.26, are extracted from *Carrier System Design Manual*, Part 4, 1969 by Carrier Corporation, U.S.A. with permission)

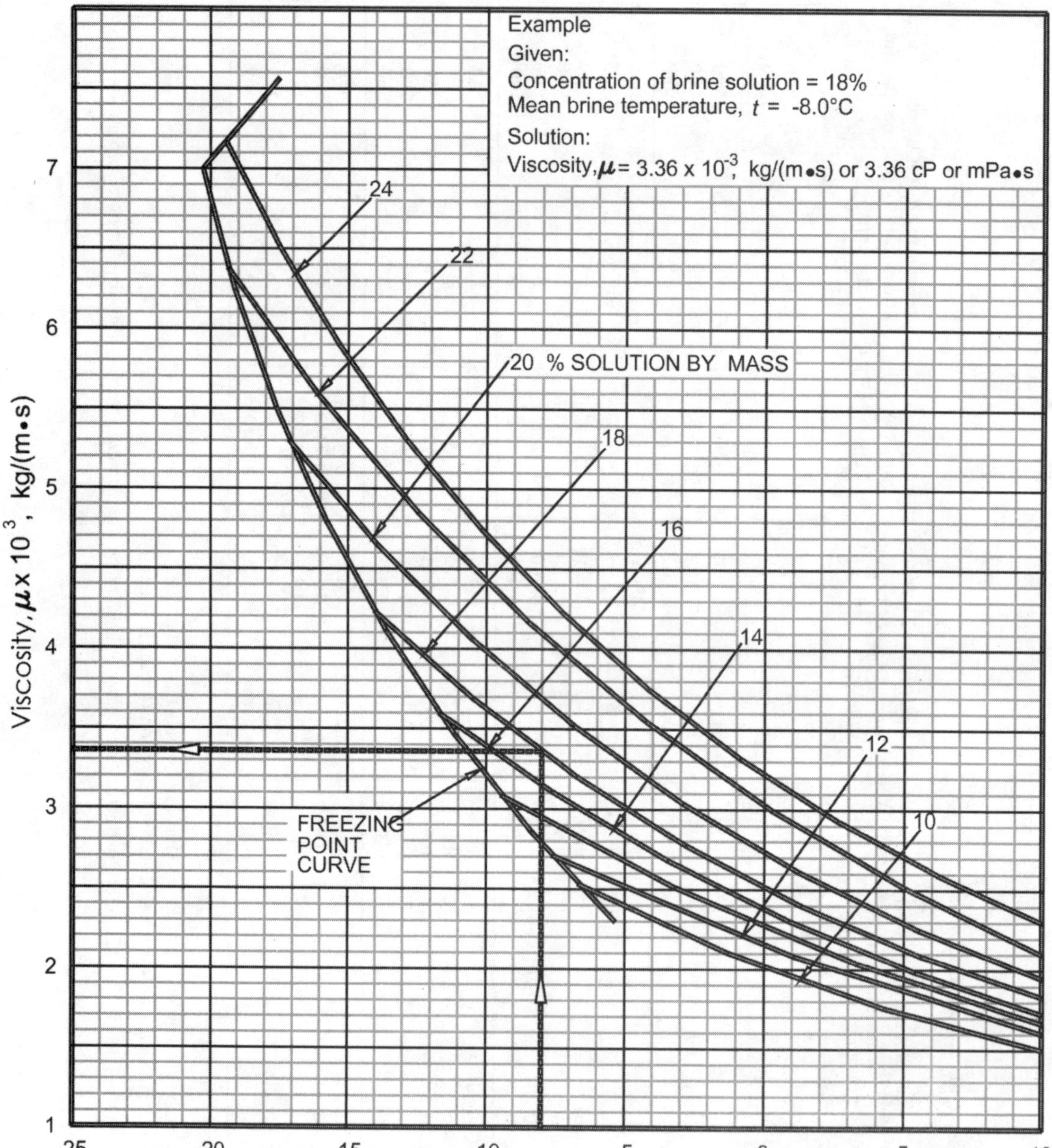

Fig. 12.1: Viscosity of Aqueous Sodium Chloride Brine

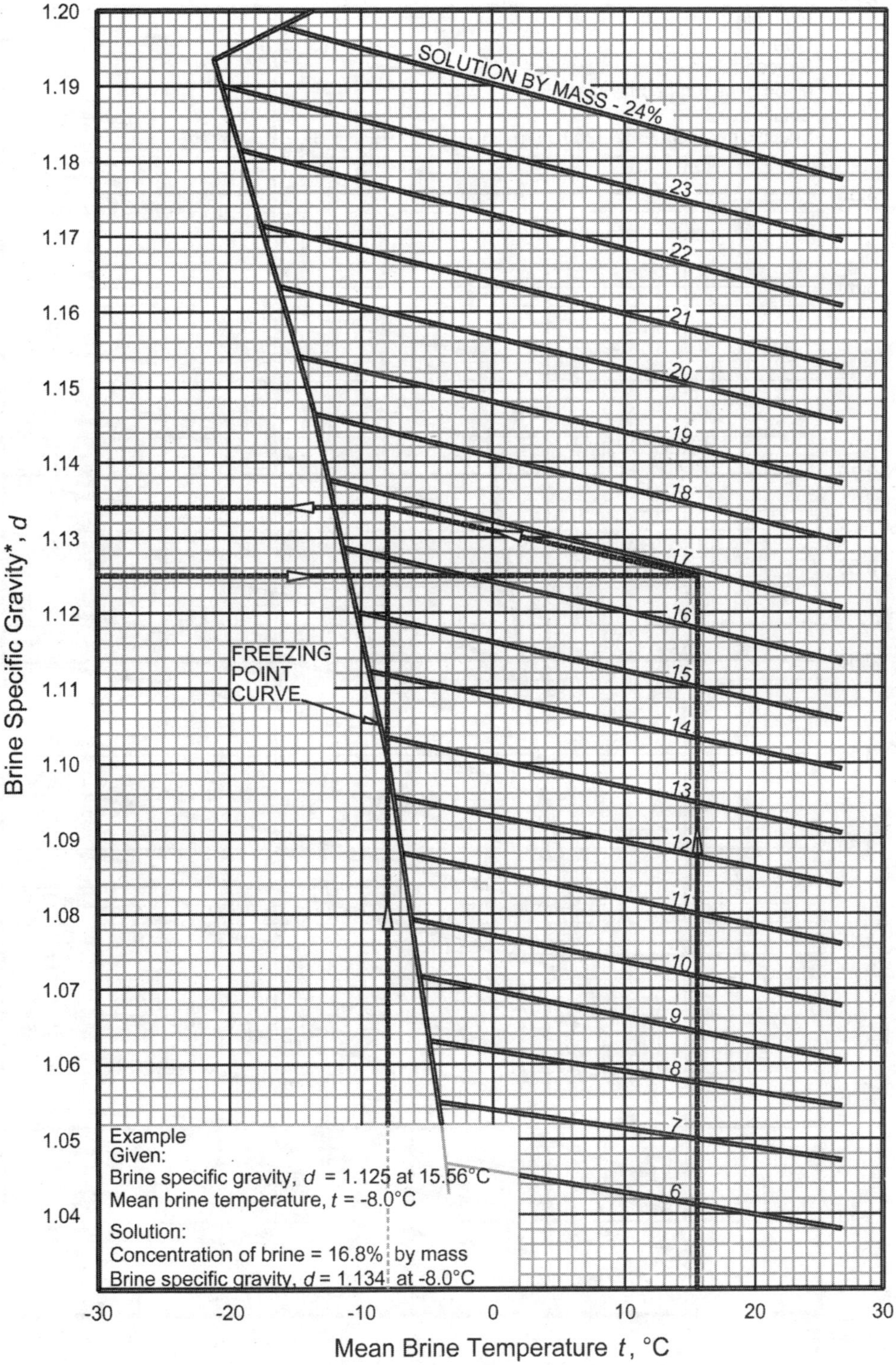

*with reference to 15.56°C (60°F) water

Fig. 12.2: Specific Gravity of Aqueous Sodium Chloride Brine

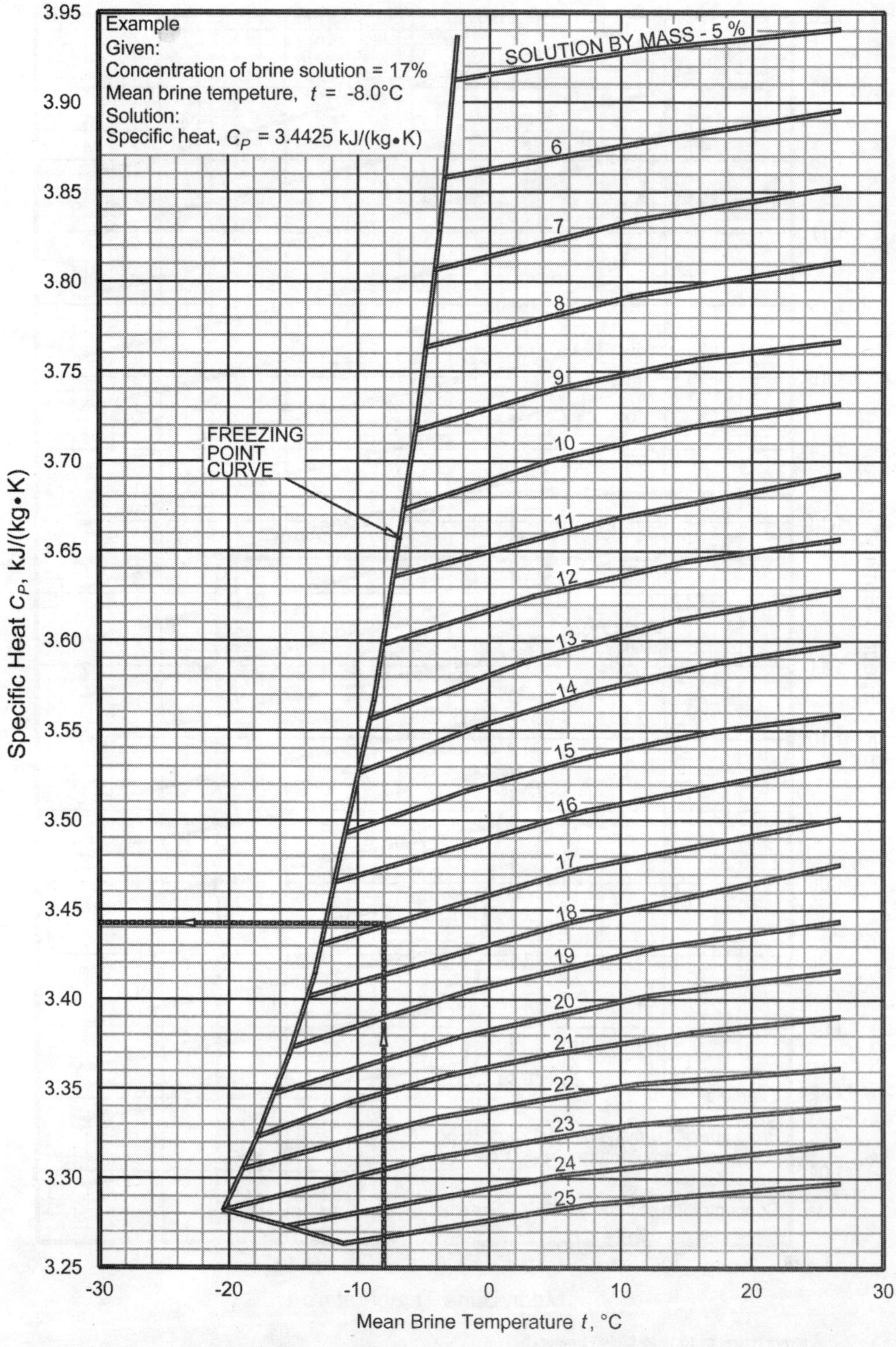

Fig. 12.3: Specific Heat of Aqueous Sodium Chloride Brine

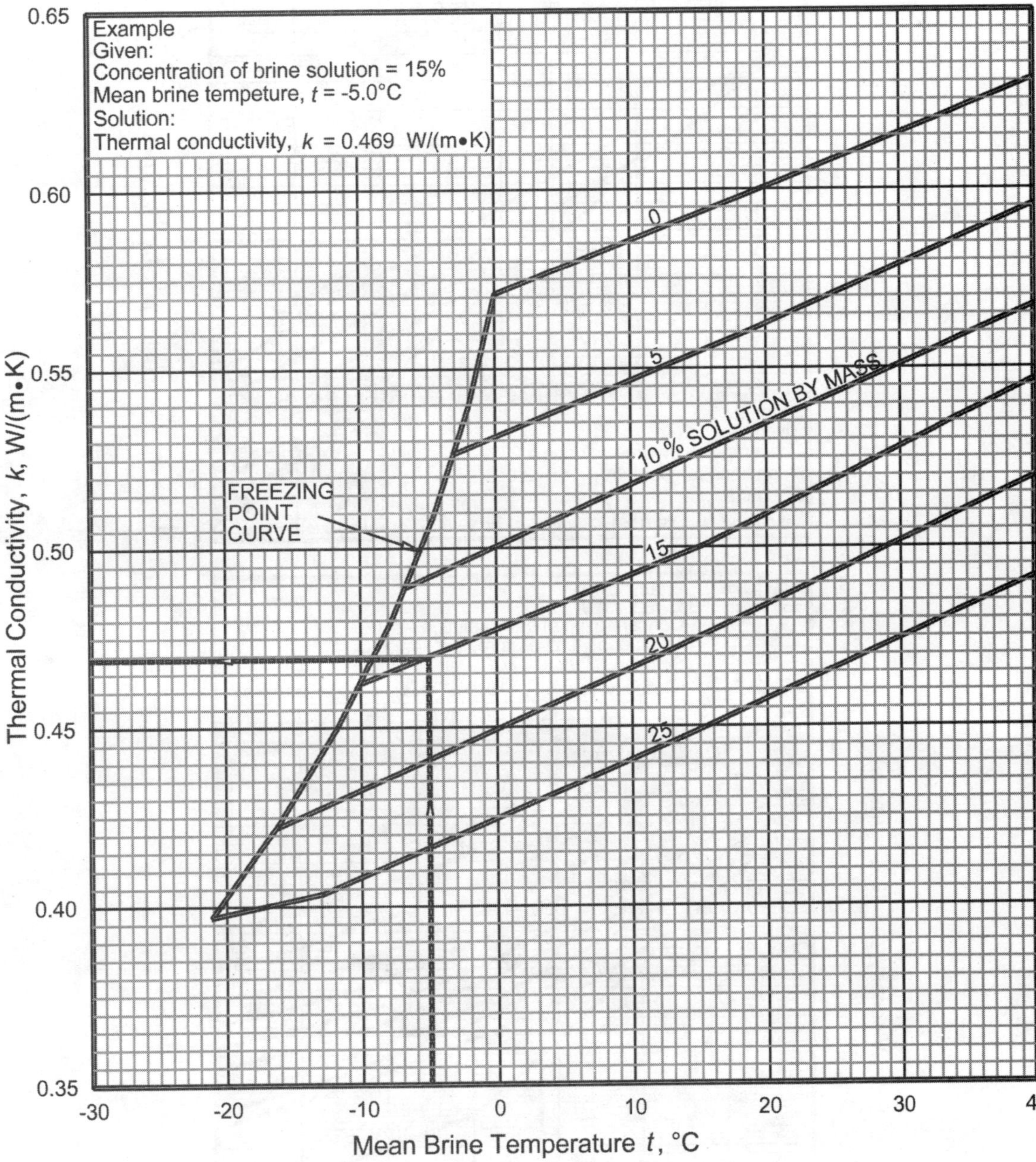

Fig. 12.4: Thermal Conductivity of Aqueous Sodium Chloride Brine

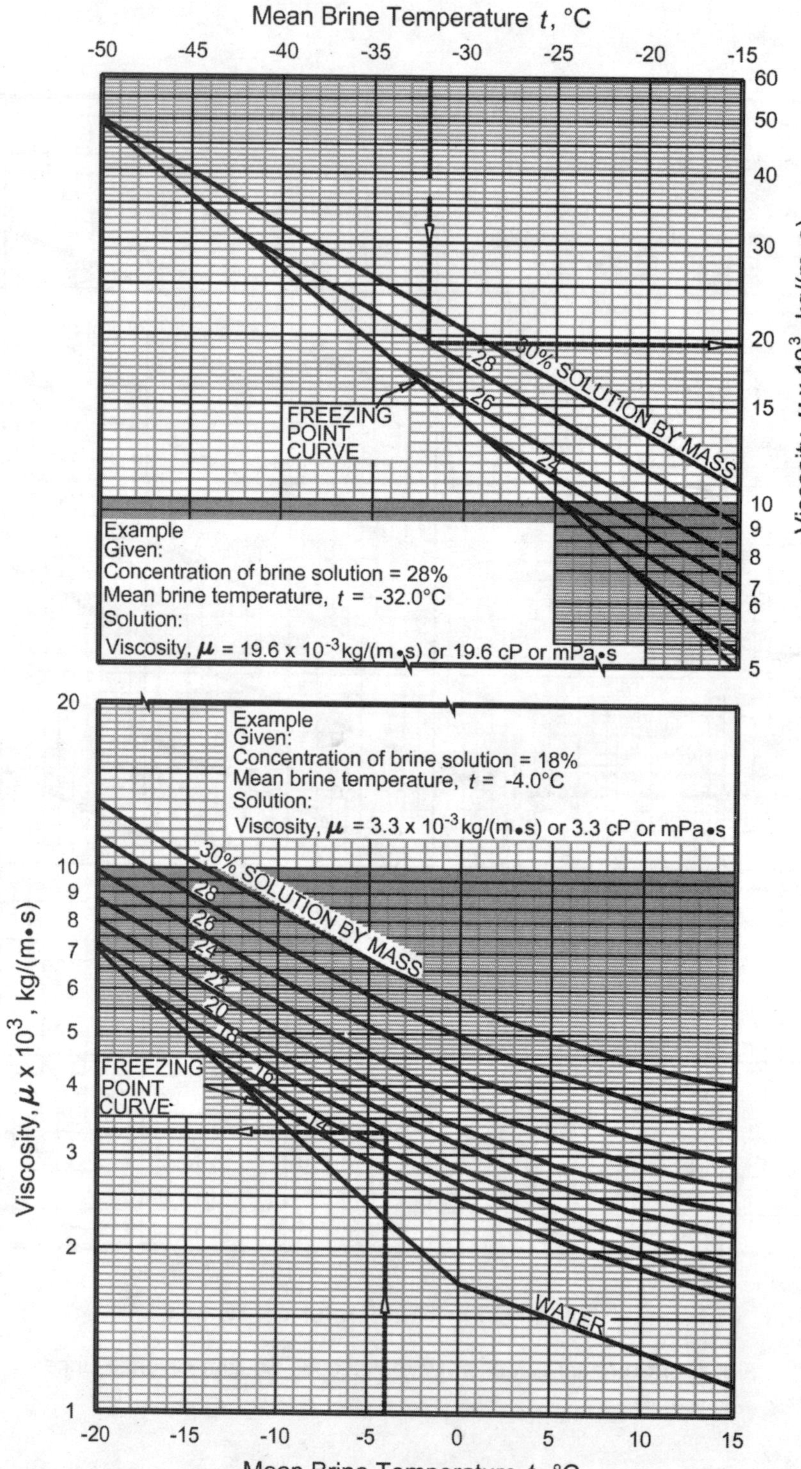

Fig. 12.5: Viscosity of Aqueous Calcium Chloride Brine

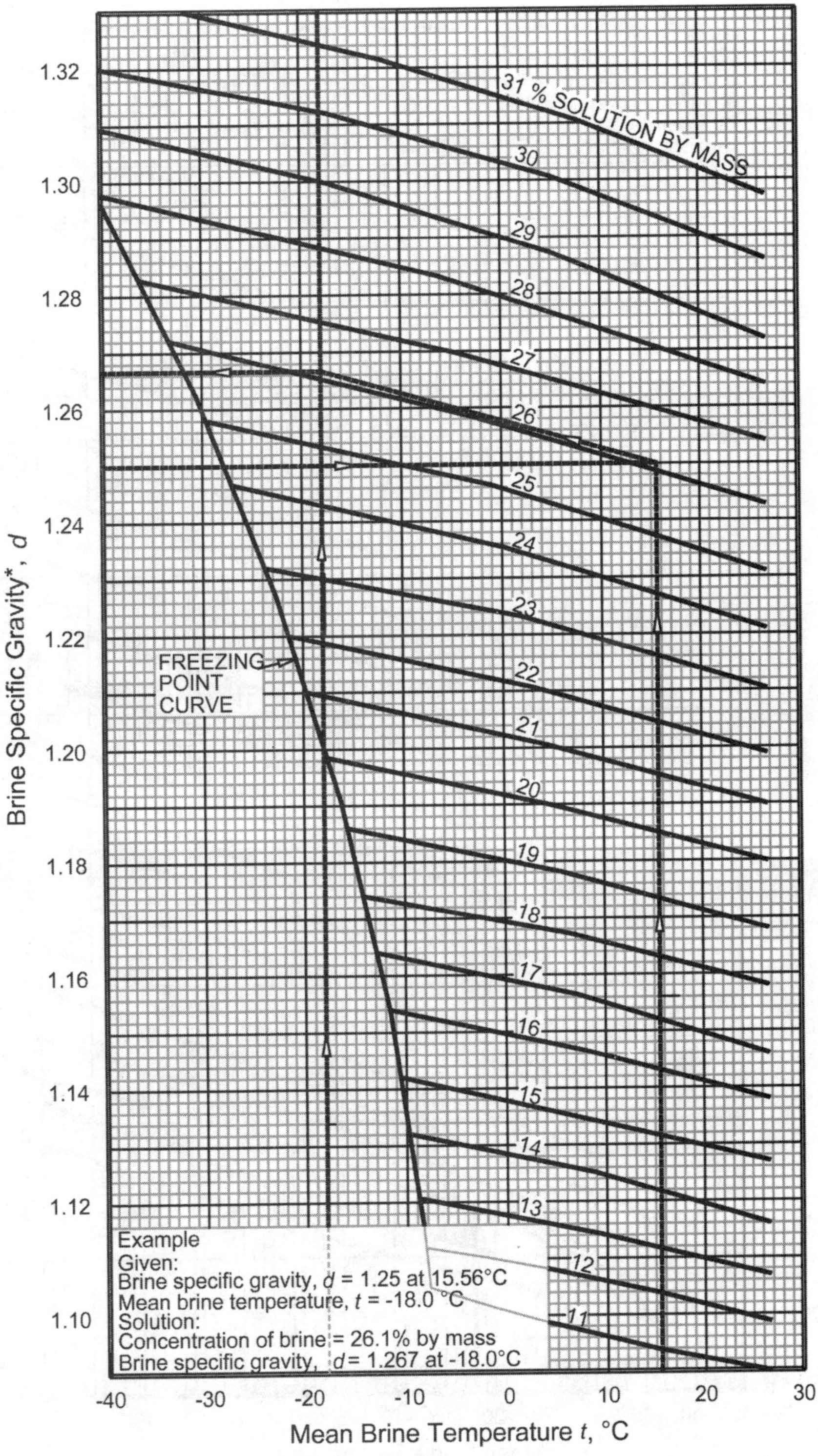

Fig. 12.6: Specific Gravity of Aqueous Calcium Chloride Brine

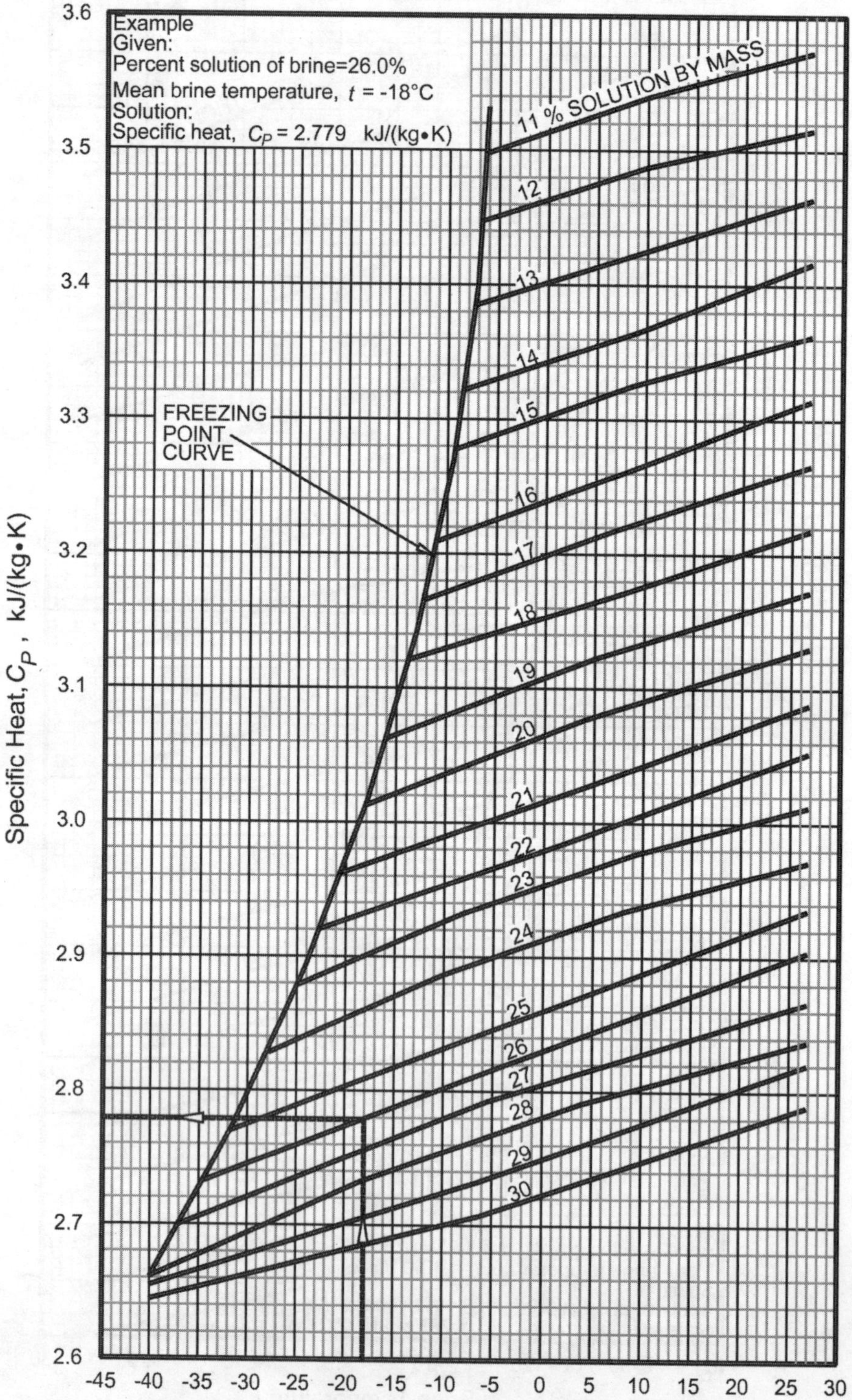

Fig. 12.7: Specific Heat of Aqueous Calcium Chloride Brine

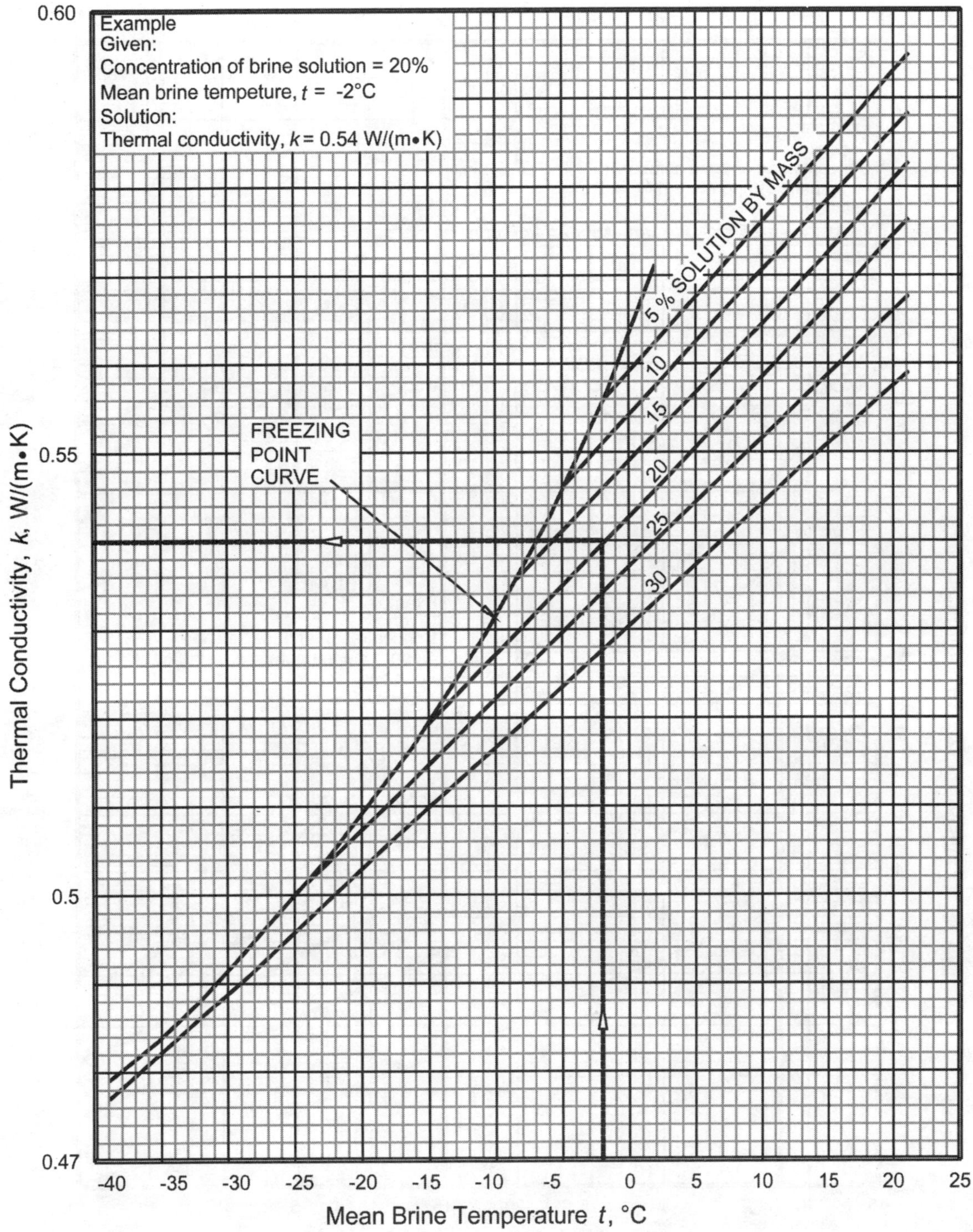

Fig. 12.8: Thermal Conductivity of Aqueous Calcium Chloride Brine

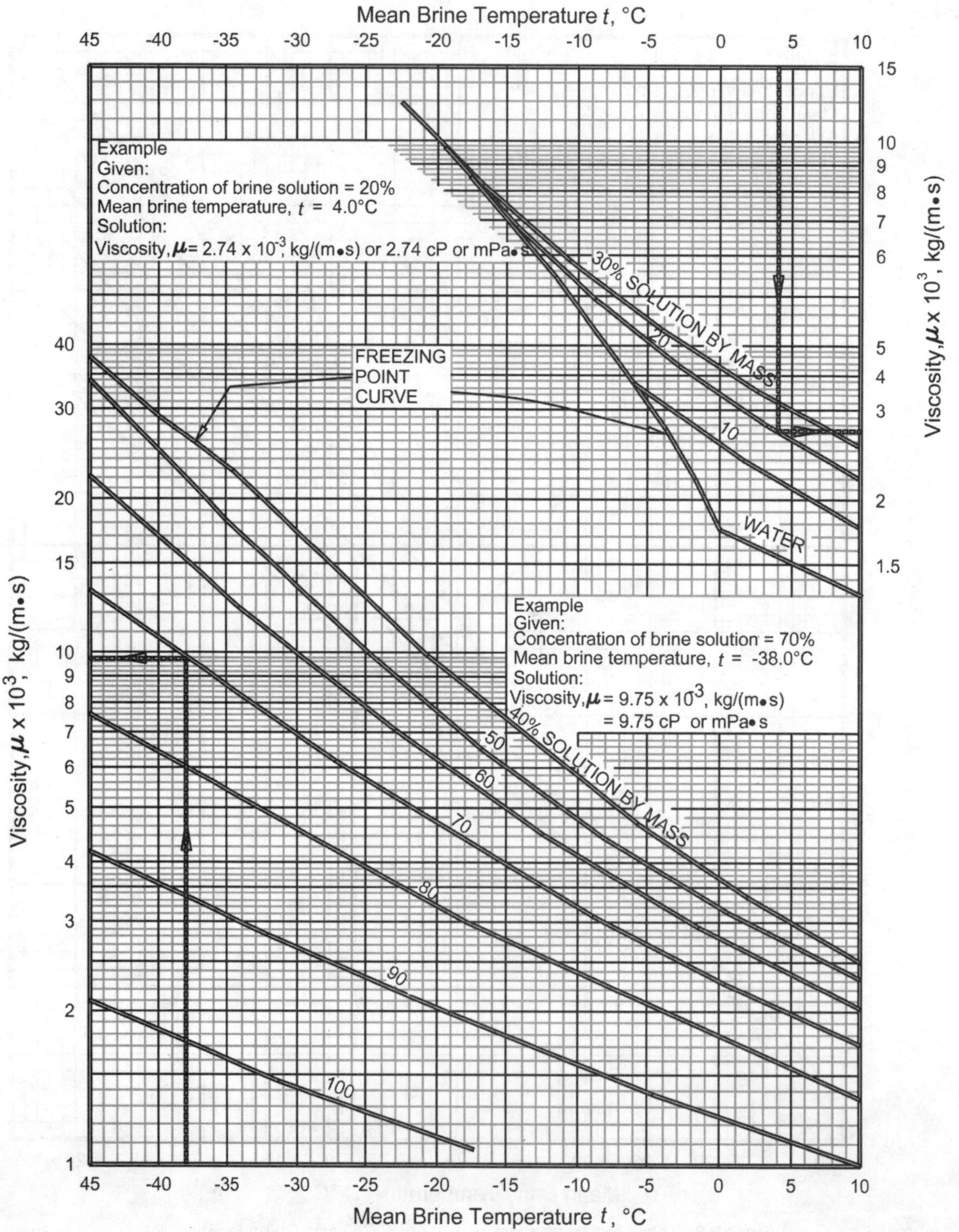

Fig. 12.9: Viscosity of Aqueous Methanol Brine

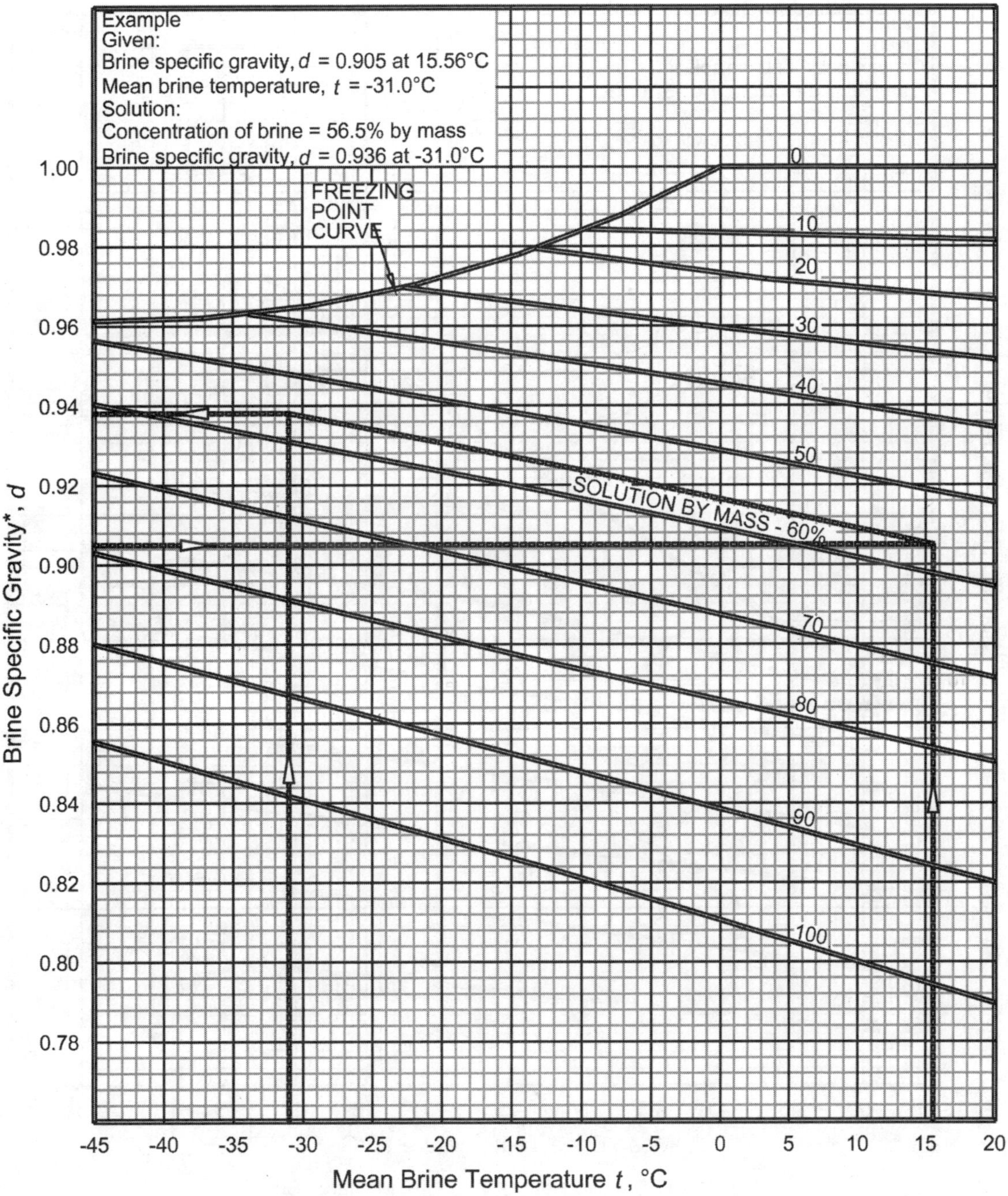

*with reference to 15.56°C (60°F) water

Fig. 12.10: Specific Gravity of Aqueous Methanol Brine

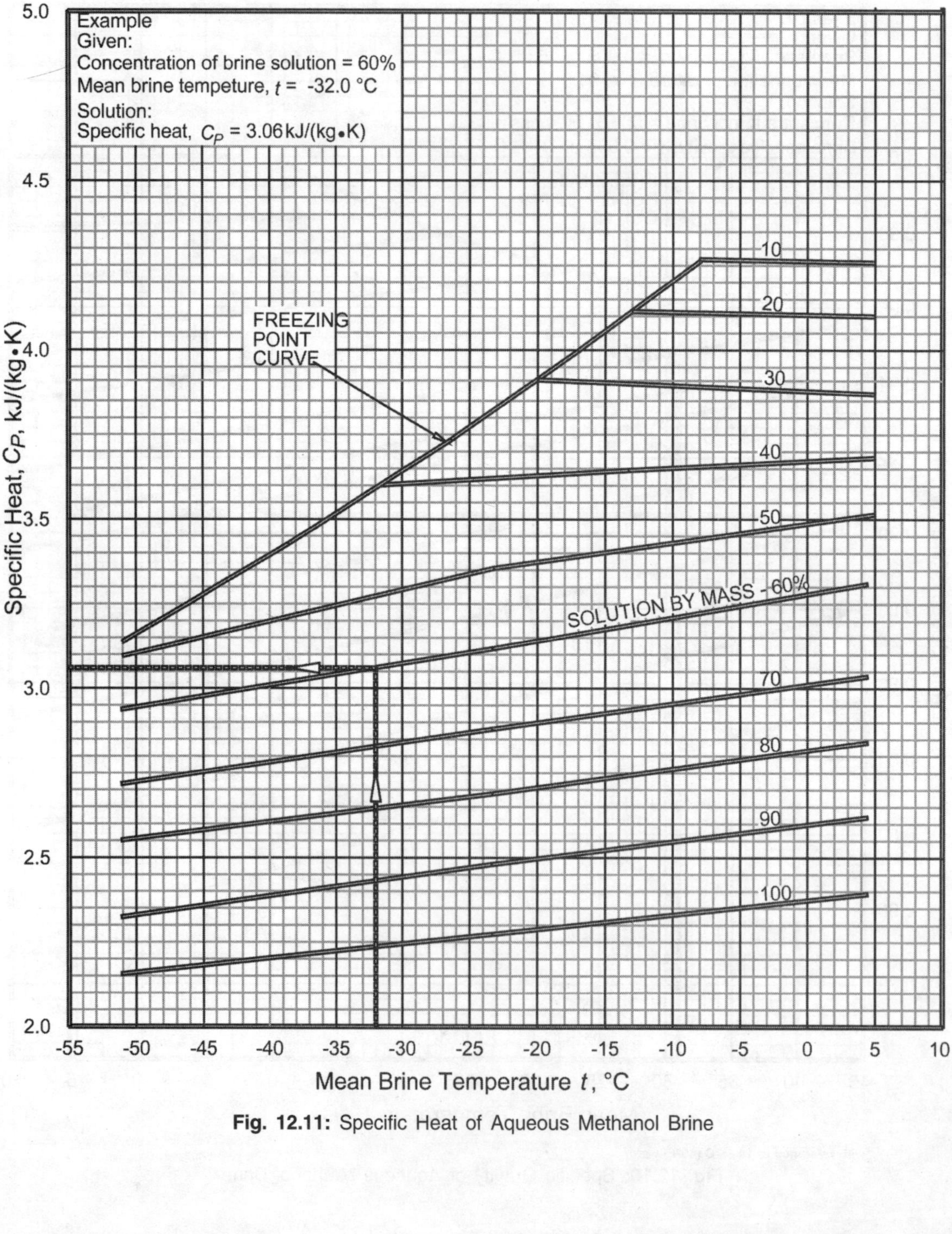

Fig. 12.11: Specific Heat of Aqueous Methanol Brine

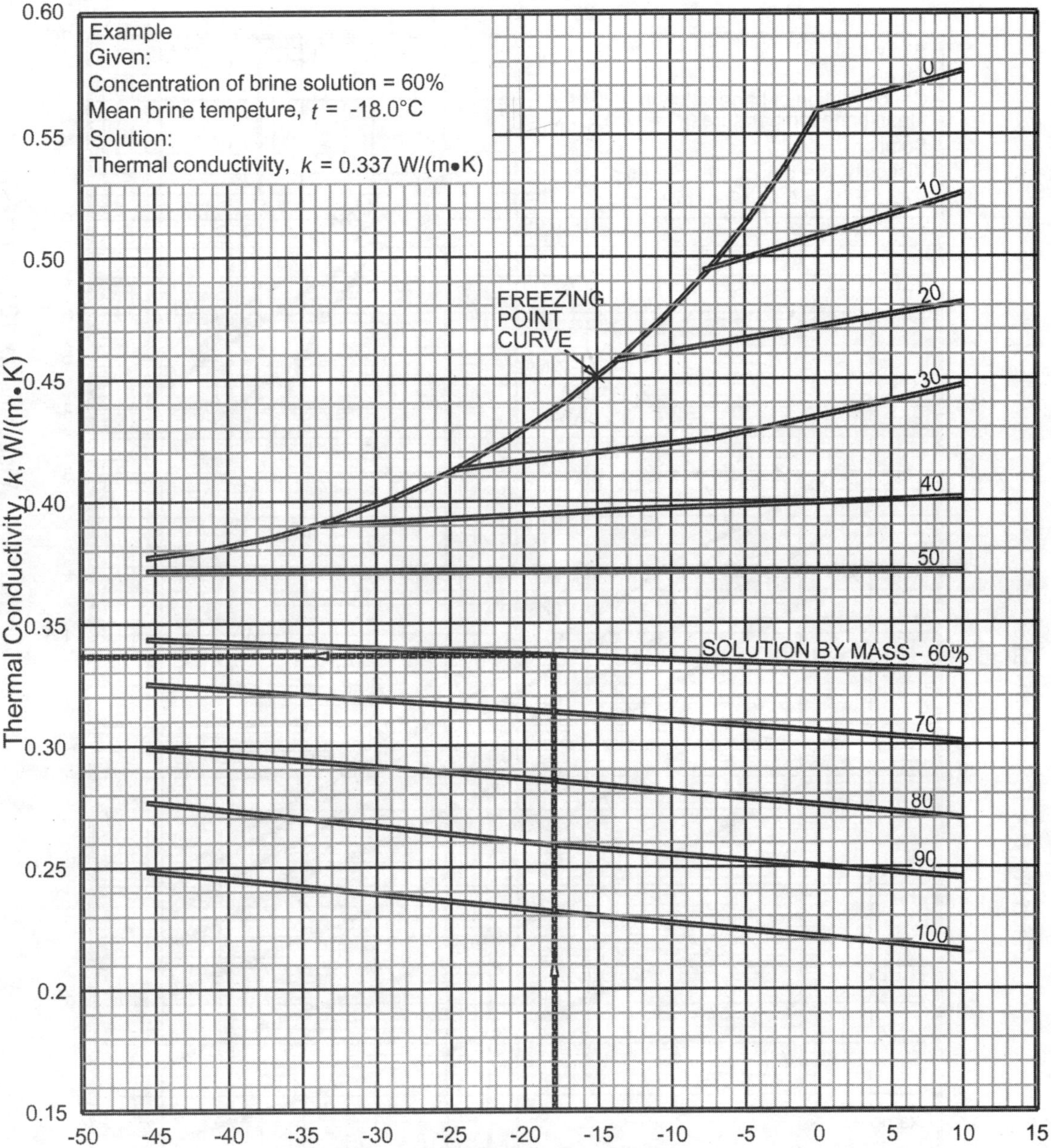

Fig. 12.12: Thermal Conductivity of Aqueous Methanol Brine

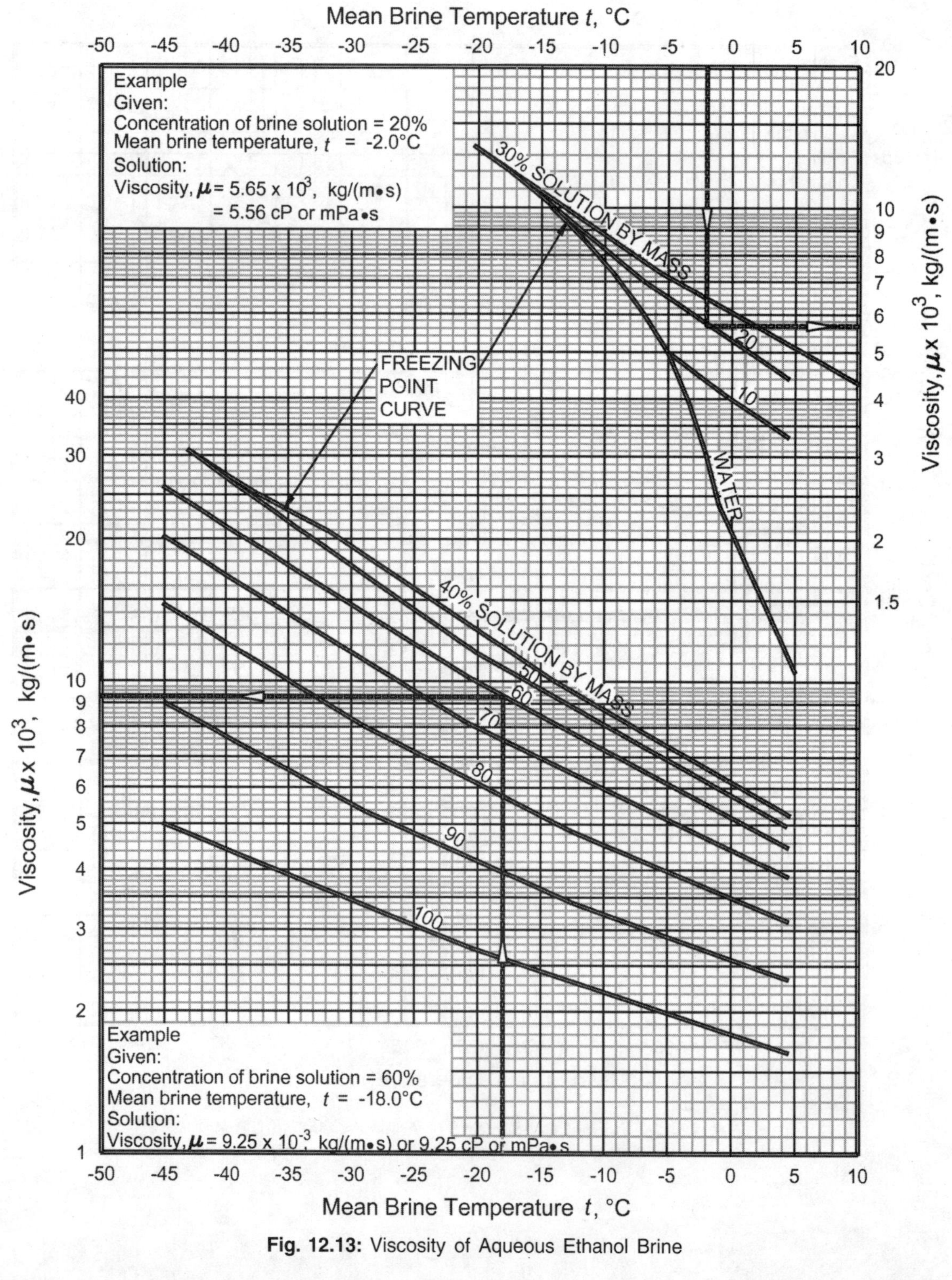

Fig. 12.13: Viscosity of Aqueous Ethanol Brine

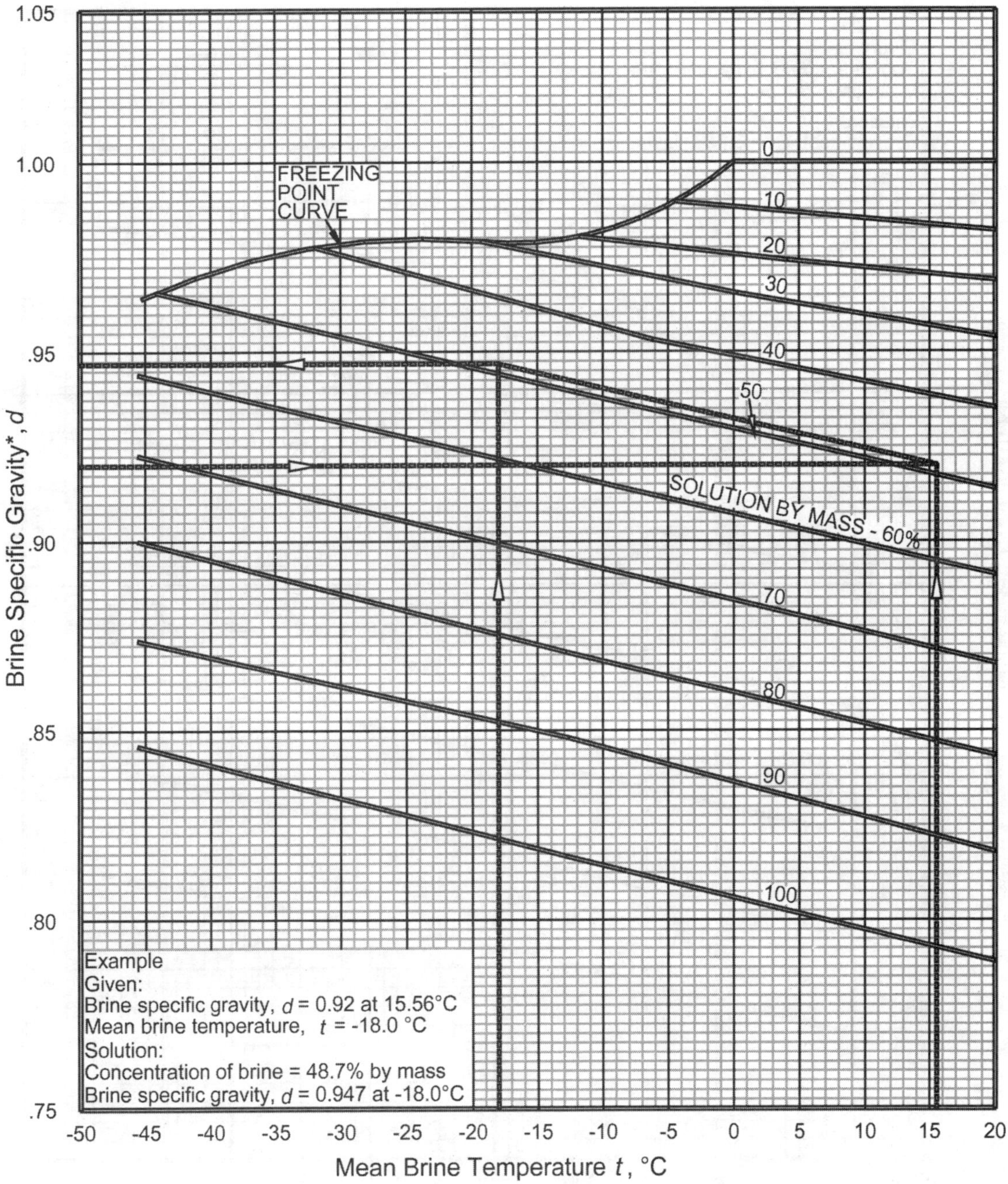

Fig. 12.14: Specific Gravity of Aqueous Ethanol Brine

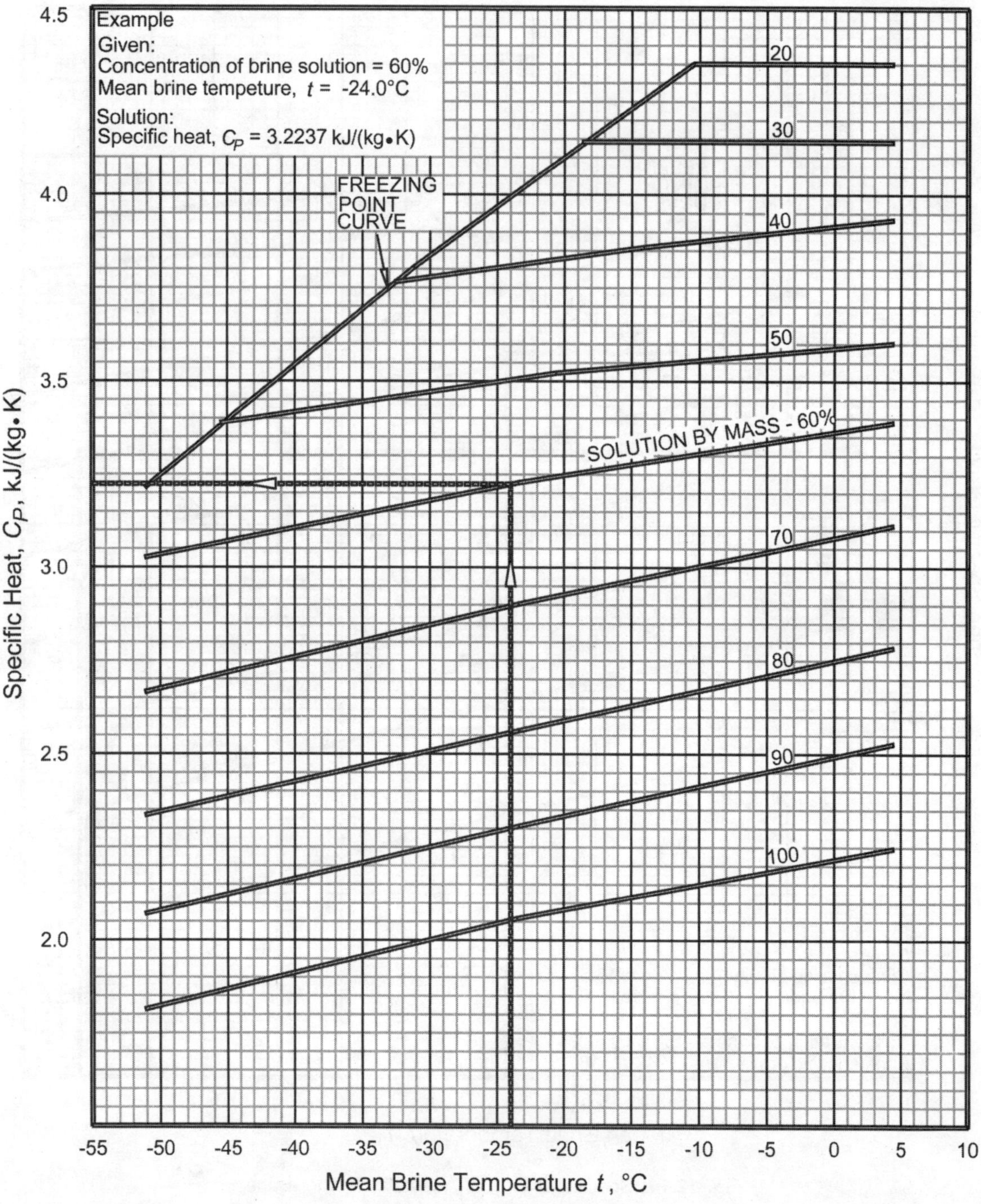

Fig. 12.15: Specific Heat of Aqueous Ethanol Brine

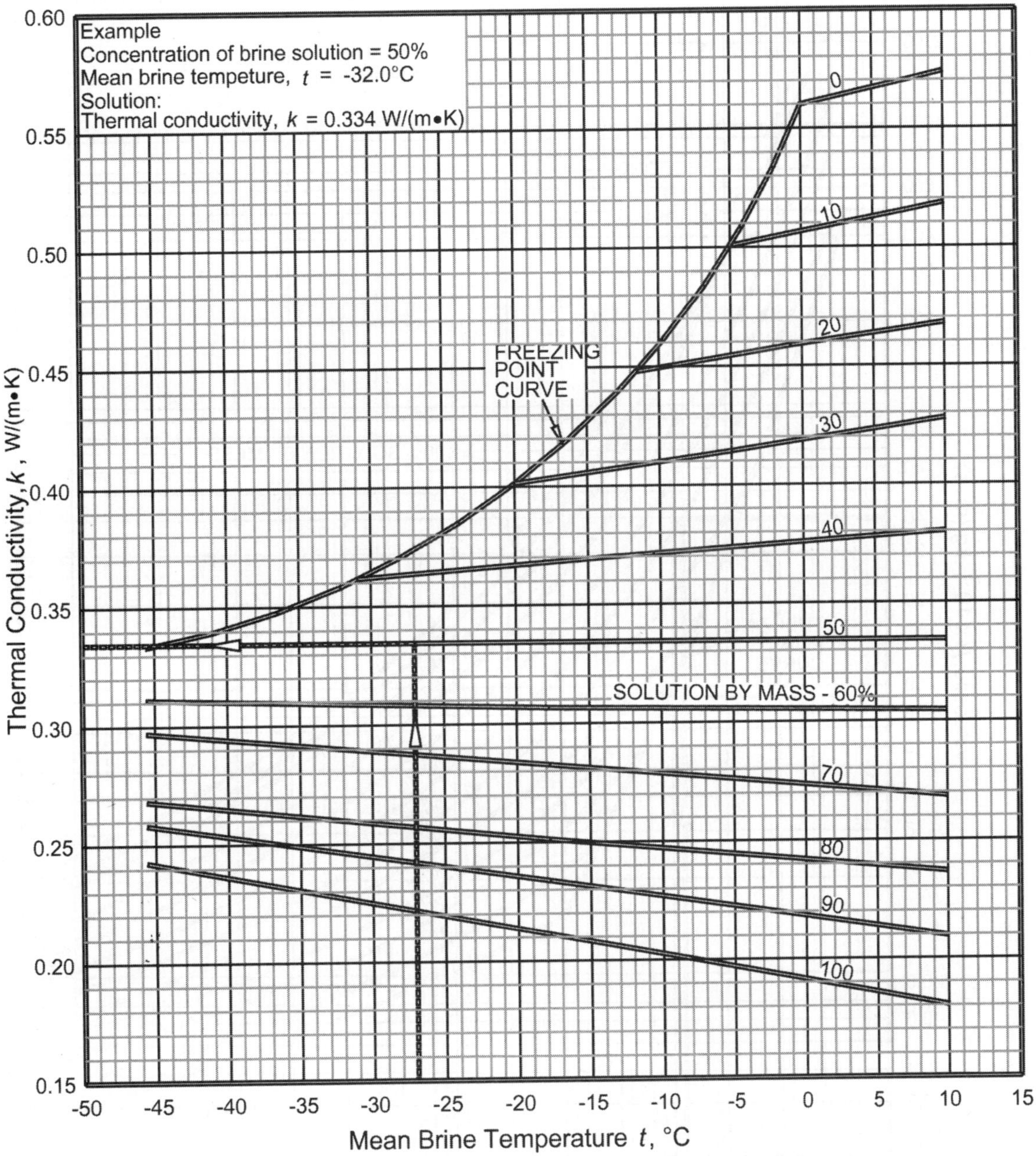

Fig. 12.16: Thermal Conductivity of Aqueous Ethanol Brine

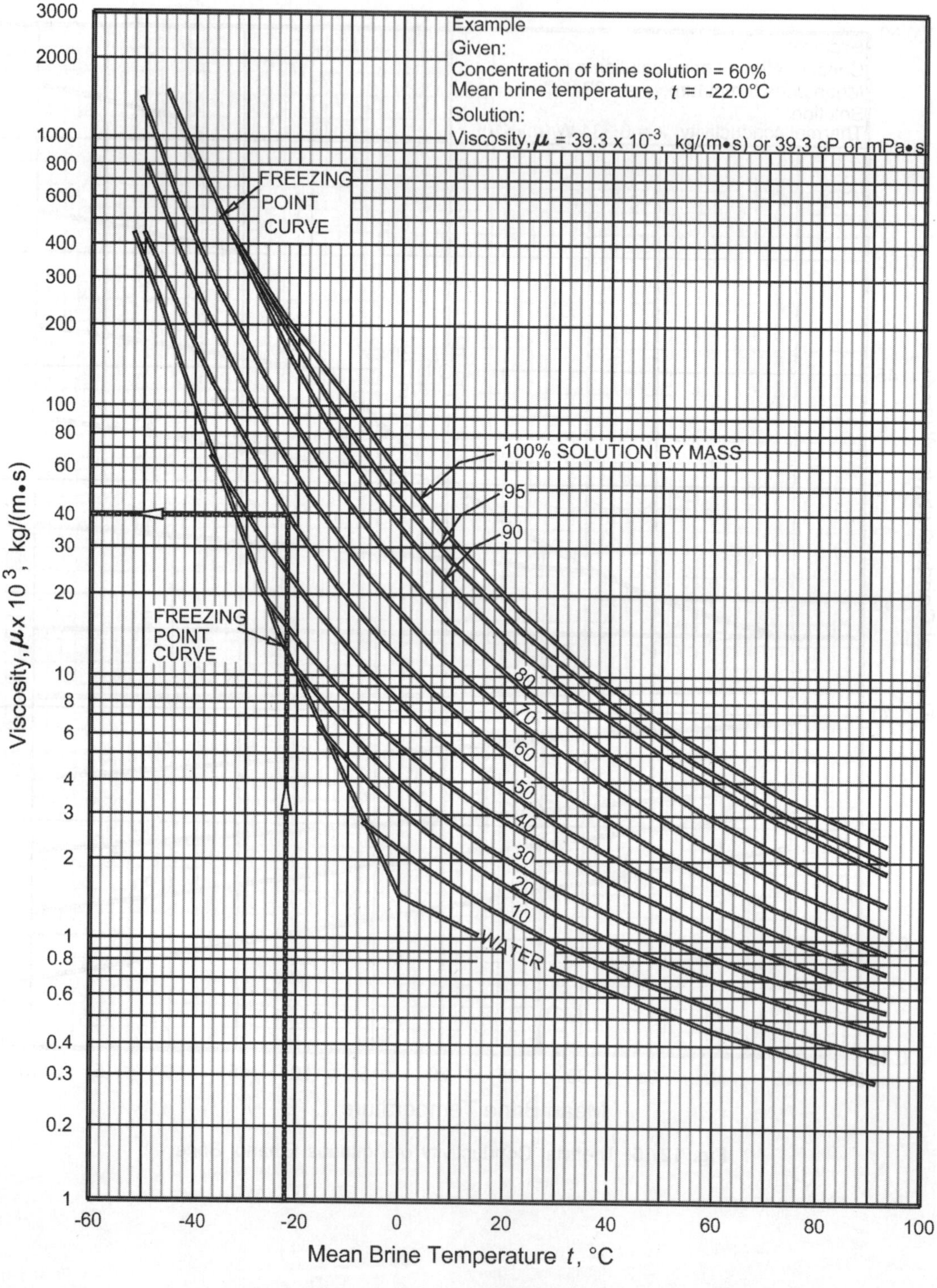

Fig. 12.17: Viscosity of Aqueous Ethylene Glycol Brine

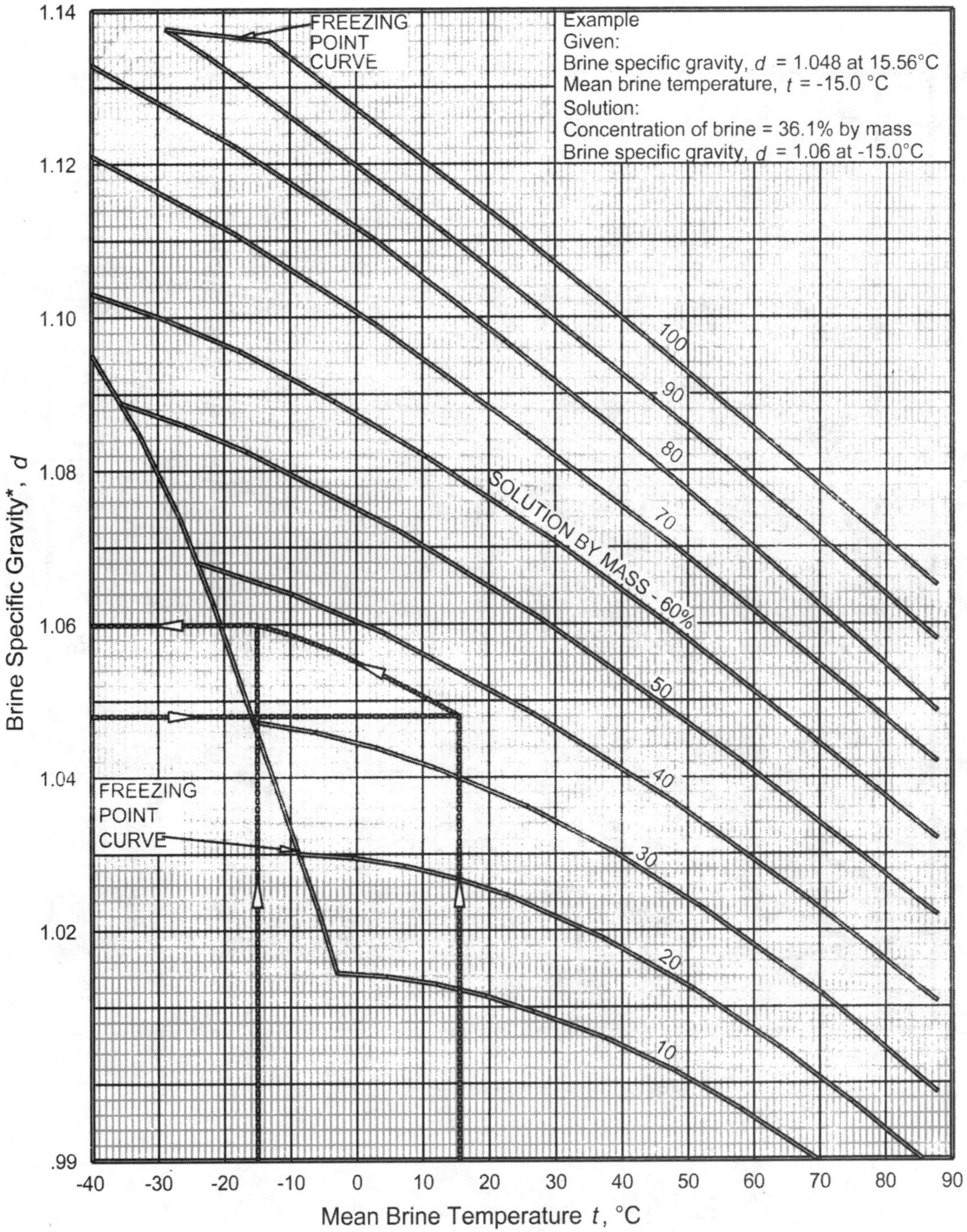

Fig. 12.18: Specific Gravity of Aqueous Ethylene Glycol Brine

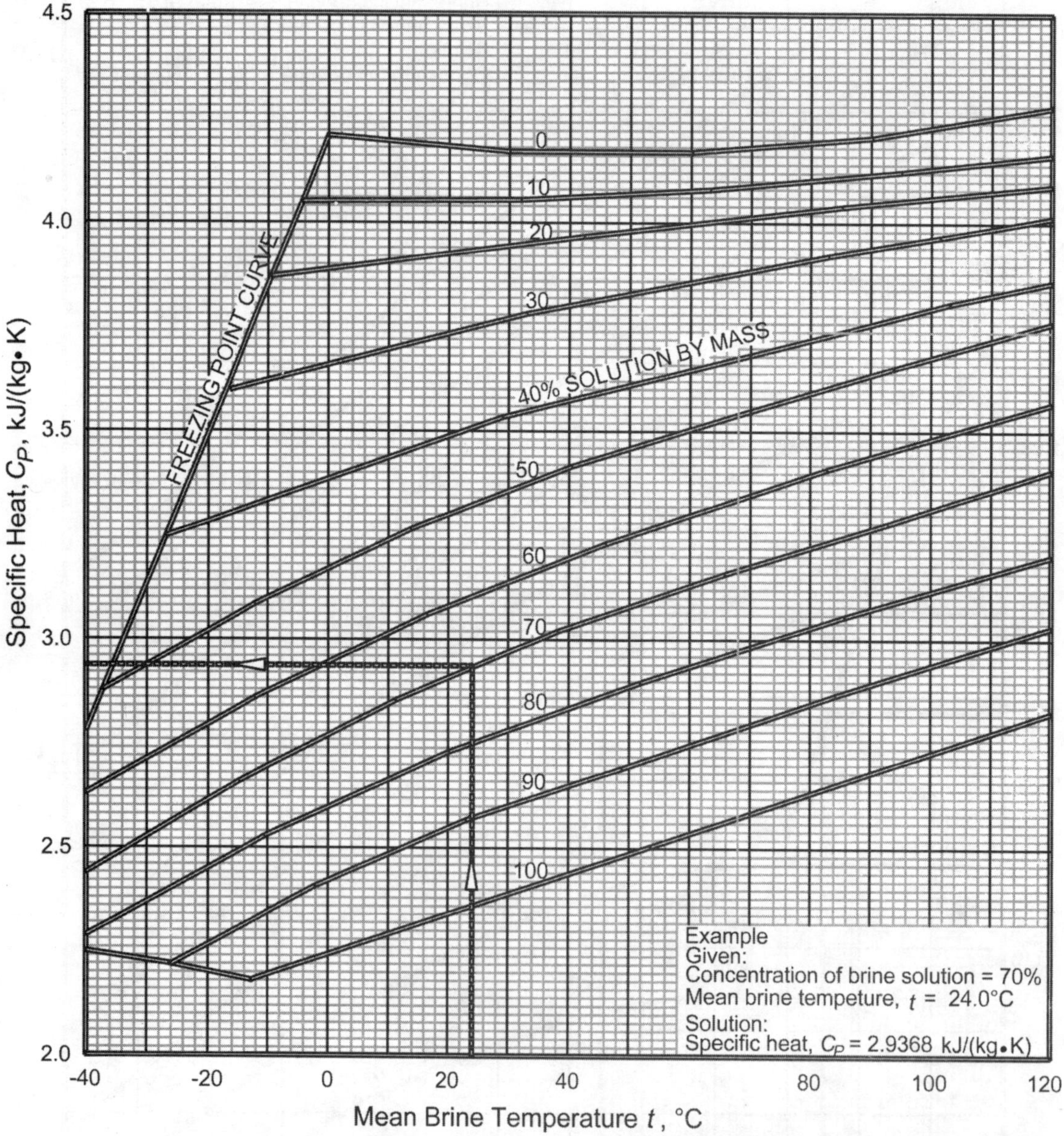

Fig. 12.19: Specific Heat of Aqueous Ethylene Glycol Brine

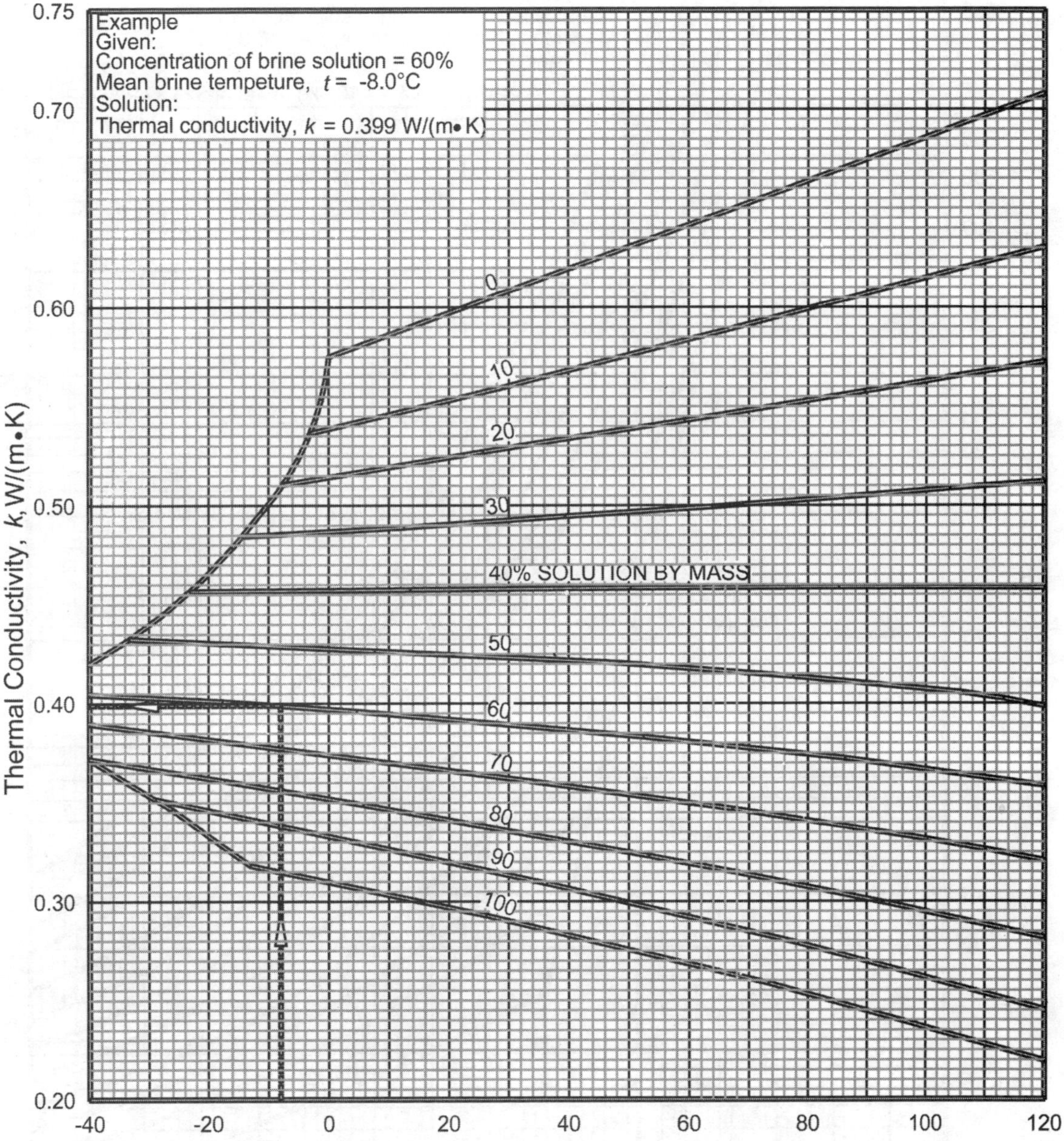

Fig. 12.20: Thermal Conductivity of Aqueous Ethylene Glycol Brine

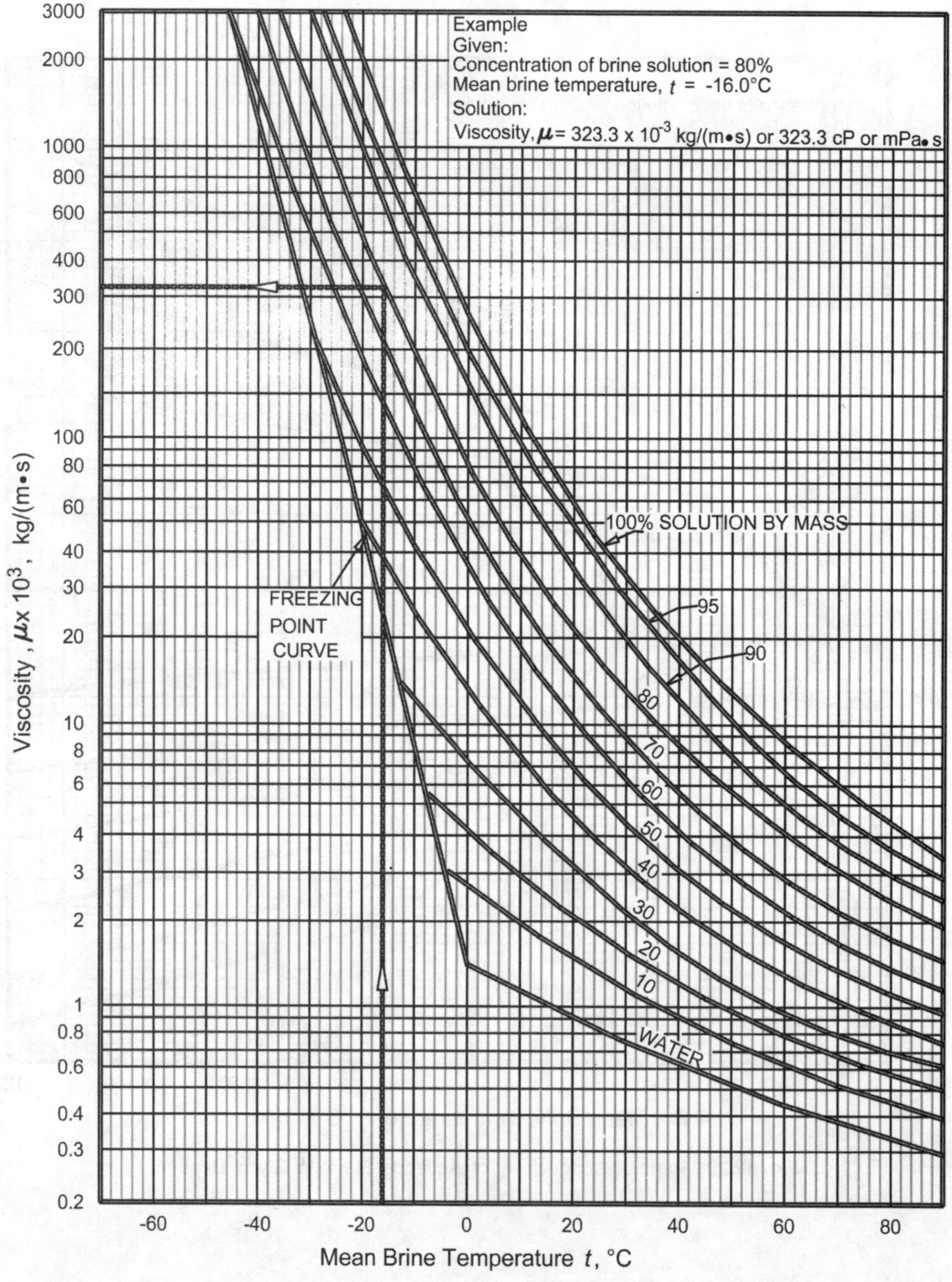

Fig. 12.21: Viscosity of Aqueous Propylene Glycol Brine

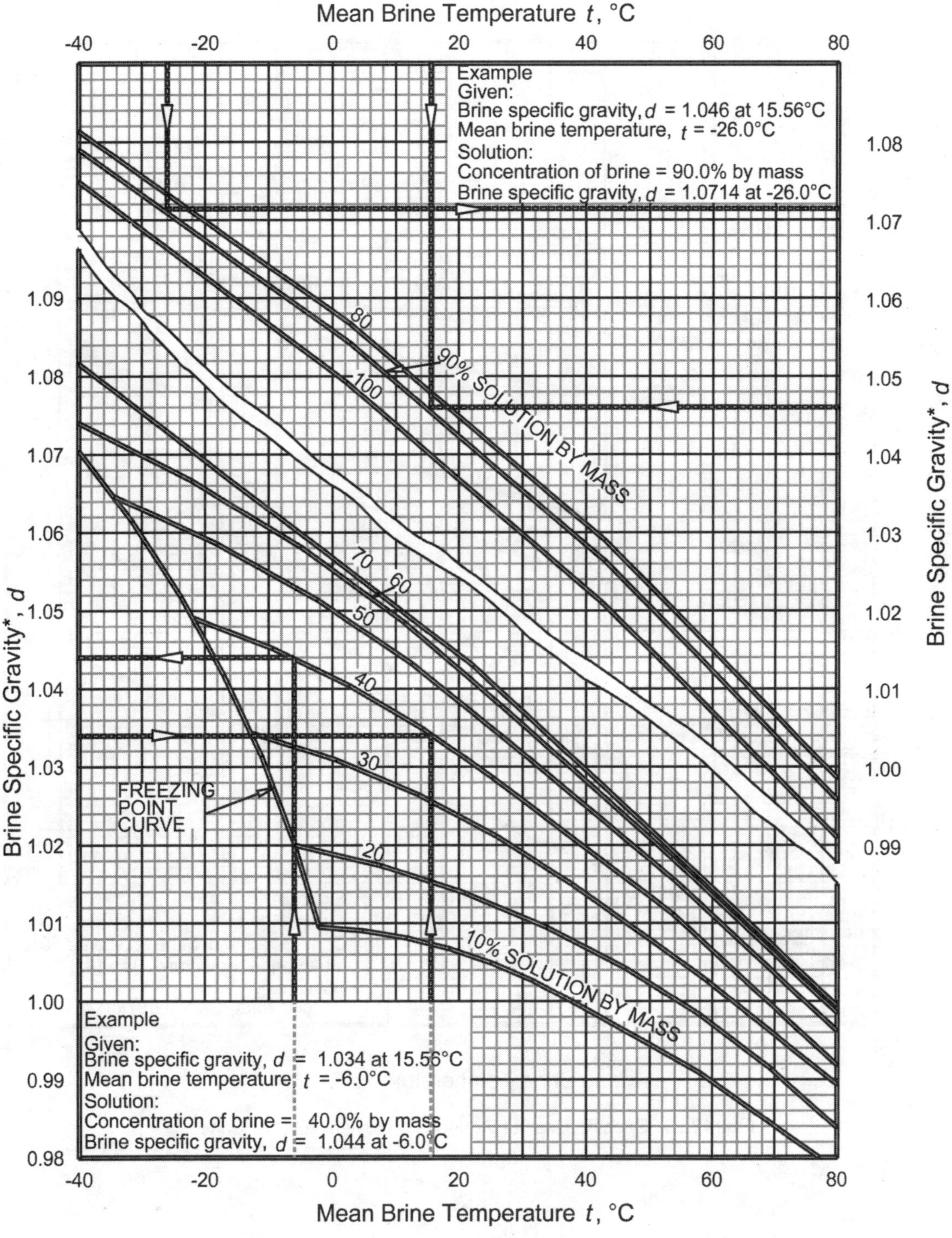

Fig. 12.22: Specific Gravity of Aqueous Propylene Glycol Brine

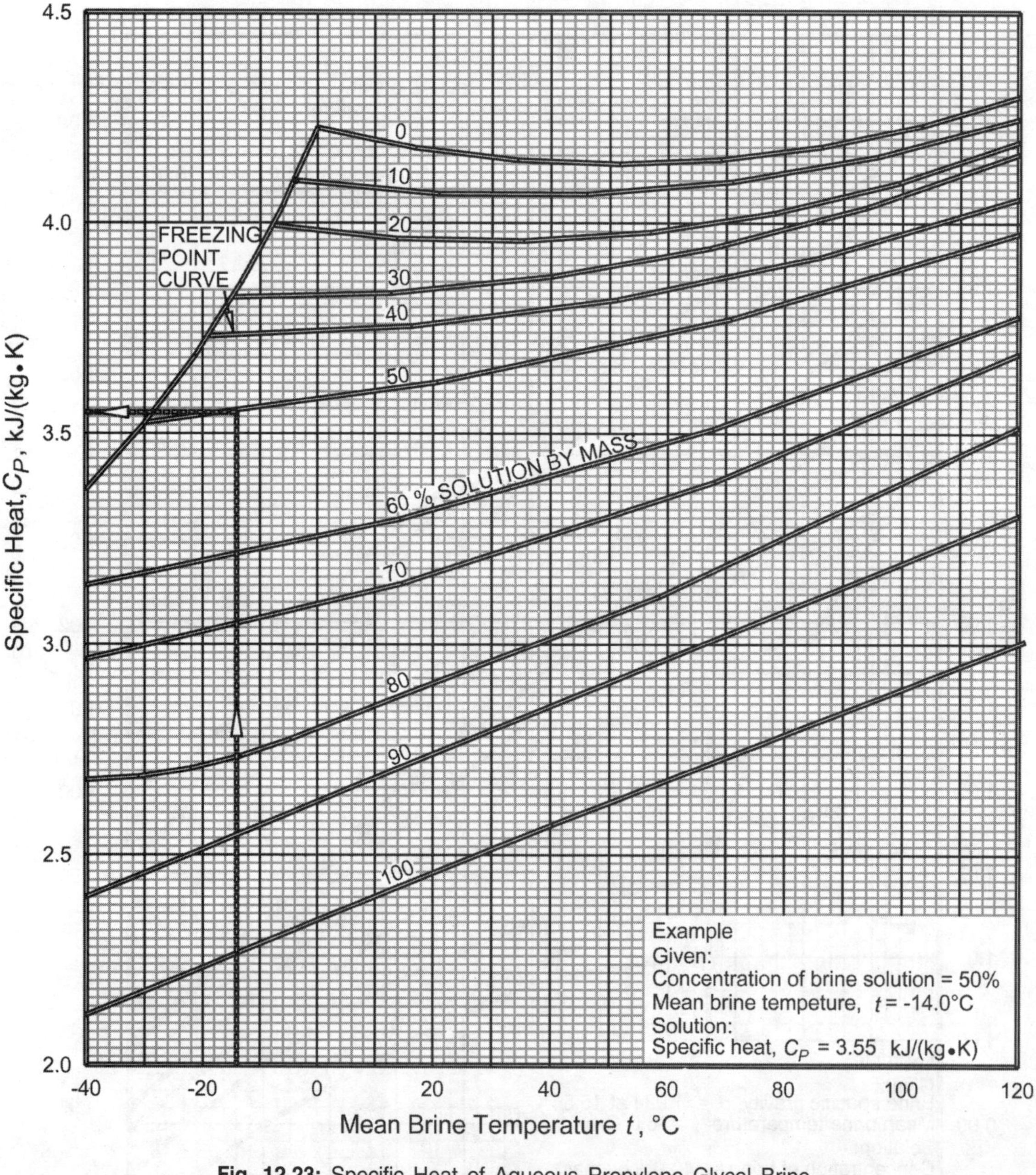

Fig. 12.23: Specific Heat of Aqueous Propylene Glycol Brine

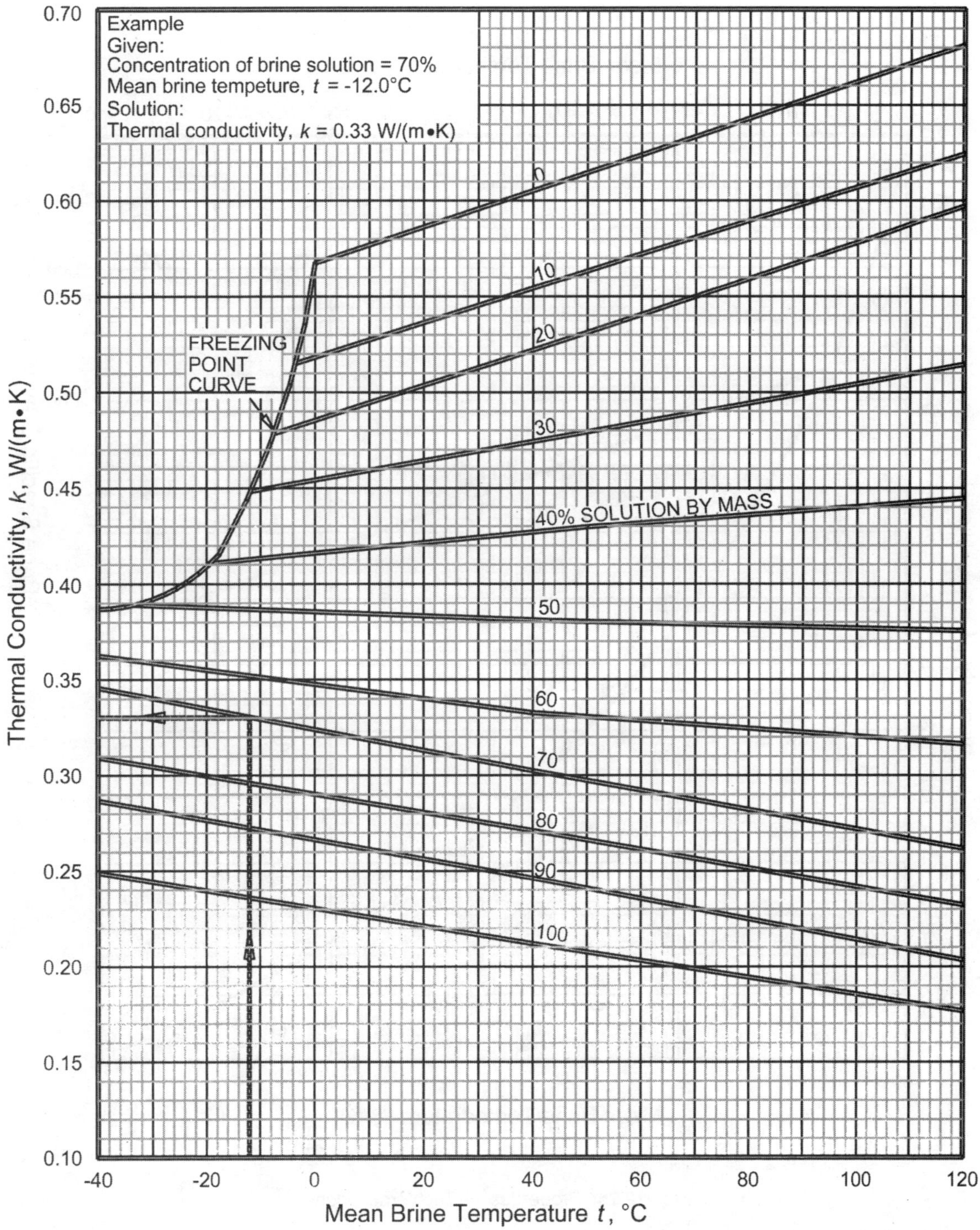

Fig. 12.24: Thermal Conductivity of Aqueous Propylene Glycol Brine

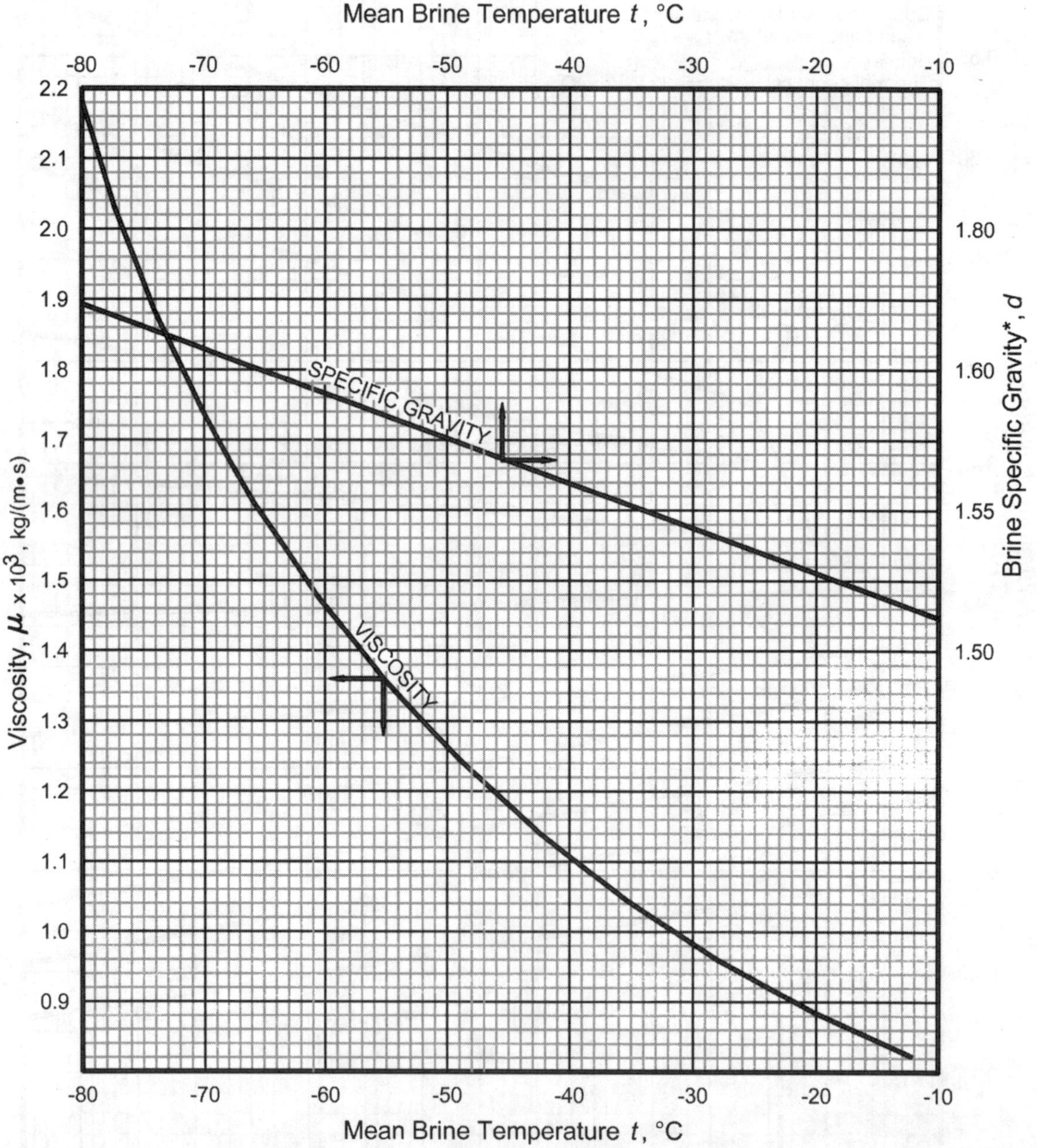

*with reference to 15.56°C (60°F) water

Fig. 12.25: Properties of Trichloroethylene

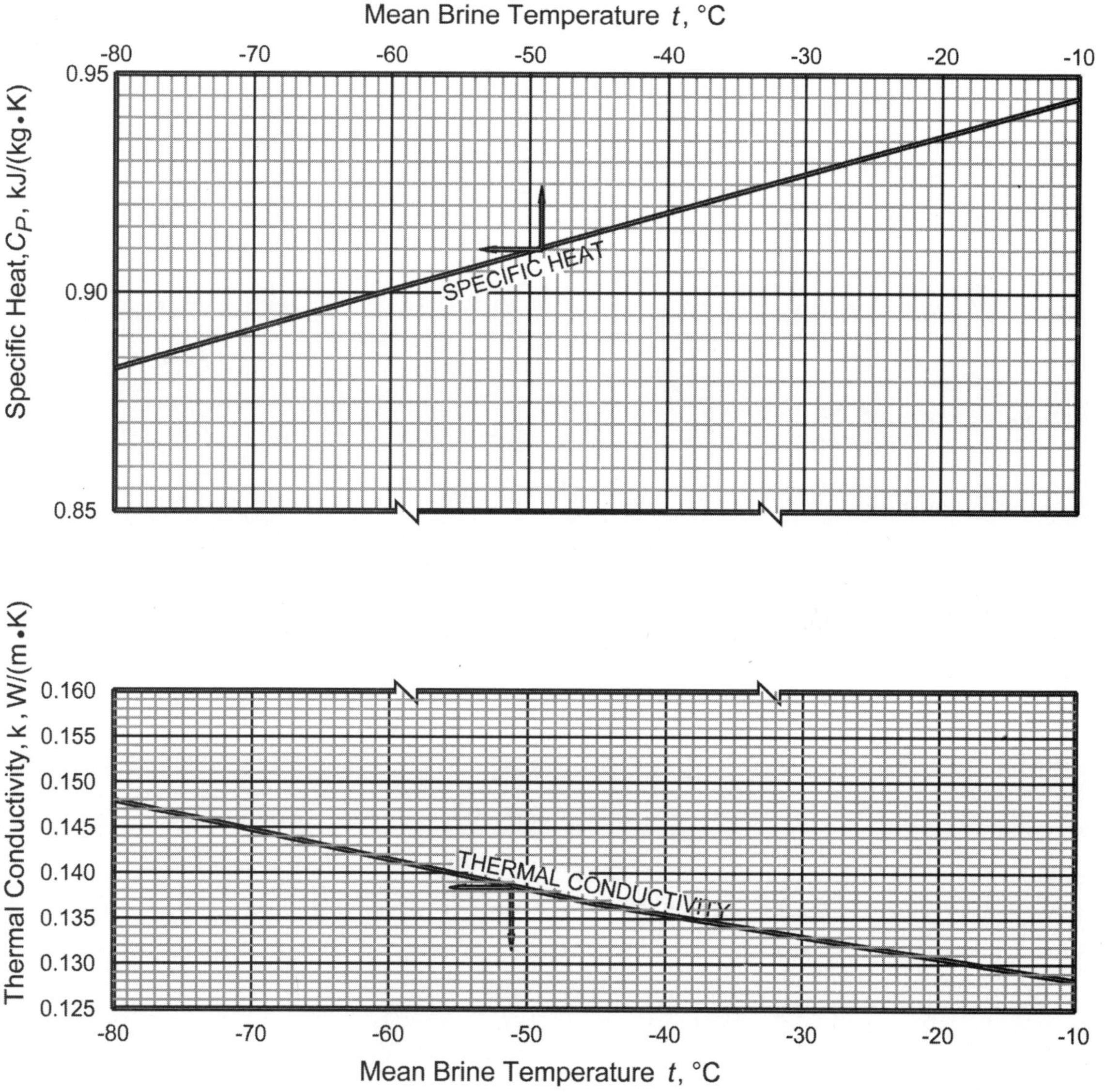

Fig. 12.26: Properties of Trichloroethylene